# The Asian Elephant
# in
# Captivity
## A Field Study

Fred Kurt
and
Marion E. Garaï

FOUNDATION® BOOKS

Delhi • Bangalore • Mumbai • Kolkata • Chennai • Hyderabad

Published by:
**Cambridge University Press India Pvt. Ltd.**
under the imprint of Foundation Books
Cambridge House
4381/4 Ansari Road
Daryaganj
New Delhi- 110 002

C-22, C-Block, Brigade M.M., K.R. Road, Jayanagar, **Bangalore**- 560 070
Plot No. 80, Service Industries, Shirvane, Sector-1, Nerul, **Navi Mumbai**- 400 706
10, Raja Subodh Mullick Square, 2nd Floor, **Kolkata**- 700 013
21/1 (New No. 49), 1st Floor, Model School Road, Thousand Lights, **Chennai**- 600 006
House No. 3-5-874/6/4, (Near Apollo Hospital), Hyderguda, **Hyderabad**- 500 029

© Cambridge University Press India Pvt. Ltd.
First published 2007



ISBN: 978-81-7596-358-0 (Paperback)

Typeset by Amrit Graphics, Shahdara, Delhi- 110 032

Published by Manas Saikia for Cambridge University Press India Pvt. Ltd.

# Contents

# Preface

Once upon a time the Raja of Savatthi called his servant and said: "Go and gather together in one place all the men of the capital who were born blind and show them an elephant". The servant did as he was told. To one blind man he presented one ear of the elephant, to another its trunk, to another a foreleg, to the other three either the belly, the tail or the tuft of the tail. When the six blind men had felt the elephant, the Raja asked each of them: "Tell me, what sort of thing is an elephant?" Thereupon the man who was still holding the ear answered: "Sir, an elephant is like a fan". The others said that the elephant was like a tube, a pillar, a wall, a rope or a brush. Now illuminated, the blind men no longer needed to beg. With such specialised knowledge they were qualified to establish a school of elephantology. The whole world would beat a path. But how to present to that world the arcane laws of elephantology? A fan, a tube, a pillar, a wall, a rope, a brush? They meditated, they discussed. Chaos resulted. Co-operation was impossible.

Based on the idea of the Canadian writer, Patricia Kathleen Page, this is a new version of an old Buddhist canon about human ignorance. Each of the sight-challenged experts, mistakes the parts for the whole, just as today adherents of different elephant management practices, students of different scientific disciplines engaged with the life of captive elephants or representatives of animal welfare organisations claim a firm hold on truth, while disparaging or ignoring other findings and opinions. In a parallel way, most elephant experts are blind to the fact that Asian elephants living in jungle camps, temples, zoos or circuses are not domesticated animals like dogs, camels or cattle, but wild animals living in captivity similarly to the dancing bears of the gypsies, the cobras of the snake charmers or, once upon a time, the cheetah, the 'hunting leopard' of the North Indian princes.

For several decades we have studied captive elephants in various facilities. Fifty years ago, Fred Kurt started his research in a circus before concentrating on wild and captive elephants in South Asia. Marion Garaï began her studies in three European zoos, before embarking on a long-term study of the behaviour of translocated African elephants and joined the studies on captive elephants in Sri Lanka. Both of us were interested in a holistic approach to captive elephants, which resulted in the finding that the ecological and social conditions in captivity and the well-being of elephants are measurably linked to physical, physiological and behavioural parameters. Through this book we have tried to create an awareness for a better life of captive elephants, well knowing that we have opened only a little window for a better understanding of the elephants in captivity, but looking positively forward to overcome one day, the simile of the blind men and the elephant.

# Acknowledgments

The idea to write the first book about behaviour and ecology of captive elephants based on research in the countries of origin are rooted on many, often night-long discourses with Professor Panicker Chandrasekheran and Professor Jacob V. Cheeran (Veterinary College, Trichur) Dr. Krishnan Easwaran (Forest Veterinary Officer, Kerala), Professor John Eisenberg (Smithsonian Institution, Washington) Professor Heini Hediger (Zoo, Zurich), Dr. V. Krishnamurthy (Forest Veterinary Officer, Tamil Nadu), Professor Hans Kummer (University of Zurich), Professor Gunther Nogge (Zoo Cologne), Professor Charles Santiapillai (University of Peradeniya), Professor Raman Sukumar and Mr. Surendra Varma (Asian Elephant Research Centre, Bangalore). A number of statistical data were provided by Mr. Alexander Haufellner and Mr. Jürgen Schilfarth (European Elephant Group) and many stem from the project 'Elephant in Sri Lanka'.

Without the help of many people the project 'Elephant in Sri Lanka' would not have been possible. First of all, we would like to thank the 111 participants from 10 universities and zoos, who spent several weeks of their vacation working in an exemplary way in the various research projects on Asian elephants. The University of Veterinary Sciences in Vienna financially supported the field periods of its students. In Austria, the organisation of the project was made possible by Professor Walter Arnold, Professor Frieda Tataruch and Ms. Helena Lederer of the Institute of Wildlife Biology. In Vienna, the project was supported by H.E., the Ambassador of the Republic of Sri Lanka, Mr. C.S. Poolokasingham, who showed his great interest also during regular meetings with the participants in the Tiergarten Schönbrunn. The Director and the Vice Director of the Vienna Zoo, Professor Helmut Pechlaner and Dr. Harald Schwammer invited the team to these regular meetings.

In Sri Lanka, important contacts were made possible by Mr. Ravindra Anthony (Hemtours), Mr. Jayantha Jayewardene (Biodiversity and Elephant Conservation Trust, Sri Lanka), Professor W.D. Ratnasooriya (University of Colombo) and Professor Charles Santiapillai (University of Peradeniya). Studies on captive elephants in and around Kandy were made possible by Mr. Nerajan Wijeyeratna, the Diyawadana Nilame of the Temple of the Tooth, Mr. Palitha W.B. Udurawana, Mr. Henry Gunasekera and several other elephant owners. The studies in the Pinnawela Elephant Orphanage were made possible by Mr. K.D.R.N. Wijesinghe and Mr. S. Gunasena (Directors of the Department of Zoological Gardens, Dehiwala), Dr. M Rajapaksa (Veterinarian), Mr. F. S. M Seelarathna (Curator) and his Assistant Mr. S. M Ratnayake, as well as the mahouts, especially the chief mahout Mr. Sumanbanda. Many of the research details were discussed with Dr. Jayanthi Alahakoon, Professor Mangala de Silva, Professor V.Y. Kuruwita, Professor W D. Ratnasooriya and Professor Charles Santiapillai. Mr. Udo Zimmermann of the ARD - studios in Munich provided us with video-copies of films on wild and captive elephants in Sri Lanka. These TV-films were made by Mr. Felix Heidinger and his team in 1997. Dr. Peter Bardehle, VIDICOM Hamburg, made further films on captive elephants in Kandy and Pinnawela and on wild elephants in Uda Walawe. We received copies of the complete videos for analyses. Dr. Barbara Herzig, Museum of Natural History in Vienna, allowed measurements of skulls of Asian elephants to be taken. The first version of the text has been reviewed by Ms. Irmgard Hofstetter – Collin (Munich). Furthermore, we thank Mr. Wolfgang Weihs (Vienna) for the unique drawings and water-colour paintings.

In this book, we have included measurements of captive elephants which stem from European zoos and circuses and from Myanmar. These data were provided by Mr. Franco Knie of the Swiss National Circus, Dr. Hans Frädrich and Mr. R. Pankow (Zoo, Berlin), Dr. Claus Hagenbeck (Hagenbecks Tierpark, Hamburg), Mr. J. Adler (Zoo, Münster), Dr. Alex Rübel and René E. Honegger (Zoo, Zurich). Measurements of elephants in Myanmar were made possible by Lt.Gen. Chit Swe, Minister of Forestry, Dr. Khyne U Mar of the Myanma Timber Enterprise and Mr. Herwig Pucher (Vienna). Material for genetic research could be collected in European zoos and circuses as well as in Myanmar, Sri Lanka, Kerala and Thailand. In collecting these data we have been assisted by Dr. Jayanthi Alahakoon (Colombo), Mr. Lyn de Alwis (Colombo), Dr. Visit Arsithamkul (Bangkok), Professor Dr. J. V. Cheeran (Trichur), Mr. K. Kock (Hamburg), Dr. S. K. Karunaratne (Colombo), Dr. Alongkorn Mahannop (Bangkok). Dr. Khyne U Mar (Yangon), Mr. Myat Win (Yangon), Dr. G. Ruempler (Munster), Professor Dr. KC. Panicker (Trichur) and Dr. Wisid Wichsilpa (Bangkok).

Genetic studies were financially supported by the Foundation of Ruth and Rudolf Schlageter (Zurich) and Mr. Alexander Haufellner (European Elephant Group, Munich). The ecological and ethological studies in Sri Lanka received financial support from the Universities Federation of Animal Welfare (UFAW, England), the International Fund for Animal Welfare (IFAW, South Africa), the Förderungsverein des Tiergarten Schönbrunn (Vienna), the Swiss National Circus Knie (Rapperswil) and VIDICOM (Hamburg).

Last but not the least, Ms. Suparna Baksi-Ganguly (CUPA, Bangalore), who introduced us to Cambridge University Press India Pvt. Ltd.

Fred Kurt<br>Marion E. Garaï

# 1

# Introduction

According to recent estimates about 15,000 Asian elephants live in captivity. Their high number represents a quarter to a third of the total population of *Elephas maximus* according to the prevailing estimates of the number of surviving wild conspecifics. About 14,000 captive Asian elephants live in the countries of origin and the rest in western zoos and circuses.

Numerous reports have considered captive Asian elephants as 'domesticated' animals. However, most of the elephants living in captivity have been wild caught and a large number of elephants that have been bred in the range countries are offsprings of mating between captive females and wild males. Although captive elephants have existed since 3,500 years in close contact with man, they have never been the subject of sustained captive breeding programmes, nor have they been selected for particular characteristics as it was, and still is, the case with truly domesticated animals such as cattle, horses or dogs. The genetic make up of all timber and temple elephants, zoo and circus elephants has therefore not been changed and they are wild animals in captivity and should be treated accordingly.

For numerous animal welfare organisations the suffering of captive elephants has become the main subject for their campaigns, and sophisticated minds conclude that 'elephants do not belong to captivity', well knowing that 15,000 elephants cannot be set free and brought back to the remaining pockets of jungle, where the last wild elephants survive in habitats which have become increasingly smaller due to encroachment by a fast growing human population. To protect humans and their crops from elephants, and elephants from poachers, their last refuge must be fenced. It leads to the fact that wild elephants will soon have to be managed like captive populations by employing methods of translocation and population control.

During the last two decades an undesirable trend has been observed. Thousands of elephants, traditionally kept in jungle camps by tribal people, have lost their work as a consequence of the protection of the remaining last pieces of the forest. The elephants have been brought to temples, cities and tourist resorts, where their lives often become more miserable than it was while they lived in their natural habitat. Animal welfare activists have done their utmost to treat the wounds and diseases of these elephants. However, this is not sufficient, and in order to improve the situation the living conditions of these captive elephants must be changed fundamentally, i.e. they should lead life under more natural conditions.

Very often animal welfare activists and elephant owners, trainers and keepers or mahouts share a common characteristic – their poor knowledge of  wild elephants. This lack of fundamental knowledge induces anthropocentric actions and argumentation, but is of little help to the captive elephants. With this book we hope to provide some data on ecology and behaviour of captive elephants in relation to their wild conspecifics. A large amount of data published here stems from a recent research project in Sri Lanka, but also from a number of our studies on wild elephants in Sri Lanka, South India and Africa, captive elephants in Sri Lanka, South India and Myanmar as well as in several European zoos and circuses.

Between 1997 to 1999, i.e. 30 years after the legendary Elephant Survey of the Smithsonian Institution of Washington, a team of 111 young veterinarians and biologists from 10 universities and zoological gardens studied wild and captive Asian elephants in Sri Lanka. The total field work encompassed 120 'man – months', and it can be considered as one of the most extensive studies on the species. The project was organised by the Institute of Wildlife Biology of the University of Veterinary Sciences, Vienna (Austria), the Zoological Institute of the University of Colombo and the Zoological Institute of the University of Peradeniya, the Department of Zoological Gardens in Dehiwala, the Department of Wildlife Conservation in Colombo, as well as members of the Asian Elephant Specialist Group (AsESG) of the IUCN.

The project had the following aims: (1) The participants were to obtain information on the life of wild and captive elephants as well as on different aspects of traditional and alternative elephant-keeping systems. (2) The participants were to carry out simple research projects without using highly specialised technical equipment. (3) The data collected was to give some guidelines for improvement of the living conditions of captive elephants. (4) The results of the studies were to add to the present knowledge of the Asian elephant.

We attempted to find answers to problems which could be solved by relatively inexperienced, but highly motivated students with simple methods within a short time of a few weeks or a few months. The financial resources and the technical means were limited. Furthermore, we endeavoured to create methods which could easily be taken over by Asian students, who had not yet received special training in wildlife management, ecology or ethology. For the captive elephants, we concentrated on the often cruel keeping systems and attempted to make proposals for improved keeping systems. Furthermore, we tried to find answers to simple questions, such as: "How much food is eaten per day?"; "How is the food prepared?"; "How do elephants grow under different living conditions?"; "How and how long do elephants sleep?"; "How does stereotypical behaviour, such as 'weaving', evolve?" In 2001, the first report on our studies was published in German (c.f. Kurt, 2001). In addition, some summaries appeared in German and English in several journals as well as in the proceedings of the International Elephant and Rhino Research Symposium in Vienna (2001). In this English version original reports have been compiled and discussed, new data and recent knowledge on elephants have also been added.

This book gives an overview of the history of the Sri Lankan elephant. The various types of keeping conditions are described. Furthermore, patterns of body growth as well as feeding behaviour are given of 123 captive elephants living either in private ownership or in the Pinnawela Elephant Orphanage. Some aspects of social behaviour, reproduction and musth as well as stereotypical behaviour, sleep and tool use of wild and captive elephants have been described. The studies on population genetics deal with inheritance of tusks. Some practical advice on estimates of age and body weight is given and patterns of body condition and signs of physical and social disorders are described. Recommendations have also been made to improve the living conditions of captive elephants.

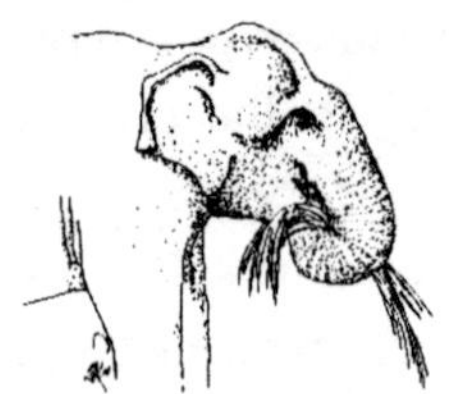

# 2

# The Sri Lankan Elephant

## 2.1  Tuskers and Tuskless Bulls in Asian Elephants

History and population genetics of a man-made phenomenon

In Asian elephants tuskless bulls or maknas (called 'aliyas' or 'pussas' in Sri Lanka) are generally rare and their frequency hardly exceeded 10% of the sub-adult and adult male population in the nineteenth century. The only exception is Sri Lanka, where between 90% to 95% of the sub-adult and adult bulls have been reported to be maknas since the late sixteenth century. In his *Manual of the Mammals of Sri Lanka*, Major W.W.A. Phillips has put forward three hypotheses to explain this phenomenon: (1) Tuskers and maknas belong to two different races present in Sri Lanka (2) Elephants from Sri Lanka and those on the Indian subcontinent belong to different races or subspecies. According to this hypothesis, all tuskers in Sri Lanka are supposed to be descendants of escaped or liberated tuskers imported from the Asian mainland. (3) The high frequency of maknas in Sri Lanka is a result of massive capture of tusk bearing bulls.

Modern studies have proven the first hypothesis to be invalid. In Sri Lanka there exists, if at all, only one separate subspecies, namely *Elephas maximus*. The second hypothesis that tuskers were introduced by man into a population originally characterised by a natural lack of tuskers is highly questionable. Tuskers from Myanmar and the Indian subcontinent have been donated to Sri Lankan kings and temples at least since the third century BC (e.g. Geiger, 1973; Digby, 1971). These tuskers were extremely rare and well cared for, practically permanently chained when not used, and led under strict commands, when brought forward as royal mounts. Bulls kept in such a restricted manner hardly breed and cannot be expected to have had a considerable genetic impact on the local elephant population. The third hypothesis, which considers the predominance of maknas in Sri Lanka to be the consequence of selective persecution by man and first put forward by Jean Baptiste Tavernier, a French nobleman who visited the Indian subcontinent in the seventeenth century six times within a period of 40 years and had a deep insight in culture and politics. Tarvernier's hypothesis has not been convincingly refuted so far.

If the high frequency of maknas in Sri Lanka is the result of selective capturing of tuskers, maknas should generally be more abundant in populations subjected to those management practices than in other populations. It should be expected that, once selective removal of tuskers from a population has been abandoned, there should be no further increase in the frequency of maknas apart from minor fluctuations due to genetic drift. In both these considerations it is implicitly assumed that tuskers and maknas do not differ as to their reproductive success, a question that also needs to be addressed in this context.

In 1995 Fred Kurt, Günther B. Hartl and Ralph Tiedemann described the development of tusker-makna ratio in several wild-living and captive populations of South Asia and examined the results in comparison to the respective management practices applied. Furthermore, Tiedemann and Kurt investigated by a stochastic mathematical model various modes of inheritance of tusks and their consequences on the impact of selective hunting and capturing as described in the following sections.

## Definitions and methods of data collection

In Asian elephants only males possess tusks, which are elongated second upper incisors of different length and form. In Sri Lanka only 17% of the maknas lack upper incisors totally (Ratnasooriya & Fernando, 1992). The remaining bulls (like many females) possess tushes, which are incisors that are much smaller and thinner than the poorest tusks. These tushes may be either circular or oval in cross-section and sometimes protrude 20 cm or more beyond the lips. The oval type, which is generally 15 to

30 cm long and decurved (curved outward / down, as opposed to the recurved, i.e. curved inward / up like tusks) seems to occur only in males. The circular type, which is often broken at the gum line, is found in both sexes (McKay, 1973).

Most data presented here were collected with no particular interest in the form of tusks, the type of tushes or complete absence of tushes. Only two types of bulls were distinguished: tuskers and maknas. Data stem from records on hunting and capture operations, from estimates of poaching, from stud books, and from investigations on wild-living populations. The demography of wild-living elephants is well-known only from studies on relatively small populations of elephants individually recognised by researchers. Population censuses carried out in large populations within a short time tend to underestimate the actual population size, the results are skewed towards adults and sub-adults, and amongst these skewed in favour of females, since unskilled observers are not always able to distinguish maknas from females (For details, see Appendix I).

In the range countries two systems are used in the management of elephants: (1) An extensive management method is used for working elephants, which are employed for transport and timber hauling and live normally in forest camps. When unemployed, they spend their time with hobbled forefeet in the nearby forests where they find natural food and meet with other tame and wild conspecifics. If desired by the owners, breeding can occur regularly and the offsprings are fathered mainly by wild bulls. (2) Temple elephants, which are mainly used for processions, are kept by an intensive management method. When not working, these elephants are chained and fed by man, like war and state elephants in former times. In this case intraspecific social contacts are minimal or absent, and breeding occurs very rarely.

The annual mortality rate for the total captive population in Asia is estimated to be about 7.5%. This estimate is based on an annual mortality rate of 11% in British elephant establishments in Bengal (Sanderson, 1907) and Sri Lanka (Tennent, 1861), as well as on recent reports from Thailand (Santiapillai and Jackson, 1990). The Delhi Sultanate seems to have lost between 5% to 10% of its tame elephants per year (Digby, 1971).

Since some techniques of capturing are highly selective and prefer mainly young males, captive born animals are distinguished from captured wild ones throughout the chapter. Furthermore, different techniques of capturing and taming lead to different mortality rates (see Kurt, 1992, 1993, for reviews). Especially the use of pitfalls caused mortality rates rise upto 90% (Sanderson, 1907), while 86% to 97% of hand-noosed elephants survived the capture operation (Jayewardene, 1994). In *kheddas*, i.e. large enclosures, by the end of the ninteenth century the mortality rate could be as high as 70%. Later this was reduced to between 5% to 30% (Stracey, 1963). The *mela-shikar*, i.e. noosing selected animals by sitting on the back of tame elephants, produced mortality rates of some 14% (Toke Gale, 1974). The various methods of capture are well described in the *Arthashastra* (330 BC–300 AD), a manual of Buddhist statecraft, and were well-known throughout antique and medieval South Asia (Meyer, 1929). The *Arthashastra* and other old sources of elephant lore provide a qualification of each of these methods according to their effectiveness in selective capture and to the survival rate of the captives.

## Are tusker-maknas  ratios in captive populations different from those in wild populations?

The least skewed sex ratio in an undisturbed population of wild-living elephants at reproductive age is 65 males:100 females (see Appendix I). In some captive elephant populations a much higher proportion of bulls at reproductive age is observed. In Kerala (South India) only tuskers are used for temple festivals. In Sri Lanka, bulls were preferred as long as selective capturing operations were permitted, i.e. before 1965. In Myanmar, where a relatively high number of working elephants are bred in captivity, the proportion of bulls is similar to that in an undisturbed wild population as mentioned above. For the time before 1980 the maximum proportion of maknas in the wild-living south Indian elephant population is estimated to be about 10%. By contrast, the frequency of maknas among privately owned elephants in Kerala is as low as 2.1%. Since capture operations were prohibited, tuskers for temple festivals in Kerala were purchased at the elephant fair in Sonpur, and they originate from captive populations or poaching operations in north-eastern India.

## Do selective hunting and capture result in increased frequencies of maknas?

In the ninteenth century maknas were rare in north-eastern India (Sanderson, 1907), but their frequency increased significantly from about 8% to 48.7% of the male population in the period between 1937 to 1950. Unselective kraaling as practiced in the late nineteenth and the early twentieth century certainly had an influence on population size, but not on the proportion of maknas present. However, in recent years tuskers have been heavily poached by tribal people for ivory and meat, which obviously resulted in a considerable increase in the proportion of maknas. In Myanmar and Thailand ivory poaching has increased in the last years to such an extent that even the tusks of living captive bulls are sawed off and stolen. In Myanmar maknas had been rare in the last century (Evans, 1910). However, by 1950 a total of 10% of tame bulls were maknas (Deraniyagala, 1955), and by 1993 their frequency was as high as 43.5% . In Tamil Nadu 217 bulls, including 4 (1.8%) maknas were caught between 1926 to 1980. From 1981 to 1983, selective poaching reduced the frequency of males to 20% of the total population, and a decline in the proportion of tuskers was also observed. In Kerala heavy poaching for ivory has taken its toll in recent years. Censuses between 1983 to 1989 revealed a strong decline in the number of tuskers, and the sex ratio in sub-adults and adults was found to be skewed towards females. The reduction of the sub-adult and adult tusk bearing male population was significantly related to an increase in the frequency of maknas (Fig. 2.1.1).

## Why did man affect tusker-makna ratios in Sri Lanka, but not on the Asian mainland?

With the exception of the remote regions in the highlands of north-eastern India, Myanmar, Thailand, Laos and Vietnam, where elephants are still hunted, captured and tamed by tribal people, hunting and capturing were always a strictly enforced monopoly of rulers and later of colonial forces. Under the Buddhist and Hindu kings of southern Asia, management of wild elephants was carried out according to Kautiliya's *Arthashastra*. It advised people to eliminate elephants from settled and cultivated valleys,

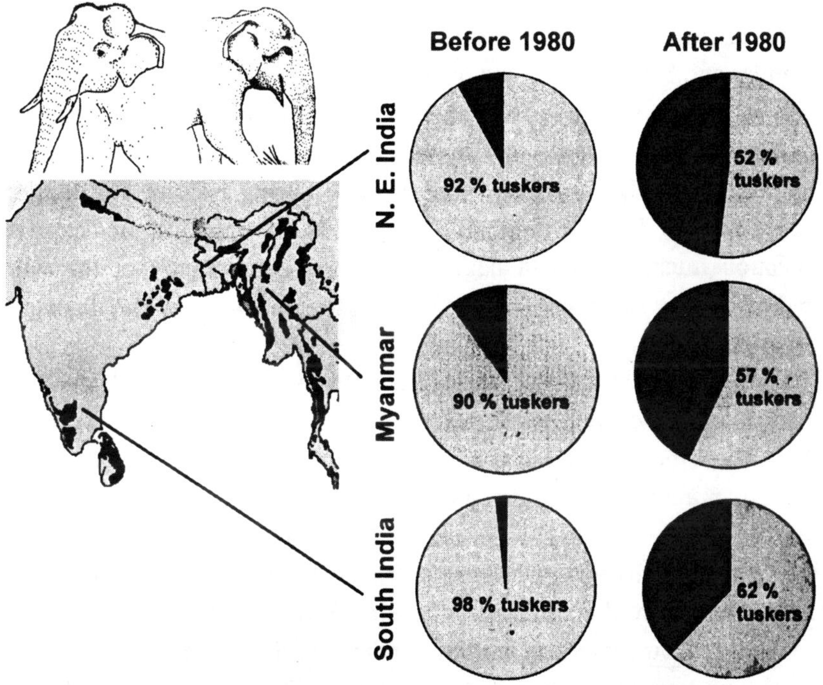


**Fig. 2.1.1** Frequencies of tuskers and maknas before and after 1980 in three mainland populations

but to preserve them in guarded elephant sanctuaries at the periphery of the respective kingdoms. Furthermore, the *Arthashastra* instructed people to capture only sub-adult and adult tuskers and to disregard sick, young and female elephants as well as maknas and bulls with small tusks. These recommendations could not be strictly followed. In each elephant - riding culture, females were kept either as docile mounts and as indispensable decoys for capture operations, or just because too few males were available.

Numbers of elephants kept by rulers in antiquity and in the middle ages have often been extremely exaggerated by contemporary writers. During its largest extension, the Delhi Sultanate (1192–1398) controlled practically the whole Indian peninsula and, thus, at least some 1.5 million km² of elephant land. If a low density of one elephant per four km² is assumed, the wild-living population consisted of at least 375,000 individuals.

Most of the captive elephants were tributes from subordinate rulers or plundered from southern India, some had been imported from Bengal, Myanmar and Sri Lanka. At the peak of its power the Sultanate possessed 3,000 elephants (Digby, 1971). However, even if one assumes a captive population with a sex ratio of 2 males : 1 female, an annual mortality rate of 7.5%, and a high mortality rate of 75% during capture and taming operations, the annual demand for animals to maintain the captive population must have been negligible to the demography of the wild population. The same applies to the South Indian kingdom of Vijaynagar (Appendix I).

During the peak of the Mughal empire in the seventeenth century, the population of captive elephants was estimated to be 30,000 (Tavernier, 1984). Even the maintenance of such a large population would hardly have had an impact on the wild-living resources. However, this only holds if the Mughals had caught the elephants in all parts of their immense reign. But they satisfied their demand primarily through unselective large-scale *kheddas* in the vicinity of their capitals. Therefore, elephant populations disappeared in northern India. On the Indian peninsula the captive population has never been larger than 8% of the wild population.

North-eastern India and South-east Asia harbour relatively large populations of tame elephants (13% to 200% of the numbers of the respective wild populations). In most of these regions elephants are regularly bred in captivity. The economics of the Khmer empire (802–1431) depended on a tremendous system of irrigation and therefore on a large-scale use of elephant power. Nevertheless, in the middle of the twentieth century the frequency of maknas in Cambodia was as low as in other regions of the Asian mainland (Bazé, 1955). The Khmer drew tributes from as far away as present day Laos, Myanmar, and Malaysia as well as regions which were later to become Thai kingdoms to the west. The large demand for working elephants could easily be satisfied from practically unlimited resources, which was in sharp contrast to the situation in Sri Lanka.

Sri Lanka always possessed a considerable tame population (20% to 33% of the wild population). Since Sri Lankan kings only exceptionally controlled territories on the Asian mainland, these populations must have always been restocked from the resources of the island itself. Import as well as breeding of captive elephants were practically absent. During the Portuguese period (1505–1658) and the Dutch period (1602–1670), high frequencies of maknas were well documented. Contrary to this, paleontological records and rock paintings indicate the presence of an elephant population with tusk-bearing males in the Neolithic and even later periods (Deraniyagala, 1955). Assuming that the high rate of maknas is man-made, it is likely that the Sri Lankan kingdoms have had a profound impact on the demography of wild-living elephant populations.

## How could Sri Lanka harbour large captive populations without captive propagation?

Contrary to the contemporary Indian or Myanma armies, the elephant units of antique and medieval Sri Lankan forces were very small (Geiger, 1973). For example, by the end of the sixteenth century, amongst 2067 captive elephants only 122 (5.9%) were war elephants (Baldaeus, 1960), while the contemporary Turko-Afghan or Mughal rulers kept 25% to 33% of their elephants fit for war (Digby, 1971). The bulk of tame elephants must have been indispensable for the construction of monumental religious and public buildings. First of all, however, they must have played a major role in the creation of a multiplicity of irrigation works, mainly in the dry zone, for nowhere in the pre-modern world was such a dense concentration of irrigation facilities observed. Elephants were used to draw earth ploughs, to haul heavy weights, and to work on levers (Murphy, 1957; Brohier, 1965; Boner & Sarma, 1972; De Silva, 1991).

Generally, captive male elephants were preferred to captive females since, on the one hand, they are substantially stronger, and, on the other hand, the wild-living herds had to be kept as large and productive as possible. A relatively large captive population had been maintained on the long run only through strictly controlled selective capture at low mortality rates. Until the middle of the seventeenth century, elephants were practically omnipresent on the island; bulls threatened people and caused crop damages (e.g. Schweitzer, 1953; Brohier, 1965). As 75% of all crop raids and 80% of all manslaughter were done by the highly sought out bulls, it was obvious to protect men and crops with selective noosing of these bulls (Jayewardene, 1993).

## What is the price for achieving a large captive stock without exhausting the wild-living population?

Elephants are polygamous, and their genetic variation is low compared to other large ungulates (Nozawa & Shotake, 1990; Hartl *et al.,* 1995). There is no seasonality in reproduction or no pronounced one, and birth intervals are some 4.5 years or more. Bulls have the largest testes amongst ungulates. Testes size is directly related to daily sperm production, sperm density, and absolute number of sperm per ejaculation (Moller, 1989). Thus, with respect to the social, genetic and physiological capacities of bulls, there does not seem to be any limiting biological factor for the maintenance of a skewed sex ratio in the reproducing age classes of a population.

This assumption is wrong. Heavy poaching of adult bulls has been found to reduce the fecundity of females (Poole, 1987, 1989; Chandran, 1990). The proportions of infants and juveniles in the population drop sharply, already at a sex ratio of one full grown bull per 6 to 10 reproducing females (Kurt, 2001). Hence, in Asian elephant the availability of adult males is a limiting factor for female fecundity as it is the case in other gregarious ungulates (Ginsberg & Milner-Gulland, 1994).

As a consequence of the former, the rulers of Sri Lanka had to take into account not only the negative effects of capturing too many females, but also those of capturing too many bulls. This could be accomplished by the very selective, but highly protective techniques of hand-noosing. One may ask the question: was it actually possible to maintain on the long run 2500 captive elephants (Pieris, 1920; Baldaeus, 1960) by permanently restocking from a wild-living population of only 12,000? (McKay, 1973). Kurt, Hartl and Tiedemann have addressed this problem by using a computer simulation program developed by Sukumar (1992) with the following basic assumptions: in the captive population sex ratio was 2 males: 1 female, and annual mortality was 7.5%. In the wild-living population there were 5.9% of juvenile males (6–10-year-old), 8.3% of adult and sub-adult males, and 42% of adult and sub-adult females. Mortality during capturing operations was 10%.

Based on these data, annual losses in the wild-living populations would have been 125 bulls and 62 females, corresponding to 8.1% and 1.5%, respectively. Such losses can be considered medium for bulls and low for females. After 50 years with constant medium off-take of males and low off- take of females the population had a sex ratio in reproducing age classes of 1 male: 8 females, and still grew at a rate of 1.5%. High off-take of bulls (15%) would have skewed the sex ratio to 1 male: 20 females, ultimately resulting in a reduced fecundity. However, the permanent and highly selective off-take of tusk bearing males must have resulted in an increasing proportion of maknas in the wild-living population, as demonstrated in the following comparison.

## Why did sacred places save the tuskers?

In the southeast of Sri Lanka tuskers are more abundant than in all other parts of the country. This area, today represented by the Ruhuna (or Yala) National Park and its surroundings, was part of the kingdom of Rohana, and until 1250 AD it was subjected to intensive wet rice cultivation. Abandoned fields and silting water reservoirs were later changed into secondary habitat with high carrying capacity for elephants. Kataragama, the heart of this area, has been considered one of the most sacred places both by Hindus and Buddhists since ancient times (Wirz, 1966). Therefore, it is not surprising that the first traceable elephant sanctuary was established in its vicinity already in 34 AD (Geiger, 1960). Ptolomy's map (100 AD) quotes the area of the present national park as 'Pasqua elephantorum'. Later maps indicated the southeast of Sri Lanka as a promising capturing ground for elephants (Baldaeus, 1960).

By the end of the seventeenth century, colonial powers introduced unselective kraaling, which resulted in a massive reduction of the wild-living population (Fig. 2.1.2) but did not affect the tusker-makna ratios. This is reflected by temporary accounts of their bags: between 1690 to 1697 (De Vos, 1872; Van Goens, 1932; Schreuder, 1946) and in 1805 (Cordiner, 1983), some 15% of all captive bulls were tuskers. In the Ruhuna National Park the frequency of wild-living tuskers was 25.4% in 1967–68 (Kurt, 1969), 16.1% in 1991–92 (Dissananyake, 1992) and 15.3% in 1993 (Santiapillai, 1993).

A completely different situation prevails in the Mahaweli basin in the east of Sri Lanka. Here antique and medieval wet rice culture was most intensive and lasted for at least 1800 years (Brohier, 1965; De Silva, 1991), and it can be assumed that the rulers of Pollonaruwa removed

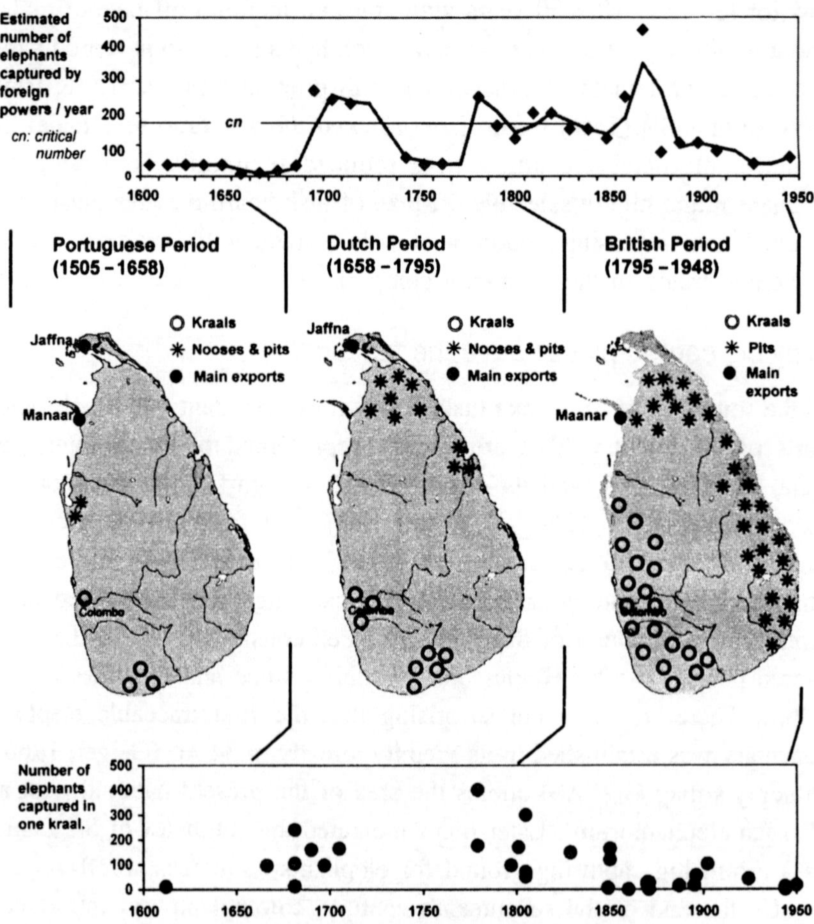



**Fig. 2.1.2** Estimated number of elephants captured by foreign powers in Sri Lanka, capturing grounds according to methods, ports of export and number of elephant captured in single kraals. C.n.: critical number

considerable numbers of tuskers from the wild population. To a certain extent selective noosing continued to be practiced here until the twentieth century (e.g. Spittel, 1951), not without an enduring effect on the proportion of tuskers present in the wild. In fact, almost all contemporary reports agree that today tuskers are completely absent in the Mahaweli population (cf Deraniyagala, 1955; Jayewardene, 1994).

## A historical approach to the inheritance of tusks

The rate of increase in the proportion of maknas in a population can be expected to depend to a certain extent on the mode of inheritance of tusks, which is as yet unknown. To provide some predictions as to the further development of tusker-makna ratios in extant wild-living elephant populations, Tiedemann and Kurt (1995) developed a stochastic simulation model. This model puts strong emphasis on the inheritance of tusks, which are assumed to be controlled by a single autosomal gene or a major gene in a polygene system. From applications to a fictive population they concluded that the tusk allele is most likely dominant under the assumption of a slightly higher (40–50%) reproductive success of tuskers. Based on real data, Kurt, Tiedemann and Hartl (1995) simulated frequencies of the tusk allele for two distinct Sri Lankan populations, differing as to human impact of their development over a certain period of time.

The most drastic impact on tuskers has been reported from Mahaweli basin. The mortality scheme (Appendix I) was simulated with an initial population size of 500. Under the assumption of an extremely high mortality, proportion of tuskers due to selective hunting, the model predicts that in about 1500 AD the proportion of tuskers should have declined from about 95% to zero (Fig. 2.1.3). This result is in agreement with the historical record for Mahaweli. However, for this case the prediction of the model is similar, when a recessive inheritance of the tusk allele is assumed.

A more moderate mortality scheme is reported for the Ruhuna area in the southeast of Sri Lanka where the initial size was about twice as high as in the Mahaweli basin. In the Ruhuna area a period of slightly higher mortality for tuskers (9–10%) than for maknas (7–7.5%) had lasted from about 300 BC until 1600 AD. The available census data suggest a reduction of tuskers from an initial proportion of 95% to a mere 10% within this period. However, the highest human impact on males, causing a mortality of about 15% between 1600–1900, had been unselective with respect to tusks. The simulation is again in agreement with these data, if a dominant inheritance of tusks and a selective advantage of tuskers during mating of about 40% to 50% are assumed (Fig.2.1.3). On the contrary, if sexual selection is excluded, the model predicts the extinction of the tusk allele after about 500 years, which

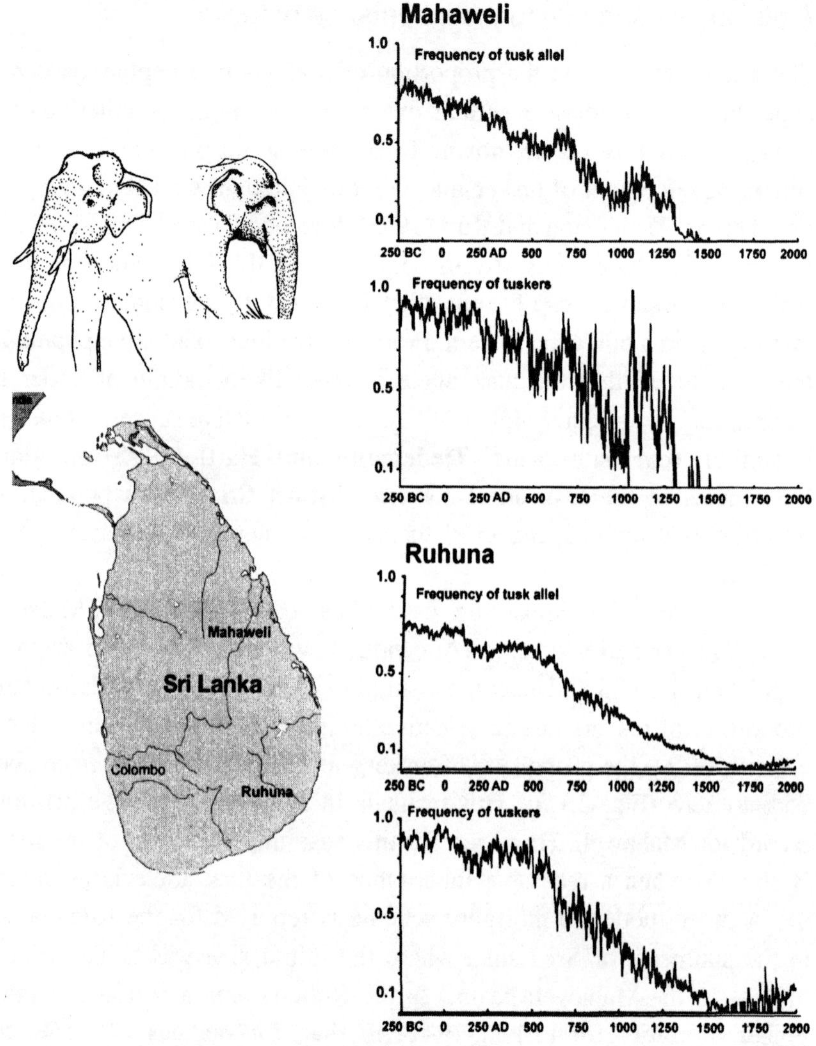


**Fig. 2.1.3**   Frequencies of tusk allele and tuskers in two populations of Sri Lanka between 250 BC to 2000 AD
(Source: Kurt *et al.* 1995; Tiedemann *et al.* 1995)

does not correspond with the census data concerning tuskers. Given a recessive inheritance of the tusk allele without sexual selection, the model predicts a survival of the tusk allele at such low frequencies that tuskers are predicted to be absent or very rare, which is again in contrast to the historical data.

But how can the remaining possibility of recessive inheritance of the tusk allele in combination with sexual selection towards tuskers be excluded?  To evaluate this, the population from southern India, before man started to hunt selectively was simulated. The model predicts that the selective advantage of tuskers will lead to the extinction of the makna allele, if the latter is dominant. On the contrary, when the tusk allele is dominant and abundant, sexual selection against maknas is overridden by drift processes due to the low frequency of the makna allele, and the makna allele may survive for thousands of years.

In summary, the results of all applications of the simulation model support the assumption of a dominant inheritance of the tusk allele in conjunction with sexual selection in favour of tuskers. However, this only holds for a selective advantage of about 40–50% in mating. A higher selective advantage would no longer cause a decline of tuskers in the simulated population of southern Sri Lanka. A lower selective advantage, however, will lead to the extinction of the tusk allele in that population, and cause a decline of tuskers in the simulated population of southern India. Both cases are contrary to the census data.

## Are tuskers doomed to extinction?

In the foregoing sections we have shown that (1) when using elephants, man has always preferred tuskers, (2) selective hunting and capturing frequently led to a decrease of tuskers in wild populations, (3) the impact of selective hunting and capturing was highest in isolated populations (e.g. Sri Lanka and recently in most isolated elephant areas on the South Asian mainland), (4) the loss of tuskers is not necessarily associated with the decline in population size, (5) selective removal of tuskers for protecting a maximum wild male population resulted in an increase of maknas, (6) unselective capturing of large numbers of elephants did not change the tusker-makna  ratio, (7) the assumption of a dominant inheritance of tusks in combination with slight sexual selection is corroborated by historical data.  Reasons for a decline of tuskers in wild populations can thus be well documented. Based on this information, it should be possible to provide some predictions as to the future development of tusker-makna ratio in particular populations.

The example from Mahaweli has shown that continuous reduction of tuskers finally resulted in the complete absence of tuskers. Given this situation, the population practically protects itself from further poaching. From a less fatalistic point of view concerning the biological significance of tuskers in present-day populations of Asian elephant, they should be protected by all means.

The mainland population still harbours enough tuskers to regain this character even in populations heavily affected by poaching. Even a complete lack of tuskers does not necessarily mean that the relevant allele is also exterminated. In fact the tusk allele has survived in the females, which are thus also shown to play an important role for the development of a particular phenotype in males.

## 2.2  Body Growth and Phenotypes

From the classical Indian Antiquity and the Middle Ages up to the present time, Asian elephants have been classified into 4 to 8 different casts (e.g. Abul'1 – Fazl' Alami, 1927; Deraniyagala, 1955; Trautmann, 1982). These traditional types were characterized by features such as body size, tail length and spinal configuration, length and form of tusks, number and colour of toe nails, colour of eyes, pigmentation of penis and depigmentation and shine of the grey skin. Rare combinations are still venerated as 'white' elephants, and certain phenotypes considered as auspicious or inauspicious respectively. But these different casts also make practical sense, as not all phenotypes are equally suitable for carrying saddles and heavy loads, hauling timber or heaving logs onto vehicles. Since antiquity certain phenotypes, mainly those with large body size, have been assigned to special geographical regions (Trautmann, 1982; Panicker, 1990). This concept has led to the definition of subspecies (Deraniyagala, 1955), which is however questionable due to genetic considerations (c.f. Hartl *et al.* 1995b, 1996) as well as to practical experience, with captive elephants, which are kept under different management systems.

The efforts of the Asian Elephant Specialist Group (AsESG) of the IUCN to integrate a relatively large captive population of some 15,000 captive elephants in its conservation programs are paralleled by a revival of studies on physical characteristics and morphometry of Asian elephants (e.g. Wemmer & Krishnamurthy, 1992; Godagama, 1996; Kurt &

Kumarasinghe, 1998). But many physical features are compared to shoulder height and body weight, i.e. exactly with those two characteristics which seem to vary strongly according to prevailing ecological conditions.

It can be expected that physical features like the number of fully developed toes, manifested in visible nails or hooves, or the presence or absence of upper incisors have a predominant hereditary component. Upper incisors can be enlarged (tusks, in males), rudimentary present as so-called tushes (in males and females) or completely absent (in males and females). Concerning these configurations 5 phenotypes (I-types) can be distinguished in a population. These phenotypes are well-known in traditional elephant lore, but no information is available on other characteristics linked to different I-types. These may be either physical features such as shoulder height and body weight, tail length, extent of depigmentation, eye colour and ear form, or profiles in body growth. For example, if certain I-types differ in the rate of body growth, different socio-ecological roles could be suspected, as it is also indicated by different strategies of tuskers and tuskless males in search of and preparation of food (Kurt, 2001). A definition of phenotypes may also serve as baseline for future genetic studies on different I-types and their population-genetic significance, so far only dealt with in models (see Section 2.1). Data on the 5 different I-types stem from captive Sri Lankan elephants and have been published before by Kurt and Kumarsinghe (1998).

They studied three populations: (1) Captive elephants from Sri Lanka (Pinnawela Elephant Orphanage or privately owned), which are kept in an intensive management system, i.e. fettered when not working and fed by man, and which are used for hauling timber, in parades or for tourism. Some 97% of the assessment of their numbers stem from animals of either sex captured in the wild areas. (2) Timber elephants from Myanmar and Thailand living under an extensive management system in forest areas, where they find their own food and meet with tame and wild conspecifics when not working. Some 95% of the assessment of their numbers stem from captive born animals of either sex. Data from Thailand and Myanmar have been pooled since keeping systems are very similar. (3) Female elephants from four European zoos. The two Asian populations stem from autochthonous sources. In some 60% of zoo elephants, the countries of origin are not known (Haufellner *et al.*, 1993), but animals from Sri Lanka and Myanmar are present in the sample. In the Sri Lankan population 5 phenotypes (or I-types) have been distinguished according to growth forms and presence of upper incisors: In males: (1) tusks-bearing 'ethas', (2) tush-bearing 'aliyas', (3) tuskless and tushless 'pussas'. In females: (4) tush-bearing 'etinnas' and (5) tushless 'alidenas'.

In the following pages a number of characteristics are presented such as shoulder height, thorax circumference, circumference of forefeet, body weight, depigmentation of the skin, form of ears, spinal configuration, length of tail and form of tail tuft, eye colour, and distribution of hair. Data from Evans (1910), Burne (1942), Deraniyagala (1955), Toke Gale (1974), Gerbet (1994) and Godagama (1996) were included where appropriate. Four questions will be discussed: (1) Which distinguishable features are related to the type of management system? (2) Which features can be assigned to other physical conditions (e.g. climate) or the overall environment? (3) Which are connected to the 5 I–types? (4) Are there explanations for these particular linkages?

## Shoulder height and thorax circumference

Shoulder height (S–in cm) is the vertical distance between the ground and the highest point of the scapula in a standing animal. Altogether, 834 S – values from 750 animals were collected. For certain animals measurements were taken in different years. Forefoot circumferences were measured in 170 animals. Shoulder heights were given for 113 Sri Lankan males. From 190 bulls from Myanmar and Thailand 211 measurements of shoulder height were available (Fig. 2.2.1). In both groups, bulls of about 20 years of age reach a shoulder height of some 220 cm. At an older age, Sri Lankan males grow taller, reaching a mean shoulder height of $265 \pm 17$ cm (n = 16) between 31 to 40 years of age, i.e. 10 cm more than bulls of the same age from Myanmar and Thailand ($255 \pm 15$ cm, n = 4). Sri Lankan bulls older than 40 years show a mean shoulder height of $262 \pm 15$ cm (n = 15), but 2 bulls from Myanmar were only 240 cm.

A total of 181 measurements of shoulder height from 152 females from Myanmar and Thailand were compared with 170 measurements from Sri Lanka. Furthermore, 159 measurements of 125 different females from zoos were available (Fig. 2.2.2). Already at the age between 11 to

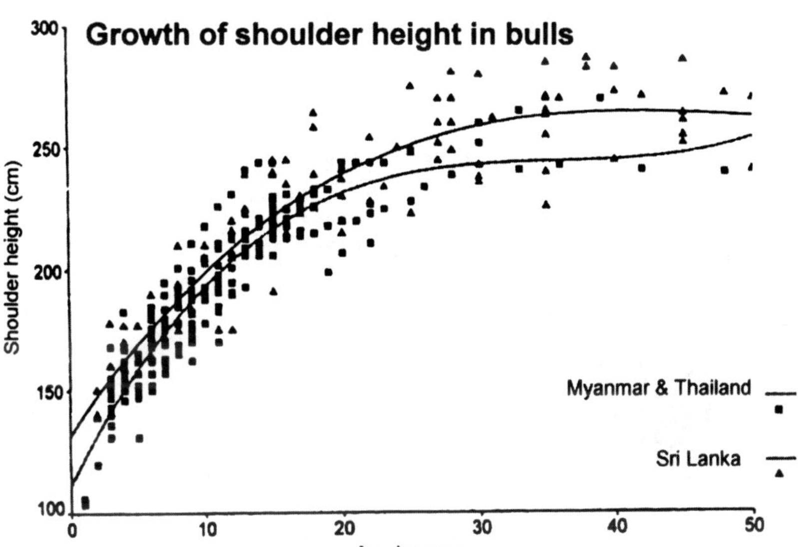


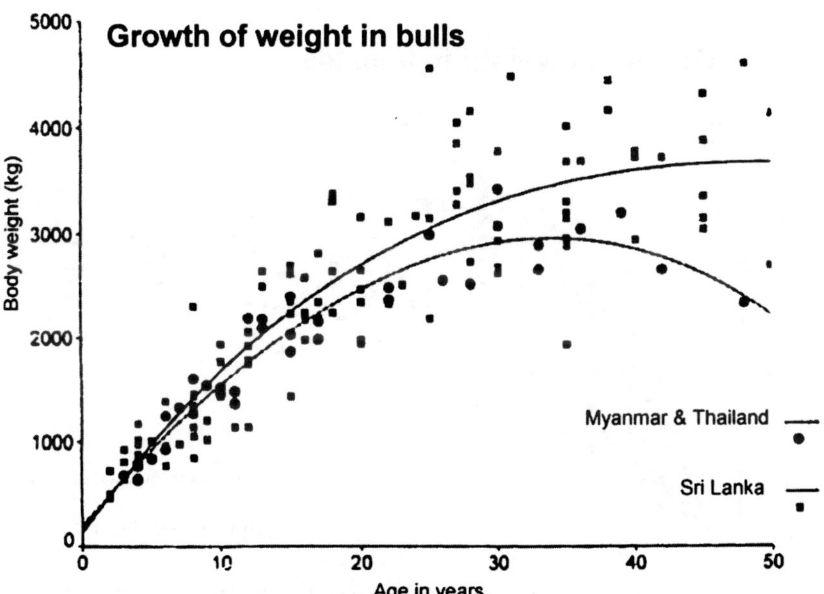


**Fig. 2.2.1**   Growth of shoulder height and body weight in two captive bull populations
(Source: Kurt & Kumarasinghe, 1998)

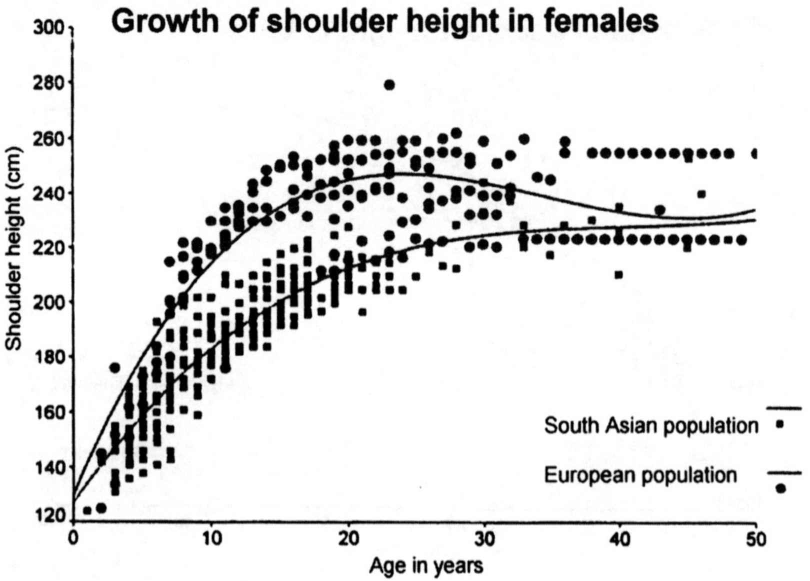


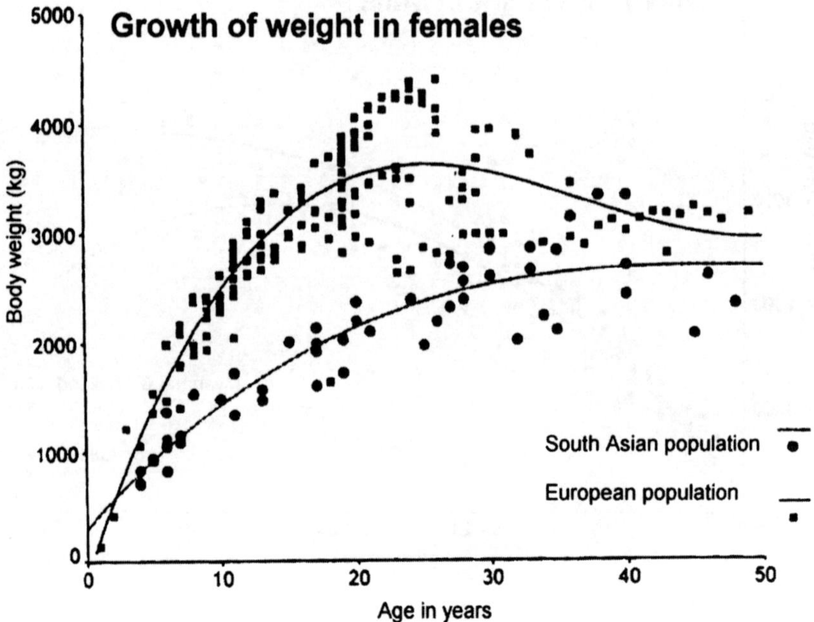


**Fig. 2.2.2**  Growth of shoulder height and body weight in two captive female populations
(Source: Kurt & Kumarasinghe, 1998)

15 years, female zoo elephants are significantly taller (241 $\pm$ 13 cm, n = 29) than those from Myanmar and Thailand (208 $\pm$ 9 cm, n = 52), and Sri Lanka (221 $\pm$ 14 cm, n = 22). But Sri Lankan females of this age group are still significantly taller than those from Myanmar and Thailand. In animals older than 30 years shoulder height in the three sample groups becomes more equal. In age classes older than 40 years, females from Myanmar and Thailand show a mean shoulder height of 236 $\pm$ 11 cm (n = 8). In Sri Lankan and female zoo elephants the corresponding values read 235 $\pm$ 11 cm and 237 $\pm$ 16 cm, respectively.

In all Asian elephants shoulder height equals twice the circumference of one forefoot or the sum of the circumferences of both the right and the left forefoot (Fig. 2.2.3). But only in males, kept under various living conditions in Myanmar and Sri Lanka, shoulder height correlates positively in a more or less linear manner with thorax circumference.

Thorax circumferences or chest girth (T–in cm) is measured immediately behind the forelegs by wrapping the tape around the torso. A total of 293 T-values was included in the study by Kurt and Kumarasinghe; 77 thorax measurements were available from Sri Lankan, and 42 from Myanma and Thai males. In the range of 240 to 260 cm shoulder height, 18 Sri Lankan males had a mean thorax circumference of 364 $\pm$ 19 cm, i.e. insignificantly larger than the mean thorax circumference of 352 $\pm$ 18 cm of 8 males from Myanmar or Thailand. Thorax measurements of 109 Sri Lankan, 39 Mynama and Thai as well as from 26 females kept in zoos were available.

In females the relation between shoulder height and thorax circumference follows an exponential function, indicating that it depends on living conditions. In the range from 220 to 240 cm shoulder height, mean thorax circumference of 46 Sri Lankan and 14 Myanma and Thai females amounts to 332 $\pm$ 17 cm and 342 $\pm$ 24 cm, respectively, but in 6 females from western zoos the mean is 373 $\pm$ 36 cm, i.e. significantly larger than the thorax circumference of Sri Lankan and Myanma and Thai females. This finding indicates that female zoo elephants of the same shoulder height as females living in traditional Asian establishments are more bulky than the latter. This difference is best expressed in body weight.

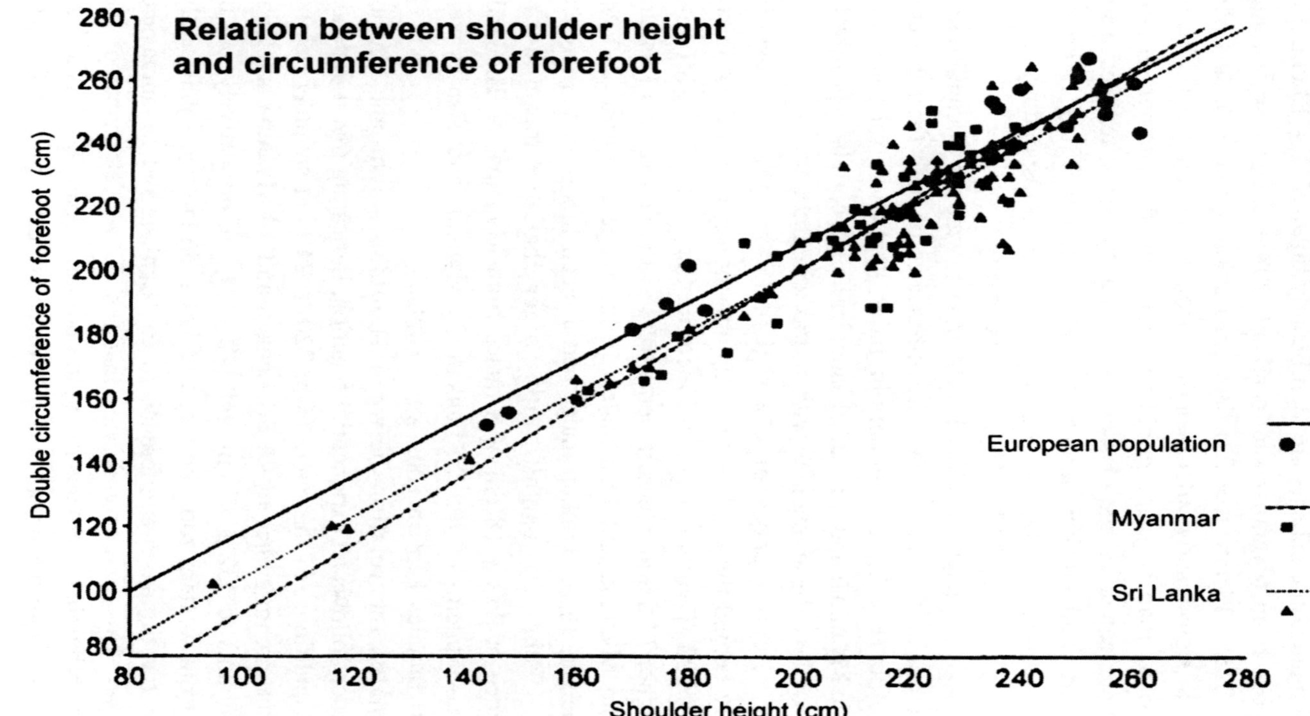


Fig. 2.2.3  Relation between shoulder height and circumference of forefoot (Source: Kurt & Kumarasinghe, 1998)

## A simple but accurate method to estimate body weight

Although about 14,000 Asian elephants live in captivity in the range countries and about 1000 are kept in western zoos and circuses, there are only few data available on body weight. The available data stem from Kerala (Ananthasubramaniam *et al.*, 1992), Tamil Nadu (Sukumar *et al.*, 1988) and Sri Lanka (Kurt and Nettasinghe, 1968), and only few western zoos and circuses regularly carry out biometrical studies such as the Swiss National Circus or Hagensbecks Tierpark in Hamburg. Only a few elephant keeping establishments possess the facilities to weigh elephants, or have elephants tractable enough to be led to a close-by weigh-bridge and to be kept under control during weighing. Consequently, several methods have been developed to calculate body weight from body measurements, such as shoulder height (Kurt & Nettasinghe, 1968; Sukumar *et al.*, 1988), chest girth or a combination of body length, neck girth and chest girth. All these methods are more or less demanding mathematical procedures which could be considered as too complicated to be carried out by ground staff such as elephant managers or veterinarians.

In wild elephants the two methods which estimate body weight from shoulder height can be considered as a useful tool, since shoulder height can be considered as valid age criterion. Furthermore, shoulder height equals quite accurately the double of the circumference of a forefoot. However, in numerous captive elephants, shoulder height correlate neither with age nor with body weight (Kurt & Kumarasinghe, 1998). Some animals, mainly females, seem to grow, if they grow at all, rather in width than in height. Consequently, shoulder height (S–in cm) as well as chest girth (T–in cm) should be taken as base-lines to establish a formula to calculate body weight (W–in kg). Considering several mathematical and biometrical theories, (c.f. Batschelet, 1970; Owen-Smith, 1988) a number of methods were tested by comparing effective weights and calculated weights, and the comparatively simple equation $W = S*T^2 / 10000$ provided the best results.

During the Elephant Survey of the Smithsonian Institution of Washington, body measurements of 77 captive elephants were taken in Sri Lanka in 1967,42 of these animals could be weighed on the weigh-bridge of the railway station in Kandy (Kurt & Nettasinghe, 1968).

Exactly 30 years later, a team of the University of Veterinary Medicine of Vienna measured 128 captive elephants in Sri Lanka and had the chance to weigh 22 of them on the weigh-bridge in Kandy (Kurt & Kumarasinghe, 1998; Weihs, 2002). In 1967 and 1997, altogether 64 data sets on body weight, shoulder height and chest girth of 21 bulls older than 6 years, 34 females older than 6 years, and 9 juveniles younger than 6 years were collected.

First, the above given formula was improved and defined for 3 classes of elephants (males and females older than 6 years, as well as juveniles below 6 years). Effective body weight was compared to body weight computed by the method presented here and the methods given earlier by Kurt and Nettasinghe (1968) and Sukumar *et al.* (1988). The standard deviation of calculated weight from actual weight as well as the maximum and minimum figures were given in per cents of the actual weight (Appendix II).

The formulae found to best assess body weight read:

For elephants upto 6 years      : $W = S*T^2/10000$
For males older than 6 years    : $W = 0.915*S*T^2/10,000$
For females older than 6 years: $W = 0.982*S*T^2/10,000$

Compared to other methods, working with shoulder height and/or chest girth the standard deviation is between 4% to 5% and the maximum and minimum figures are between 6% to 8% (Appendix II). In the two older methods standard deviations rank between 6% to 14% of the actual body weight and minimum and maximum figures between 7% to 46%. The method presented here is not only quite accurate but simple. Measurements of chest girth and shoulder height are easily obtained; and instead of shoulder height, circumference of forefeet can be used since double the forefoot circumference  quite accurately equals the shoulder height. An elephant with an estimated weight of 1000 kg may weigh in reality most probably somewhere between 950 kg to 1050 kg and in extreme cases somewhere between 920 kg to 1080 kg. For an elephant with an estimated body weight of 3000 kg the actual weight is most probably between 2850 kg to 3150 kg and in extreme cases between 2760 kg to 3240 kg.

However, it must be stressed that the weight of an elephant can change within a short time depending on feeding, defecation and urination. Furthermore, elephants hardly keep quite when standing on a weighbridge. Hence the measured weight may not be very accurate. The same applies for such simple body measurements as shoulder height or chest girth. Numerous of the elephants measured in Sri Lanka were afraid of measuring tapes and gauges and measurements could only be taken with the help of ropes which were normally carried by the animals. This procedure definitely did not improve the quality of data. Hence, the figures given in the checklists for assessing the body weight of Sri Lankan elephants (Appendix II) must be considered as crude estimation. Nevertheless, they may be helpful in practical work with elephants.

## Body weight and body growth

Elephants were weighed at heavy-weight weigh-bridges. When shoulder height and thorax circumference of an animal with unknown effective body weight were known, body weight was estimated according to the methods given above. A total of 230 measurements of effective body weight from 191 elephants and 156 calculated body weights from 155 animals were included in the study. A comparison between 224 body weights of 201 females from zoos (only effective weights) and 59 females from Myanmar and Thailand showed that differences in body weight increase. Female zoo elephants have a sharp weight increase after their fifth year, which reaches a climax at around 25 years with peaks of around 4 tons. Then the weight decreases until at the age of 40 years it almost attains the weight value of the females from Southeast Asian timber camps (Fig. 2.2.2). Among 4 age classes between 11 to 50 years, covering 10 years, respectively, the differences are significant. Body weight of Sri Lankan females are more or less intermediate and read $1986 \pm 541$ kg (n = 28) for females between 11 to 20 years, $2593 \pm 298$ kg (n = 16) between 21 to 30 years, $2564 \pm 724$ kg (n = 13) between 31 to 40 years, and $2588 \pm 475$ kg (n = 25) in females older than 40 years. Comparisons of body weight between males from Sri Lanka and males from Myanmar and Thailand show that after the age of 30 years the former become heavier due to increased height (Fig. 2.2.1).

Kurt and Kumarasinghe compared the shoulder height and body weight of the 3 different I-types of Sri Lankan elephant bulls. The tusk-bearing 'ethas' (n = 31) seem to grow faster than the tush-bearing 'aliyas' (n = 75). The mean shoulder height of 4 'ethas' between 31 to 40 years is with 275 $\pm$ 5 cm significantly more than the mean shoulder height of 12 'aliyas' of the same age class with 257 $\pm$ 16 cm. The mean body weight of the 4 'ethas' is 3800 $\pm$ 111 kg and that of 12 'aliyas' 3232 $\pm$ 641 kg. In bulls older than 40 years the mean shoulder height and the mean body weight of 8 'ethas' were 254 $\pm$ 13 cm and 3302 $\pm$ 430 kg, respectively. The corresponding values for 6 'aliyas' were 264 $\pm$ 14 cm and 3618 $\pm$ 538 kg. That means that, although 'ethas' grow faster between the ages of 30 to 40 years, thereafter, the 'aliyas' become slightly taller and heavier than the 'ethas'. The tallest and heaviest bulls are the tuskless and tushless 'pussas' (n = 12). The mean shoulder height of two 'pussas' between 31 to 40 years is 285 $\pm$ 3 cm, and that of four 'pussas' older than 40 years is 289 $\pm$ 19 cm. These values are significantly larger than those of 'aliyas' within the same age classes, and significantly larger than the values of 'ethas' over 40 years. Similarly, the mean body weight of 'pussas' between 31 to 40 years of age is 4324 $\pm$ 202 kg, and that of 'pussas' older than 40 years is 4589 $\pm$ 451 kg. A similar comparison of body weight and shoulder height between the 2 I-types of females, the tush bearing 'ethinnas' and the tushless 'alidenas', was not found.

## Obesity and reproduction potential

In general, regression curves of shoulder height versus body weight indicate that older animals are slightly smaller and lighter. In female zoo elephants this reduction is so extreme that it cannot be explained by physiological and anatomical processes in old age alone, but rather by a higher mortality of obese animals due to heart insufficiency and diseases of overtaxed limbs and organs used for locomotion (e.g. Ruthe, 1961; Salzert, 1976). Data presented in this chapter indicate that growth curves of shoulder height and body weight in females tend to show similar values at the age of 40 years and older. This possibly indicates the existence of an ideal maximum weight, and may be due to the fact that only females with a certain body configuration reach old age.

The body weight of an animal is associated with many features of physiology, ecology, and life history (Petters, 1983; Schmidt-Nielson, 1984), and it is also an indicator of overall physical condition. Hence, it is an important parameter used in making decisions concerning the suitability of specific husbandry regimes (Terranova & Coffman, 1997).

As in primates (e.g. Schaaf & Stuart, 1983), there is also a connection between body weight and reproductive performance in Asian elephants. Gestation period and neonate weight are positively correlated with relative body weight (weight / shoulder height) of the mother. The rate of still births is positively correlated with obesity of the mother, and obese females have shorter reproductive periods in their overall life span. Females with extremely low relative body weight are unable to reproduce (c.f. Mar *et al.*, 1997). Biometrical studies on captive female elephants are paramount for assessing animal welfare and reproduction success. Measurements of thorax circumference, shoulder height, and/or circumference of forefeet are easily obtained, and a simple method for estimating body weight from these data is available. But data collected in notoriously obese zoo elephants should be compared to findings from elephants living in their home countries, as overweight is otherwise not identified as such and considered to be natural (e.g. Fischer *et al.*, 1993).

In wild elephants shoulder height is considered as a valid age criterion (McKay, 1973; Kurt, 1974a, 1992), but in many sub-adult and adult captive elephants shoulder height hardly correlates with age, but rather reflects living conditions and individual life history. Female calves that grow up with their mothers grow faster, become taller and heavier than orphaned females (see Section 4.1). South Indian timber elephants born in captivity or caught at an age younger than 10 years have a lesser shoulder height in both sexes, and a lesser body weight in males, compared to wild elephants or elephants caught as sub-adults or adults (Sukumar *et al.*, 1988). A similar comparison can be made between mainly captive born animals from camps in Myanmar and Thailand with mainly wild-captured elephants from Sri Lanka. A reason for reduced growth may be stress from capturing, taming, training and working as well as qualitatively and quantitatively insufficient food. All these factors lead to a more or less pronounced lack of a secondary growth spurt which would normally take place after puberty.

These restraints are absent in zoos, and in many sterile female zoo elephants extreme growth in height and weight is recorded between 10 to 25 years, when first pregnancies take place in wild elephants. In males, growth in shoulder height may be influenced by living conditions, but not growth in body weight, they do not or hardly become obese. Therefore, in order to maximize their fitness, Asian elephants display the same tendencies in body growth as other large polygamous ungulates: under optimal nutritional conditions males invest in body size and females in storing resources (see Geist 1987, for an overview). Body size may reflect the ability of bulls to find, and process food resources, and can thus be an indication of good survival abilities. Female choice favours the tallest bull (in musth), and shoulder height correlates positively with reproductive success in wild as well as in captive Asian elephants (Kurt, 1992, 1995). Males in musth, i.e. those with the highest current breeding potential, raise their head and thorax conspicuously during social encounters, in order to demonstrate their body size.

Neonate elephants are extreme 'follower types' (e.g. Lent, 1974), as is typical for offsprings of highly mobile large ungulates. Young elephants grow rapidly and their mothers must have the ability to utilize a good portion of the energy intake necessary for their own growth and maintenance in favour of prenatal growth and the production of nutrition-rich milk.

The comparison of growth profiles of male I-types indicate that the tusk-bearing 'ethas' grow faster and reach maximum shoulder height and body weight at a younger age than do tuskless males. In cervids two phenotypes in relation to body and antler growth have been distinguished and characterized by genetic markers (Hartl *et al.*, 1991, 1995a). Studies on behaviour and ecology of several ungulates lead to the assumption that fast-growing phenotypes tend to disperse, and slow growing ones tend to spend a more sedentary life (c.f. Kurt, 1991, for overview). The hypothesis that tuskers rather represent the dispersal type and tuskless bulls the sedentary type is backed by findings from ecological and ethological studies on wild elephants: Female-offspring groups concentrate in areas of secondary growth, rich in appropriate food and open water. Males that are not in musth live at the periphery of these core areas, often in primary forests or dry areas of secondary growth, where open water is scarce or absent. The feeding activities of bulls could also be responsible for creating new core areas rich in secondary growth (Kurt, 1974a,b). Due to their tusks, 'ethas' are better equipped than tuskless males to tackle problems of procuring and processing food from woody vegetation and digging for shallow ground water. Different I-types seem to play different social and ecological roles in the population, as they can be characterized by the frequencies of certain spinal configurations and patterns of depigmentation, i.e. features which may function as optical signals.

## Spinal configurations and skin depigmentation

For a closer study of the 5 I-types in Sri Lankan elephants three different forms of spinal configurations were distinguished: (1) 'flat' back – the back line is horizontal without knobs. (2) 'Broken' back – the back line is generally sloping down from shoulders to tail but shows one or more conspicuous knobs. (3) 'Sloping' back – the back line is knobless and follows a slightly rounded sloping line from shoulder to tail base. Spinal configurations were noted from 305 Sri Lankan elephants. 'Flat' backs are significantly more common in females (35.2%) than in males (19.2%), whereas sloping backs are more rare in the former than in the latter (Appendix II). But there is also a significant difference in the frequency of spinal configurations within males. Sloping backs are more often found in 'ethas' than in 'aliyas' which more often have 'flat' backs.

So far sexual dimorphism in Proboscidea has been described mainly in terms of tusks and tusk – like structures, body size or the musth period in males. This study shows that sex dimorphism may be apparent also in body shape. For example, in bulls the spinal configuration tends to be more often sloping down from shoulder to tail and this phenomenon is most common in 'ethas'. Sloping backs underline and exaggerate body size optically. But sex dimorphism is also underlined by depigmentation, as discussed here.

The extent of skin depigmentation on trunk, temples, ears, and shoulders of 125 Sri Lankan elephants from the age of 15 or more years was noted using the following classification: 0: absent, 1: depigmentation occurs on 25% of the area concerned, 2: 26–50%, 3: 51–75%, 4: more than 75%. Depigmentation stemming from chaining or other wounds inflicted by man was not considered. Eyes and the opening of the temporal gland may be accentuated by more or less complete circles of depigmented skin. The extent of this accentuation was described as the extent of depigmentation of the trunk, temples, ears and shoulders. In all 5 I-types the extent of depigmentation of trunk, ears, temples and shoulders increases with age; and 'ethinnas' are significantly more depigmented than 'alidenas' (Appendix II). In males, eyes are significantly more accentuated by depigmentation than in females. In 'ethas' this characteristic is significantly less pronounced than in 'aliyas' and 'pussas'. However, no significant difference was found between males and females concerning the degree of depigmentation of the opening of the temporal gland. But in 'ethas' the accentuation is significantly less prominent than in 'aliyas' and 'pussas' (Appendix II).

In Asian elephants depigmented skin patches seem to play a similar role as do the conspicuous hair fields and tufts which accentuate the head and especially the eye and skin glands in ungulates (Portmann, 1948; Geist, 1971), where dark colouring of the face is linked to aggressive behaviour, and white marks seem to be associated with peaceful gregariousness (Hamilton, 1973; Kurt, 1978). In Asian elephants naturally depigmented skin patches are concentrated in the area of the head. In young animals, depigmentation starts on the trunk tip and base, on the upper distal tip of the ear and the ear lobe. In old animals depigmentation increases and predominantly accentuates the eyes, openings of the temporal gland and ears (Fig. 2.2.4 and Fig. 2.2.5). These parts are also often touched by the trunk tip of the partner in the course of social contacts (Eisenberg *et al.*, 1971; Garaï, 1992).

Although depigmentation can hardly be recognized in feeding or walking elephants, as their bodies are covered with substrate, the bright marks of depigmentation immediately appear after a shower and a bath at water holes, where most social contacts take place (Kurt, 1992). The overall extent of depigmentation is similar in males and females, and in both sexes the opening of the temporal gland is accentuated. In females the temporal gland secretes rarely, and if so only in context with the end of a mating period or during parturition. In males it is active during the musth period, when temporal glands are swollen and the dark fluid contrasts with the bright depigmentation (c.f. Kurt, 1992). In females, the eyes are

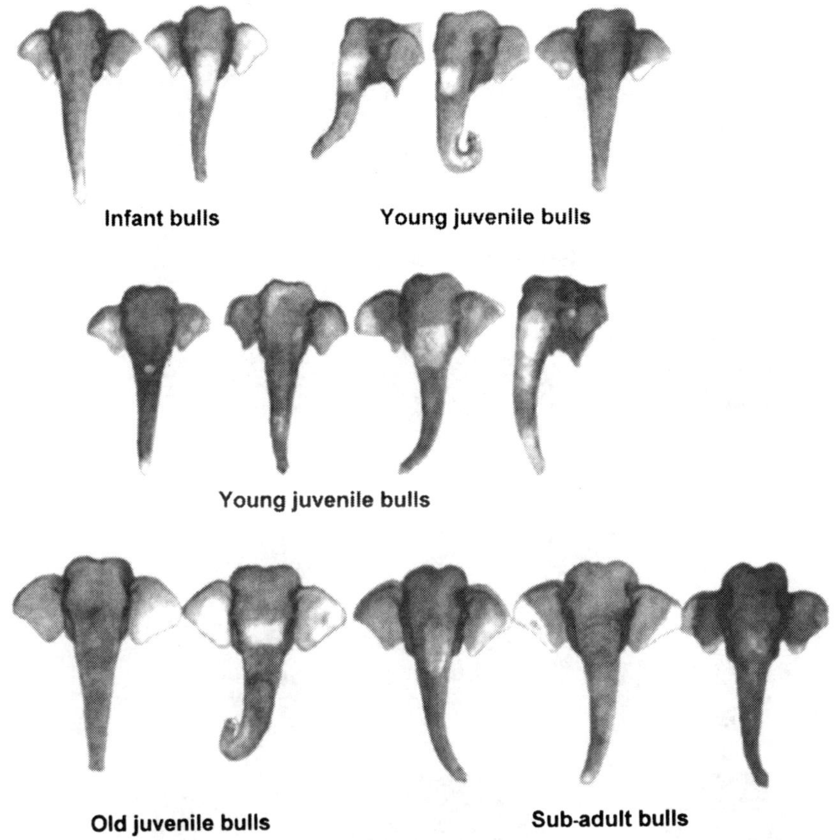



**Fig. 2.2.4**  Depigmentation and ear forms of 14 wild infant, juvenile and sub-adult bulls in the Uda Walawe National Park of Sri Lanka

significantly less accentuated than in males and the tusk-bearing 'ethas' show significantly less depigmented eye fields than do tuskless bulls. The difference could be explained by different strategies in aggressive encounters. Although tusks are unimportant in direct combats which are carried out by pushing heads, pressing together the inner trunk bases, scratching with tushes and biting, tuskers can easily threaten tuskless bulls by maneuvering the tip of one tusk under the eye of the opponent (Kurt, 1992), where the conspicuous white skin patch might signal peaceful submission. Lengths and forms of tusks as well as patterns of depigmentation vary immensely among individuals and may serve as distinctive optical signal for each elephant.

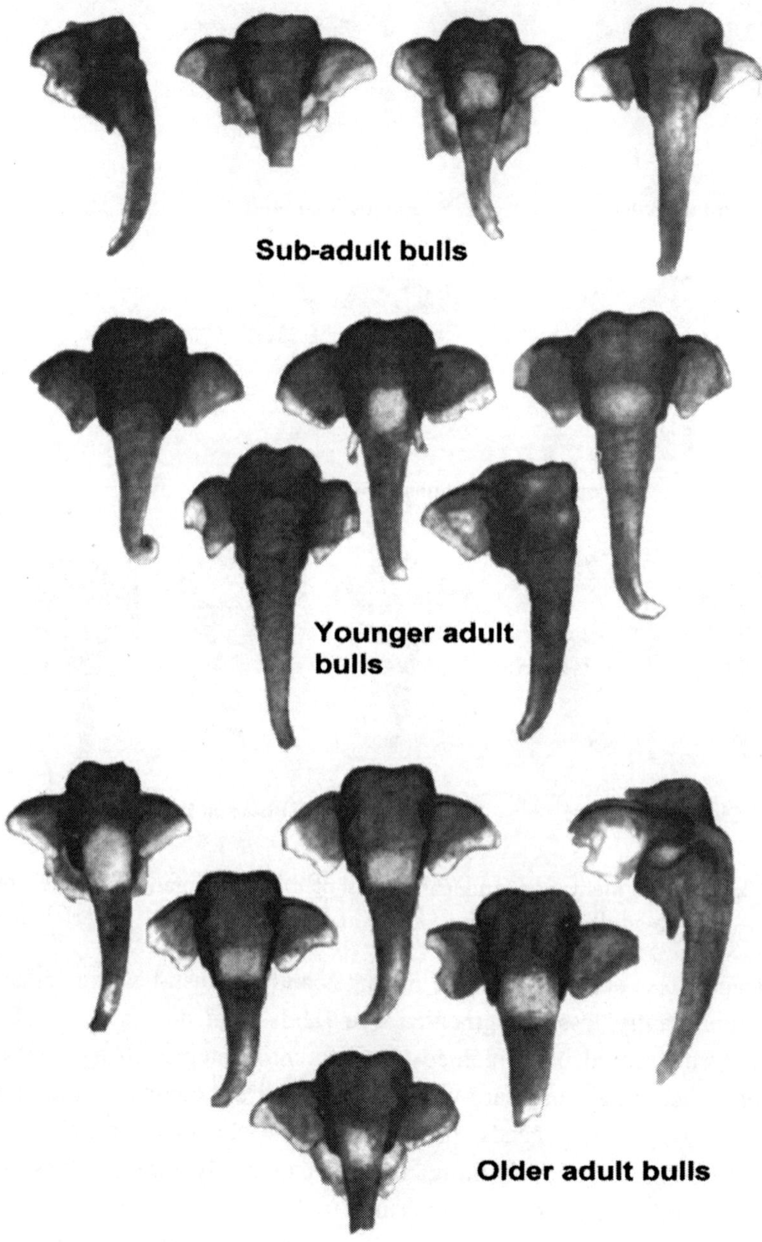


**Fig. 2.2.5** Depigmentation and ear forms of 15 wild sub-adult and adult bulls in the Uda Walawe National Park of Sri Lanka

But not all significant links between certain physical features and certain I-types could be explained. Remaining questions are for example: Why are 'alidenas' less depigmented than 'ethinnas'? One could speculate and mention that females with tushes have a weapon and need to signalize appeasement, therefore 'ethinnas' are more depigmented. Why is the opening of the temporal glands less accentuated by depigmentation in 'ethas', less pronounced than in tuskless males, or, as shown below, why are honey-coloured eyes in 'ethas' more abundant than in other elephants?

## Eyes, tails and toes

Four eye colours were distinguished: (1) grey, (2) dark brown, (3) light brown, (4) honey. In 238 elephants from Sri Lanka (116 males, 122 females), 19 (8.0%) had grey eyes, 30 (12.6%) dark brown, 142 (59.7%) light brown and 47 (19.7%) honey-coloured eyes. A significant difference was found between 'ethas' and tuskless males. In 18 (81.2%) of 22 'ethas' eyes were honey-coloured, but this characteristic was found in only 50 (53.2%) of all tuskless bulls studied.

Five forms of tail tufts were considered in 160 elephants from Sri Lanka: (1) The long bristles are organized in egg shape and a line of bristles grows on the ventral part of the tail above the tuft; (2) as (1), but the tuft is open at the tip; (3) and (4) as (1) and (2), but hair is often shorter and there is no hair line on the ventral part of the tail above the tuft, (5) tuft incomplete, consisting of a few hair only on one side of the tip of the tail. In wild-living elephants the form and extent of the tail tuft may be considered a rather weak age criterion. But in captive ones the forms correlate neither with age nor with sex or I-types. They seem to depend on food and mainly skin care by man.

In 177 Sri Lankan elephants, tail lengths were measured from the highest point of the anal folds to the tip. Tails that had been bitten off to some extent were not considered. Kinked tails were also not considered as a special form of tail, as this characteristic is a product of biting by conspecifics or an accident during or shortly after birth. Tail length of 73 males measured between 68 to 160 cm, with a mean of 117.7 $\pm$ 21.5 cm. In 104 females the mean reads 109.1 $\pm$ 20.0 cm, with extremes of 53 and 160 cm respectively. Tail length correlates positively with shoulder height. The mean value of relative tail length (tail length / shoulder height) is the same in males (51.8 ± 5.0%; min.: 40%; max.: 65%) as in females (51.5 ± 6.1%;  min.: 32%; max.: 68%). The tail length as well as the form of the ear lobe (see below) vary considerably and seem to be hereditary, as members of the same mother families in wild elephants often possess tails with similar relative tail length and ear lobes of the same shape, a fact already known since the nineteenth century (cf Tennent, 1861).

Out of a sample of 1337 elephants from Myanmar, only 11 (0.8%) had 20 toe nails or 5 nails on each foot; 1119 (83.7%) had 18 nails, 5 on each forefoot, and 4 on each of the hind feet; 16 (2.1%) had 17 nails, 5 on one forefoot and 4 on the remaining 3 feet (Evans, 1910; Toke Gale, 1974). In a sample of 313 Sri Lankan elephants, 1 (0.3%) had 20 nails, 299 (95.5%) 18.1 (0.3%) 17 and 12 (3.8%) 16 nails (Deraniyagala, 1955; Godagama, 1996). The occurrence of aberrations in the number of nails is significantly higher in Myanmar (16.4%) than in Sri Lanka (4.5%), and seems to be a hereditary characteristic of a distinct population.

## Ears and hair

Three forms of ear lobes were distinguished: (1) round; (2) vertically elongated; (3) more or less horizontally elongated. Ear forms were recorded from 170 elephants in Sri Lanka. The form of the ear lobe correlates neither with sex and age nor with body weight. In 74 males and 94 females from Sri Lanka, 57% of the ear tips were round, 24% were vertically elongated, and 19% more or less horizontally elongated.

The upper edge of the ear was classified into four forms (Fig. 2.2.6): (1) No fold along the top margin of the ear pinna; (2) fold is not wider than a finger; (3) fold is as wide as 1 or 2 fingers and forms a small distal triangle, which hangs down when ears are not moved; (4) fold is as wide as 3 or more fingers and forms a conspicuously large distal triangle. In type (3) and (4) the distal third of the pinna can be folded inwards or outwards. The form of the upper ear edge changes with age and weight. Foldless edges (form 1) are mainly found in young elephants, forms 3 and 4 correspond to old age and heavy weight. Among different forms of ear edge, paired comparisons revealed a stronger statistical correlation with weight than with age (Appendix II).

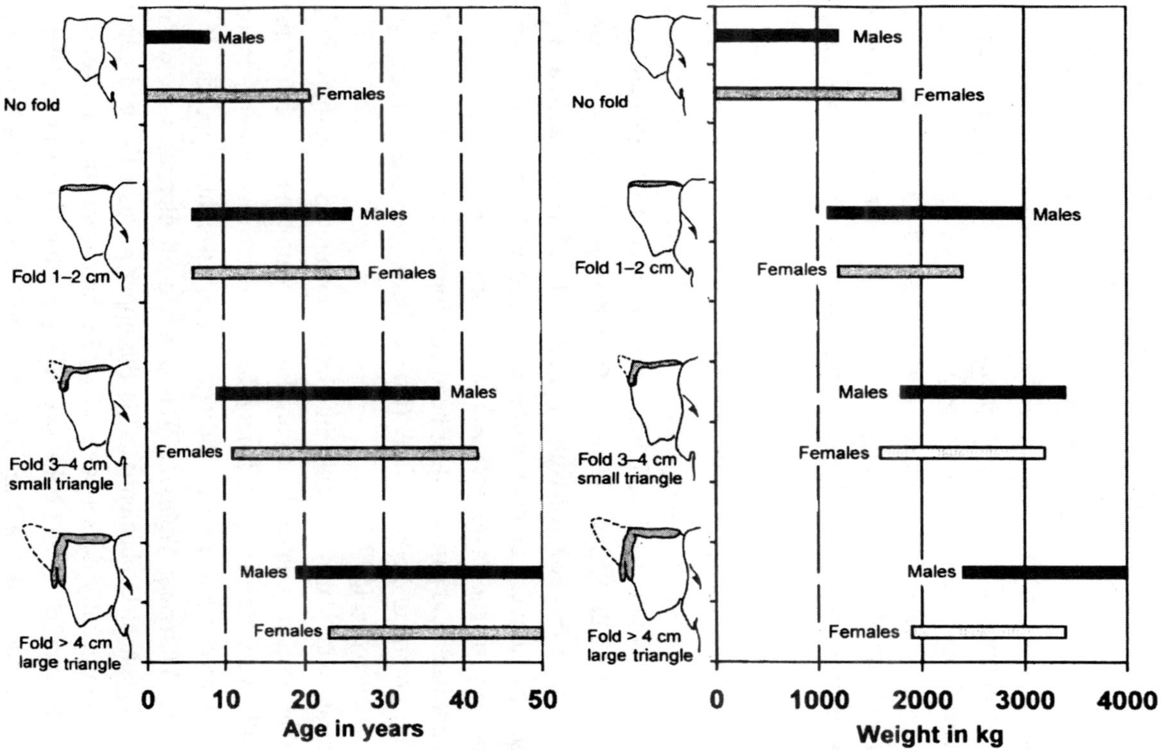

**Fig. 2.2.6** Development of ear fold according to age and body weight in 170 captive elephants (77 bulls, 93 females) (Source: Kurt, 2001)

Animals with extreme body weight living in tropical climates are faced with problems of thermoregulation, unlike African Savannah elephants which have enlarged ears containing a dense network of subcutaneous blood vessels (e.g. Shoshani, 1991). In Asian elephants, relationship between the temperature of the environment and the frequency of ear flapping was reported (McKay, 1973). The present findings show that the occurrence of more or less enlarged folds on the upper edge of the ear correlates better with body weight than with age. This indicates that the distal triangle of thin, extremely depigmented skin, which is well-supplied with blood, may have a thermoregulatory function. It is exposed and fully enlarged during ear-flapping and hangs folded down when ears are not moved.

But Asian elephants use many more methods to regulate body temperature. They spend the hottest hours of the day in shady places. They cool their bodies by taking shower and bath, and they cover it more or less completely with sand, soil, mud or plants. Out of 80 wild elephants observed in the Uda Walawe National Park between 6 a.m. to 6. p.m., 66 were more or less covered with sand and soil, 35 with plants, and 12 with mud. The amount of body surface covered with substrate as well as the density of the layer increase with body size. Neonates cover mainly their back and partially their head and forehead. But in adult bulls almost the complete body surface (frontal parts, shoulders, back and pelvis) is covered (Fig. 2.2.7).

Substrate and plants are held in place by hair, possibly in such a manner that insulating air-cushions are formed. With the exception of the occasionally very dense fur of lanugo-hair in newborns, no correlation was found between hair growth and age or body weight (Appendix II.). Furthermore, Wolfgang Weihs (2001) found in 30 elephants from the Pinnawela Elephant Orphanage that on the upper parts of the body, hair growth is significantly denser with $4.0 \pm 3.6$ hair / cm² than on ventral parts such as chin, chest and belly with $2.3 \pm 1.7$ hair / cm² or on lateral parts, such as the region of eyes, base of trunk, cheeks, upper forelegs, flanks and thighs with $1.7 \pm 3$ hair / cm² (Fig. 2.2.7). But the relatively rare hair have further functions serving as grid to hold wet substrate and plants, which cool the body by evaporation process and protect the skin

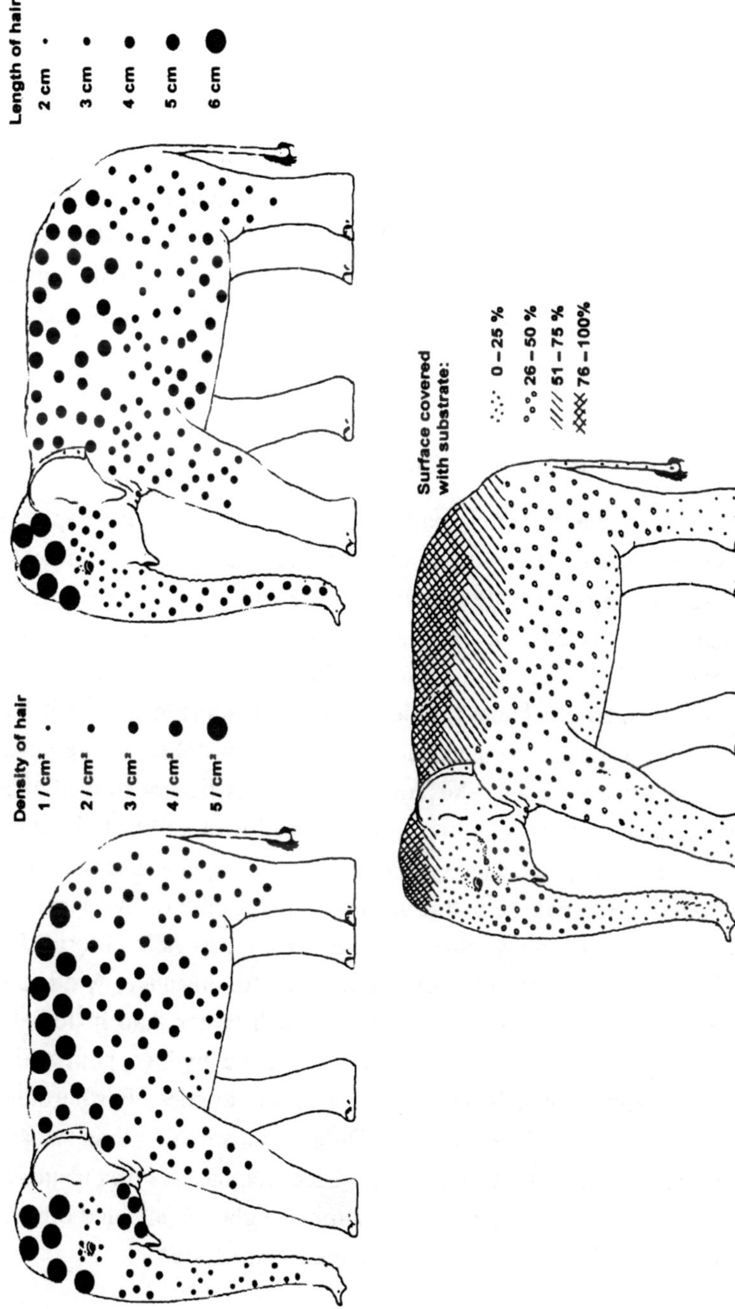


Fig. 2.2.7 Mean density and mean length of hair in 30 captive elephants and coverage of body surface by substrate in wild elephants

from sun and insects. The tuft of very strong hair at the tip of the tail is used to brush the anus, thighs and hind parts of the back with very masterfully applied tail movements. A dense cushion of thin, short hair protects the opening of the ears.

Finally, hair are also tactile sense organs of elephants (e.g. the hair on the trunk tip). This function depends, amongst other anatomical particularities, also on the length of hair. Short hair measure between 1 to 3 cm, long ones between 8 to 10 cm. The mean hair length on front, shoulder, back and pelvis measures $4.9 \pm 2.2$ cm (n = 120) and on chin, breast and belly $4.6 \pm 2.2$ cm (n = 90). These regions of the body cannot be visually easily controlled, as compared to proximate regions such as eyes, trunk base, trunk, cheeks, upper forelimbs, flanks or thighs where mean hair length measures $2.7 \pm 1.4$ cm (n = 210), i.e. significantly less.

## 2.3  Age Criteria and Social Classes

In many captive elephants age is quite known as they come into captivity at a young age, when age estimate is quite accurate. But in wild-living elephants accurate age estimate is only possible with accurate age criteria. Field researchers may also face problems in sexing the elephants, mainly younger ones. This applies particularly to Sri Lanka, where most of the bulls are tuskless. In elephants the descensus testiculorum is absent, i.e. bulls have no scrotum. Younger bulls are easily mistaken for females. Moreover, the female organs like vulva and clitoris lie not immediately below the anus, as in other mammals, but on the lower belly, at about the same place where the penis can be seen in bulls. Inexperienced observers often confound the female sexual organs with the sheath of the penis and do not know that especially older females can erect their clitoris. For trained observers easily visible sex criteria are for instance, width and size of the scull. In bulls the head appears much larger than in females of the same age. In adult bulls the head is topped by two helmet-like structures, or domes, which are rarely seen in females. Furthermore, bulls extend their penis while urinating. Older bulls are taller than females and look more athletic and well-nourished. Older females are best identified by their well-visible and well-rounded mammary glands (situated directly behind the forelegs) and their relatively lean constitution.

Elephants grow during the first three decades of their life, but they do not only increase in shoulder height, they also gradually change body proportions and physiognomy (Fig. 2.3.1). Shoulder height equals double the circumference of the forefoot, and can be measured in footprints, or can be estimated when a certain animal stays close to a tree with known height. Therefore, it is obvious that shoulder height is considered as an age criterion.

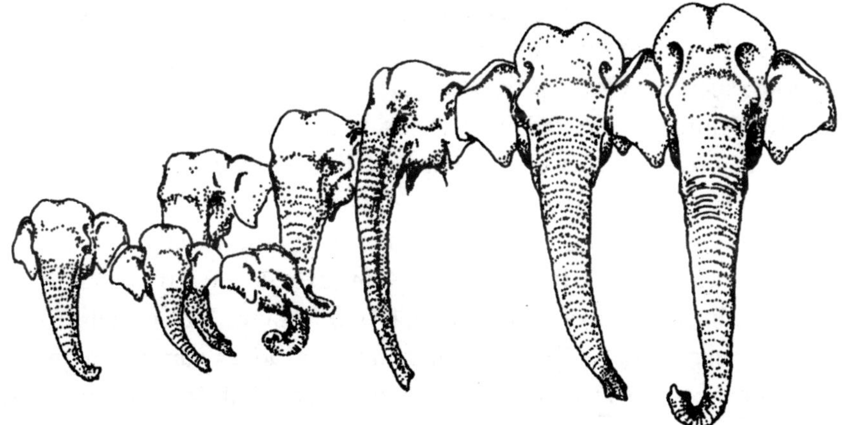

Fig. 2.3.1   Changes of physiognomy in Asian elephant
(Illustration: Wolfgang Weihs)

The method to gauge the age of elephants by their size is a useful rule of thumb. It must be followed up by other methods. As already mentioned in the foregoing chapter, the amount of depigmentation increases with age, and the upper folded edges of the ears become wider. Furthermore, the older an elephant grows the more it carries skin injuries mainly on the edges of the ears and tails. Occasionally tails are broken and tips are bitten off. As shown in the following, chewing frequency decreases with age from about 44 beats per min in neonates to about 27 beats per min in old adults, and, hence, can be considered as a further tool to estimate age (Weihs, 2002).

## Molar growths and chewing frequency

The total grinding surface of all simultaneously operative premolars and molars in all 4 jaw halves was measured in 37 skulls of Asian elephants from captive elephants in the National Museum in Colombo (Sri Lanka), the Museum of Natural History in Vienna (Austria) and from wild elephants in the Ruhuna National Park of Sri Lanka. Data on chewing frequencies were collected in 1230 samples from 48 captive elephants (27 males, 21 females) in the Pinnawela Elephant Orphanage as well as in Kandy (Sri Lanka). Data from 30 wild elephants (15 males, 15 females) of all sizes and age classes stem from the population of the Uda Walawe National Park. The frequency was given as number of chewing movements per minute. The wild elephants were eating more or less dried perennial grasses such as *Cymbopogon nardus* and *Imperata cylindrica*. The captive animals were fed with stems and leaves off the Kitul palm (*Caryota urens*), leaves of Cocos palm (*Cocos nucifera*), and branches of the Jack fruit tree (*Artocarpus inegra*).

The total grinding surface in neonates (0–2 years) was 27 to 29 cm², in infants (3–5 years) 153 to 172 cm², in juveniles (6–10 years) 247 to 435 cm², in sub-adults (11–15 years) 470 to 532 cm² and in adults (older than 15 years) 442 to 730 cm². The number of ridges varied between 16 and 20 in neonates and 60 to 84 in adults. Although the number of ridges, the sizes of cheek teeth and the grinding surfaces increase considerably with age and body size, dentition was considered as a very vague age criterion in elephants older than 6 years. In living elephants inspection of molar teeth was not possible, even in very tame captive animals.

Chewing frequency correlated negatively with body weight, grinding surface, age and size (Fig. 2.3.2). In the first two size classes (136–165 cm; 166–195 cm) no differences between males and females were found. In the 3rd (196–225 cm) and 4th (226–255 cm) size classes bulls had a significantly higher chewing frequency than females. In the 5th size class ($\geq$ 256 cm) the frequencies of males and females were not significantly different. The frequency of bulls of the 5th class was significantly lower than the frequency of bulls of the 4th class. In females, however, the results were reversed: females of the 4th size class showed lower frequencies than those of class 5. These seemingly controversial findings can be explained as follows: Males grow faster than females and with 190 cm and more shoulder height they are younger than females of comparative size. In the case that molar growth is related to age, molar growth is more advanced in females than in males of similar size, the

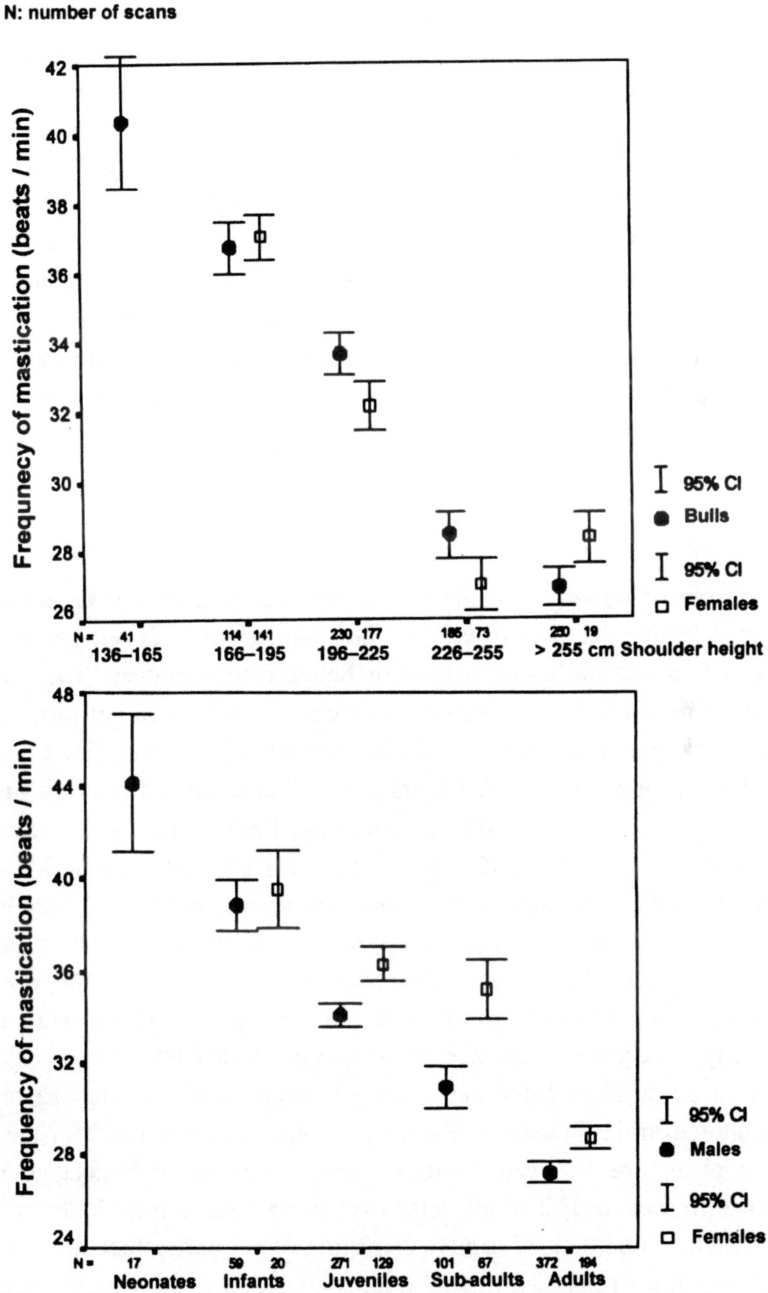

Fig. 2.3.2    Frequency of mastication according to size classes (above) and social classes (below)

grinding surface comparably larger and the chewing frequency lower. Only few females grow taller than 255 cm. It can be expected that they are of old age, and gradually lose parts of the last molar. Therefore, the grinding surface becomes smaller and the chewing frequency higher. In a comparison of chewing frequencies of juvenile, sub-adult and adult elephants (Fig. 2.3.2), females had significantly higher chewing frequencies than males. On the assumption that molar growth is related to age, animals of same age have similar dentition, but the grinding surface is smaller in females than in males because of a pronounced sexual dimorphism in skull size; hence, the chewing frequency is higher in females than in males. Data presented here advocate for the hypothesis that chewing frequency could be a valid age criterion for elephants with known sex.

## Puberty and birth interval

The time of puberty and the period between 2 consecutive births are highly influential parameters for the demographic development of a population and the social behaviour between its members. The time of puberty depends amongst other environmental factors on nutrition. Some females kept in western zoos and circuses gave birth to their first offspring at the age of 8 or 9 years, i.e. they must have reached puberty already at the age of 6 or 7 years (Haufellner *et al.*, 1993, 1996, 1999). However, in zoos and circuses hardly any mother has been older than 30 years. Female timber elephants in Myanmar can become mothers already at the age of 8 years and continue to reproduce until the age of 50 or, in few cases, 60 years (Mar *et al.*, 1996). Data from more than 1,000 working elephants in South India revealed that the youngest mothers were 18, i.e. puberty took place at the age of 16. Such a rather late puberty may be due to stress from capturing, taming, training and working. Elephants monitored in the Pinnawela Elephant Orphanage parturiated for the first time at an age between 11 to 40 years, females with delayed body growth reproduced, if at all, later than those with normal body growth.

Living conditions of captive elephants do not only affect body growth but also time of puberty. In well-nourished and physically hardly stressed females reproductive status is reached earlier than in undernourished and physically overstressed females. Therefore, captivity marks the extremes of the reproduction physiology of the species. Within a population of 88 newly captured elephants at least one female with an estimated age of 12 years had a baby and well developed mammary glands (Kurt, 1974, 1992; Sukumar, 1992). In the Ruhuna National Park females seem to reproduce for the first time at the age of 10 to 15 years (Kurt, 1992). Similar results were obtained during our studies in the Uda Walawe National Park. In Mudumalai (South India) most wild females start reproduction apparently later and give birth for the first time at the age of 17 or 18 years.

Pregnancy lasts about 20 to 22 months, but some extremes of 17 to 25 months have also been reported. Most accurate data stem from zoo elephants. These elephants are often obese and, therefore, their offsprings are heavier and pregnancy is longer than in Asian working elephants and most probably also in wild conspecifics. The interval between two consecutive births is determined by the gestation period and lactation and lasts between 4.4 to 4.8 years.

Males reach puberty at the same age as females. Male zoo elephants reproduce with young females already at the age of 8 years (Haufellner *et al.*, 1993, 1996, 1999). Wild bulls try to mate at a young age, but are prevented to do so by older males. They most probably reproduce successfully only at the age of 20 years and later, i.e. when they can establish their own musth periods. First signs of musth appear as early between the age of 15–20 years.

## Social classes

Five social classes with different social roles and functions in reproduction have been distinguished: neonates, infants, juveniles, sub-adults and adults. Other studies on Asian elephants defined fewer social classes and in some cases terms like juvenile or sub-adult were used in a contradictory way. In the following pages the 5 social classes are defined. More details on age criteria have been given above.

Neonates are newborns upto the second year of life. During their first week of life they can be identified by a red ring around the eyes, and often their bodies are covered by a dense fur of fine and long lanugo hair. Depigmentation in absent. Chewing frequency measures 42 to 47 chewing movements / min. Body size in either sex is similar. Neonates are nursed regularly.

Infants are offsprings at an age between 2.5 to 5 years. The red eye-ring and the lanugo fur are missing. Differences in body size between males and females become slightly visible. Their skin is hardly marked by depigmentation. Chewing frequency measures 38 to 40 beats (one full jaw movement) per min in females and 38 to 42 beats in males. Infants are nursed very rarely and for very short periods. Normally their mothers are pregnant.

Juveniles are elephants between 6 to 10 years. The sexual dimorphism in body size becomes easily visible. Well visible depigmentation are found on the edges of the ears and on the base and the tip of the trunk. Chewing frequency in males and females measure about 36 movements / min. Normally mothers of juveniles nurse a smaller offspring.

Sub-adults are between 11 to 15 years. They reach puberty and/or give birth for the first time, and in few cases, for the second time, but they are not yet fully-grown. Sub-adult females show a chewing frequency of about 30 beats / min and sub-adult males of about 34 beats / min. Depigmentation is present on several parts of the body and some injuries can be found on the ear edges.

Adults are older than 15 years and reach the maximum body size. In adult males chewing frequency is about 28 beats / min, and in adult females the value reads about 27 beats / min. Whenever possible, young and old adults are distinguished, i.e. elephants at an age between 16 to 20, and those older than 20, respectively. In adult bulls the two bulges or domes on the top of the head become more and more prominent with age. In either sex depigmentation and size of the triangle on the upper ear edge reach the maximum size. In very old adults cavities on the temples and cheeks become prominent and chewing frequency becomes faster.

Grouping members of a population into five social classes is a tool to characterize the structure of the population or to follow the genesis of certain behaviours such as the use of tools or the methods of food preparation. However, it must be considered that elephants grow during a long period of their lives, that they reach puberty relatively late, that they live in a highly structured social organization, and that offsprings of high-ranking females grow faster and reach puberty earlier than offsprings of low-ranking mothers as shown later. Accordingly it can be expected that criteria of physiological age is more valid for biological studies than criteria of effective age, as they are found in zoo populations or in the Pinnawela Elephant Orphanage. Elephants are not only long lived, they are known for their high intelligence and extremely good memory. Such characteristics, as well as the fact that they live in many different environments, lead to different life histories and personalities. Therefore, the following will deal, whenever possible, not only with social classes but also with individuals.

# 3

# The Captive Populations, Keeping Systems and Health

## 3.1 The Pinnawela Elephant Orphanage and its Elephants

The Pinnawela Elephant Orphanage is situated in the Kegalle District, some 80 km northeast of Colombo, north of the city of Kegalle, or about half-way between Colombo, the present capital of Sri Lanka, and Kandy, the last of several historical capitals of the country. The orphanage was established in 1975 with the aim to give young orphaned elephants, which are found on the island, a safe home. In 1978 the orphanage was taken over by the National Zoological Gardens from the Department of Wildlife Conservation. The orphanage encompasses an area of about 9 ha of mainly coconut plantations. With a population of 70 captive elephants in 2002 the Pinnawela herd is one of the largest permanent concentrations of captive Asian elephants in the world next to the camp of Maesa in Thailand's Chang Mai Province with 85 elephants in 2000 (Tipprasert, 2002) and the Temple of Guruvajoor in Kerala with about 65 elephants in 2002.

### Population and population fluctuation

In 1975 the Pinnawela population consisted of 10 elephants, in 1986 there were 32 elephants, 53 in 1993, 56 in 1997, 64 elephants in 1998 (Kurt, 2001a), and 70 in 2002 (Rajapaksa *et al.*, 2002). Between 1983 and 2003, 22 elephants were born at Pinnawela by 8 orphaned and 3 captive born females. One of the males was born in the Dehiwala Zoo in 1991. In

1998, when we completed most of our studies the Pinnawela orphanage harboured 61 elephants (26 males, 35 females). 13 of them were captive born. The others came as orphans to Pinnawela, 16 as new born babies, 7 at an estimated age of one year, 7 at the age of 2 years, 3 at the age of 3 years, 3 at the age of 4, 3 at the age of 5 and 5 as juveniles, i.e. between the age of 6 to 10 years. Anusha (estimated date of birth: 1949) came from the Dehiwala Zoo, where she had been measured in 1967 during the Smithsonian Elephant Survey (Kurt & Nettasinghe, 1968). The Tusker Raja came to Pinnawela on the 29th of March, 1994 at the age of 40 years. He was captured after poachers blinded both his eyes with gun shots. On the 14th of May, 1994 the adult bull Sanka II was brought to Pinnawela. As a former working elephant he had escaped, freed himself, lived in the wild, raided fields and crops and killed 12 people before he could be recaptured. It seems that Sanka II has died in the meantime. In 1994, the female Sama II, who lost her right forefoot in the wild area due to stepping in a land mine, had been subsequently treated in the Dehiwala Zoo and came to Pinnawela at the age of about 6 years (Alahakoon & Santipillai, 1997). In 1997 she was integrated into the herd. Two attempts in 1997 and 2001 to fit her with an artificial limb failed.

During the study, the Pinnawela herd had the following population structure: 31.1% neonates and infants (18.0% males; 13.1% females), 37.7% juveniles (13.1% males; 24.6% females), 11.5% sub-adult females and 19.7% adults (11.5% males; 8.2% females). No data is available on the mortality rate of the Pinnawela herd. According to the finding the infant male Anura died in 1998, between July 1998 to July 1999 the orphaned neonate male Mihindu and 3 orphaned neonate females (Sapumali, Surangani and Meena) died. More detailed data on mortality of animals brought from the wild into captivity are available from the Elephant Transit Home (ETH, Ath Athuru Sevana) which was established in 1995 at the edge of the Uda Walawe National Park. Between October 1995 to December 2001, 83 young elephants were brought to the ETH, and 39 (47.0%) died subsequently due to serious illness and disabilities (Daily news, Colombo, 25th January, 2002). Some of the animals from the ETH were gifts to temples, other public institutions and to the Pinnawela Elephant Orphanage.

By 1995, 11 elephants from Pinnawela had been given to temples or private owners, amongst them was Ceyla Himali that now lives in a zoo in Zurich, where she has given birth to 5 offsprings. Later, at least 2 young elephants went to Japan. In 2002 the Government gave away 6 elephants from the Orphanage (The Island, 2nd August 2003): The orphaned sub-adult males Tharaka, Choola, and Singharaja, were donated to the Kaduwela Devalaya, Palmadulla (Sabaragumawa) temple, and Polwatte temple in the Gampala district respectively. In the Polwatte temple an orphaned female from Pinnawela died in 2002 at the age of 10 years. The juvenile orphan Kumari II was sent to Croatia in 2002 but entrance was refused due to an outbreak of foot and mouth disease in Sri Lanka. After her return to Sri Lanka she was given to the Kande Viharaya. The Pinnawela born male Isuru was given as a juvenile to the Mahiyangana Devalaya, and the Pinnawela born Saliya as a four year old infant to the Kataragama Kiri Vehera. The captive born male Arjuna was given in October 2002 at the age of five as mascot to the Sri Lanka Army's Light Infantry Regiment to replace under the name Kandula VI his predecessor, Kandula V, who also came from Pinnawela (Sri Lanka Army, News Report, 30th August 2003).

There may be several reasons why the present Sri Lankan Government is willing to reduce the Pinnawela herd even further: In 2003 the elephant owners demonstrated at the residency of the prime minister and asked to be able to increase the fast dwindling numbers of privately owned elephants. Already in 2001, Jayantha Jayewardene, head of the Biodiversity and Elephant Conservation Trust, suggested at a FAO meeting on captive elephants in Bangkok: "A policy decision should be made by the government to either sell the Pinnawela elephants or allow capture from the forest. This former measure, however, should only be carried out if prospective owners meet certain criteria ...". "If the government adopts a policy of selling some of the large number of elephants at the Pinnawela Elephant Orphanage to selected persons, it will ensure that the elephants would be better looked after than they are now. They will be given better individual attention by the new owners and mahouts. This policy of selective disposal will ensure better care of the elephants that are left". This proposal may be well intended. Unfortunately 'well intended' is often the opposite of 'good'. Firstly, the living conditions of elephants in the Pinnawela Orphanage can be improved fundamentally, as we shall discuss in Chapter 7. Furthermore, the Orphanage earns considerable amount alone by visitors (see below), but the intensive keeping of elephants in Sri Lanka by temples and private owners can only be improved by better healthcare, which is doubtlessly an important step for animal welfare. Nevertheless, these intensively kept elephants still have to live under completely unnatural conditions (see Section 3.2, 3.3 and Chapter 5).

## Keeping system, daily routine, food and milk

Seen from the elephant's perspective the Pinnawela Elephant Orphanage can be divided into three sections, the lawn, the bath and the night quarters (Fig.3.1.1). Since 2001, new plans exist to modify the establishment. In 2002 the work was started. But later one shall concentrate on the living conditions of the elephants in 1997 and 1998, when most of the studies were carried out. During this time most elephants were kept in 4 buildings during the night. 2 of them were round, roofed constructions with a rugged floor made of stones. Here, mainly younger elephants were housed. Each of them had, when chained a portion of about 11m² at its disposal. Each of these rondavels could house ten elephants. But 2 stands were not used. However, this did not mean there was more room for the other elephants living in this construction. A third building was an older tin-roofed longer construction with a rugged stone floor, open to all sides, where the animals had to be fettered in a line. The average space left to a chained elephant measured 18.5 m². This stall was planned for ten animals, but during our study only three were kept here. The fourth building, similar to the third one was only open to one side. The other three sides had walls, some of them with a height of two meters, which were to work as wind-brakes. On either side of this building there were 2 'boxes' of about 20m², separated by walls from the other elephants, for the 2 females with the youngest offsprings. The space between these 'boxes' was planned to house 6 elephants. However, it was occupied by 11 animals, some of them were very young offsprings. These buildings were cleaned every morning.

During the study 20 elephants had their night quarters in the open and were chained to one or between two Coconut palms, where the

## Plan of Pinnawela Elephant Orphanage night quarters in 1997

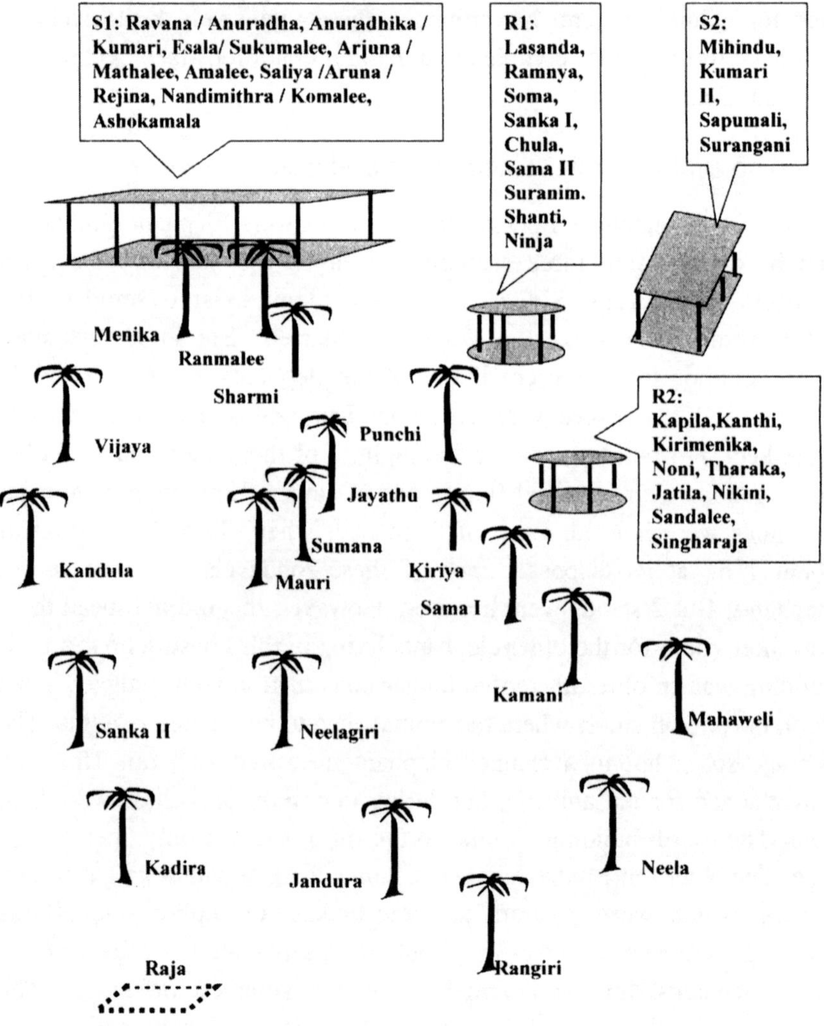


Fig. 3.1.1  Sketch of the night quarters of the Pinnawela Elephant Orphanage. S1, S2: roofed building with stone floors: R1, R2, round buildings with concrete floor. Most elephants were chained on coconut palms.

space for free movement measured between 20 to 48 m². Here, they had to stand on a mixture of dung, urine, mud and food remains. Dung was piled up in tremendous amount behind the elephants since these stands were hardly cleaned. These heaps measured about 3.5 × 2 m² with a height of 0.5 m. Most of the 20 elephants kept in the open had to stand with their hind extremities on a large stone slab or on thick wooden boards, which soon became soaked with urine. These platforms had a mean surface area of 0.8 × 1.2 m². At about 6 p.m. the elephants were brought to their night quarters and chained until about 8 a.m., when they were released, i.e. for 14 hours they had to stand chained (except for neonates and some infants) on often filthy floors.

After release in the morning the herd was driven onto a grass area (the 'lawn') of about 12 ha with a few coconut palms and a smooth depression where rain water would collect. Here, the elephants were allowed to have social contacts, to play and to wallow, especially during the rainy season when water had collected in the depression. However, they were constantly driven together in a bunch since there was no strong fence to keep elephants inside the lawn apart from a weak barbed wire. At 10 a.m. the elephants were herded over the main road of Pinnawela and down a path lined by tourist stalls to the near Maha Oya, the Big River, a shallow water body. Here, the elephants were allowed to drink, splash water and lie down if they wished, however, only a few individuals chose to lie down as the river bed is mostly covered by large boulders, making it rather uncomfortable. Most elephants just stood idly. This so-called elephant bath is foremost an entertainment for tourists, that jostle to get a front view along the main viewing area. At noon the herd was driven back to the grass area in the orphanage, and had to stand in the sun until 2 p.m., when they were again forced to stand in the Maha Oya for 2 hours, bunched in a group and only exceptionally were allowed to walk to the opposite river bank where they could find suitable opportunities for sand bathing. Occasionally one or the other elephant would be taken to the shore to be photographed by the tourists.

In elephants, skin care includes not only bathing or at least showering but also mud-wallowing, sand bathing and rubbing the skin against suitable trees or rocks. In traditionally kept Asian elephants such behaviours are normally not allowed. However, their mahouts lead them once or twice daily to a clean water body, where the animals are made to lie down on their sides and are scrubbed vigorously with special scrapers made out of coconut husks to scrub the body thoroughly to remove dirt, parasites and fungus, as mainly the areas around the nails, base of tushes and tusks, behind the ears and base of the tails are likely to harbour ectoparasites and egg packets of oestrid flies. After cleaning, the animals are often massaged with flat pebbles to increase the cutaneous blood circulation (Krishnamurthy, 1992). At the Pinnawela Elephant Orphanage only the adult and a few sub-adult bulls that were trained and used for transporting food received such skin care. The other elephants were hardly trained and there were no respective opportunities for skin care behaviour available to them.

The neglect of skin care is a small but important argument to conclude that the daily routine of the Pinnawela elephants is far from the one observed in wild herds. A more general argument advocates our opinion that they are neither kept 'semi-wild' nor 'semi-natural' as is often postulated by some authors. This argument concerns the daily activities. In Pinnawela the elephants have to stay during the hottest hours of the day either in the shallow water or in the practically shadeless grass area. Wild elephants spend this period of day resting in the shade, and drink and bath mostly during early morning and especially in the late afternoon or early evening. Furthermore, elephants feed when not resting, sleeping or bathing, i.e. most of the day. In Pinnawela they are normally prevented even of grabbing some plants along their way from the lawn to the river or on the opposite bank of the Maha Oya. They receive their food in the evening and must eat it during the night.

The Pinnawela elephants are mainly fed with the stems and leaves of Kitul palms (*Caryota urens*), leaves of Coconut palms (*Cocos nucifera*) and branches of the Jack tree (*Artocarpus integra*). Occasionally, they are additionally given branches of Bread-fruit trees (*Artocarpus nobilis*), Banyan (*Ficus bengalensis*), Bo (*Ficus religiosa*) and a few other tree species. Perennial grasses, a very important part of the daily diet of wild and many captive elephants in India, Myanmar or Northern Thailand are lacking. In Pinnawela, food is brought in daily from outside (e.g. Gampaha, Pimimatalawa, Kadugannawa, Molagoda, Rambukkana, and Kegalle) by suppliers selected through a tender process. In Pinnawela stems and branches are split and food bundles are brought by a few trained adult and sub-adult bulls to the stands of the elephants.

During our studies the Pinnawela elephants received nothing other than green fodder. But it is reported that until the middle of the 90s additional rations of 500 g of boiled rice mixed with honey was given daily to younger elephants, and older ones daily received 2 kg of a mixture made out of corn flower, rice bran, minerals and trace elements (Poole *et al.*, 1997). Captive born calves were suckled by their mothers and received no additional special diets. Orphaned neonates and infants were bottle-fed 5 times daily upto the age of 3.5 years. At each bout of feeding the young elephant was given 7 bottles of milk. So in one day each of the young orphans received an average of about 26.25 liters of milk made out of Lactogen powder (Tilekaratna & Santipillai, 2002). There are no detailed reports of administration, neither of traditional nor western medicine to the Pinnawela elephants.

## Staff, visitors, finances and local economy

During our study the Pinnawela staff was headed by a curator, his assistant and since 1998 by a veterinarian. Prior to that, the veterinary surgeon of the Dehiwala Zoo was looking after the elephants. The elephants were looked after by two head mahouts, eight mahouts and eight helpers, who had to herd the animals in the grassy area and in the river, to guide them between the lawn and river, to feed them and clean their night quarters. Furthermore, they had to break in and tame juvenile and sub-adult bulls to be used as temple or working elephants. In later years following our study, the staff was increased to 26 mahouts and helpers (Tilakaratne & Santipillai, 2002). In 2002, a second veterinarian was appointed.

Elephants kept in such an intensive system require intensive care and even first class mahouts cannot fulfil their duties adequately. A ratio of one keeper per two elephants is way below the known norm. Experts like Evans (1910), Toke Gale (1974), Krishnamurthy and Wemmer (1995) reported that traditional intensive keeping systems only work properly with at least one mahout and one assistant per elephant. More problematic animals are cared for by three men. In Maesa (Chang Mai) at least 90 mahouts are in charge of 85 elephants, and in Guruvajoor (Kerala) each elephant is looked after by two to three mahouts. In Pinnawela the situation is vice versa. There are two to three elephants per keeper, which is even less than in a modern western zoo like Emmen, Hamburg or Rotterdam, where elephants regularly breed and are kept in so-called protected-contact systems which demand less staff. Here the ratio – keeper: elephant is about 1 : 1.

It was found that the Pinnawela Elephant Orphange does not only face a quantitative personnel problem but also at least 3 fundamental problems concerning captive management: (1) The standards of knowledge about care and management of elephants amongst the decision making staff, (2) the hygienic conditions and (3) safety for visitors and elephants. Since 1999, the Biodiversity and Elephant Conservation Trust organises training programmes on improvement of veterinary care for elephants in Pinnawela and other establishments. Several of these authorities (e.g. the highly recognised South Indian experts Dr. S. Krishnamoorthy and Professor Dr. J.V. Cheeran) rightfully criticized the hygienic conditions in Pinnawela. During the study the stands on natural ground were hardly cleaned and the waste from the roofed and hard-soil stables was transported daily to a huge heap at one edge of the grass area depression, where the herd had to stay during the day when they were not in the river. With the summer rains an impressive stream of manure was flowing into this area.

Furthermore, it was found that there was (and still is) no concept of safety either for humans or elephants in Pinnawela, on the contrary, during the stay elephant handling could be considered as careless. The elephants managed to escape quite easily. At night time young bulls were repeatedly able to free themselves from their chains, and one evening, after the visitors had already left the property and the main gate had been closed, 6 young bulls tried to re-enter the orphanage. Nobody had realised that they had slipped away from the herd during their stay in the river and had reached the opposite bank of the Maha Oya in search of food. Several hours later they wandered back to the orphanage. Furthermore, the elephants are not safe from visitors. Mainly orphaned neonates and infants were continuously teased by some visitors who wanted to impress their families. Proximity between visitors and elephants could occasionally reach a point when elephants were able to hit the people with their trunk. In 2001 Mathali, one of the adult females who was more or less permanently chained became so aggressive towards tourists that she killed a young mahout in due course. To protect humans and elephants from each other modern zoos have a system of double fencing: One to keep elephants from escaping and one to keep visitors at a safe distance.

In Pinnawela, admission charges vary: Local visitors pay less, while foreigners pay much more. For local adults the ticket costs Rs. 20, children between three and 16 years are charged Rs. 15. School children as a group are charged at the rate of Rs. 7. Adult foreigners pay Rs. 150, children Rs. 75 and the charge for using a video camera is Rs. 200, while professional photographers are charged Rs. 600 per head. The number of visitors has increased tremendously in the last few years. An unofficial estimation in 1997 speaks of about 350,000 visitors per year. Official numbers from the Department of Zoological Gardens state for 1998 575,000 visitors including 155,000 foreigners and for 1999 750,000 visitors including 235,000 foreigners. According to the Director of Zoological Gardens, in 2001 the average number of visitors per day was about 3,000 with peaks on weekends of 8,000 to 10,000, i.e. the number of visitors per year is now more than a million. A rough estimation from 1999 gives a figure of an average daily income of Rs. 100,000. However, a more detailed report by Tilakaratne and Santipillai (2002) estimates a lower amount of Rs. 55,000 per day or Rs. 20,065,000 (275,453 US$) per year.

The expenditure adds up to an annual total of 216,137 US$ (food for elephants: 182,500 US$; milk for calves: 19,077 US$; mahouts' salaries: 14,560 US$). The profit ranks at 59,316 US$ (Tilakaratne & Santipillai, 2002). Instead of filling the coffers of the government the profit would urgently be needed to employ more mahouts and helpers at Pinnawela.

## 3.2 Traditionally Kept Elephants

3500 years ago the first elephant-riding cultures were established in the central Jangtsekiang and in North India (Kurt, 1992). Captured and trained elephants were first of all a symbol of power for ruling families. Furthermore, they were used for transport, hauling and levering of heavy weights and for war. Captive elephants assisted in construction of temples, palaces, walls and irrigation dams. Power over captive elephants also meant power over wild populations. With the help of specially trained hunting elephants, wild conspecifics could be noosed or driven into kraals. Captive elephants were used to help tying up newly caught elephants as well as taming them. Since antiquity and in the middle ages, the techniques on how to control captive elephants were part of the education of princes in south and southeast Asia – also in Sri Lanka. The ruler-to-be first learned to identify the 'nerve-centres' where the *ankus*, the elephant hook, had to be placed to control the animal, on bronze models. In early periods a high ranking prince had the status of 'Äth Arcaria' or 'Äth Panthiya', i.e. the master of the elephant establishments of the king. Later a high ranking officer was chosen as *Gaja Nayaka Nilame*. He was in charge of the *Pannikiyas*, the elephant catchers who noosed wild elephants, the *Kuruve*, who cared for the newly caught animals, and the *Pannayas*, who had to collect food during capturing operations. The *Gaja Nayaka Nilame* was also the head of the mahouts, the elephant doctors and the specialists who knew medical herbs for the treatment of captive elephants.

In such a way the kingdom provided employment and income to thousands of people, and the ruling families increased their reputation amongst their subjects, since all their male members were capable of controlling captive elephants. Furthermore, the ruling families would demonstrate their power by their ability to control wild herds depending on the conduct of their subjects: They would capture wild elephants that plundered gardens, fields and plantations. But also the opposite was possible: The king's specialists would control wild herds and move them to almost any place. "Here these Elephants do, and may do, great damage to the Country, by eating up their Corn, and trampling it with their broad feet, and throwing down their Coker Nut Trees, and oftentimes their houses too, and they may not resist them. It is thought this is done by the King to punish them that lay under his displeasure" wrote Captain Robert Know, a sailor in the service of the English East India Company who was captured by the King of Kandy in 1660 and detained for nearly 20 years in the interior of Sri Lanka.

## Size and social structure of the captive population in Sri Lanka

In 1588 the King of Kandy demonstrated his power to the Portuguese during the siege of Colombo by the deployment of 2200 elephants. The sizes of the captive elephant populations during the colonial period are not known. But it can be assumed that the number decreased in the same way as the number of wild elephants shrank due to clearing of forests, hunting and capturing. (Fig. 3.2.1)

In 1946 official sources mentioned a captive population of 736 captive elephants (Jayewardene, 1994). In 1955 the population consisted of 670 and in 1969 of 532 animals belonging to 378 owners (Deraniyagala, 1955; Jayasinghe and Jainudeen 1970). According to the Department of Wildlife the population dropped to 344 in 1984 (Godagama, 1996). In 1997 the population consisted of 214 captive elephants belonging to 150 owners (Jayewardene 2002a). And a very exact count in 2002 found a rest population of 186 traditionally kept captive elephants (Jayewardene, 2002b). The Pinnawela population was not included in this count.

The dwindling of the captive population in Sri Lanka has several reasons: first of all, with very few exceptions captive elephants never bred in the country. In 1950 capturing of wild elephants was stopped. However, some so-called problem elephants were caught and auctioned (Jayewardene, 1994). Since 1980 elephants from the Pinnawela Elephant Orphanage were occasionally donated or sold to temples and private owners. It is also said that a few of the captive elephants have been poached (Godagama, 1996). The dwindling of the captive population in Sri Lanka is characterized by two demographic changes. The sex ratio (males: females) shifted significantly from 1 : 0.7 in 1955 to 1 : 0.9 in 1968 and 1 : 1.1 in 1995. The age structure also changed (Fig. 3.2.2). At the Kandy *Perahera* in 1947, 37.5% of the captive elephants between 11 and 20 years of age and 24.5% were between 21 to 30 years. The rest were older (Deraniyagala, 1955). Twenty years later most of the elephants that partook in the Kandy *Perahera* were between 11 to 40 years (Kurt & Nettasinghe 1968, Kurt 1992). In 1994, 33% of the bulls and 38% of the females were aged between 41 to 50 years (Godagama, 1996).

In 2003 the Sri Lankan Elephant Owner Association stated that 80% of their elephants were older than 50 year. This can hardly be true, as our data reveal. Available data on age was summarised, which was given

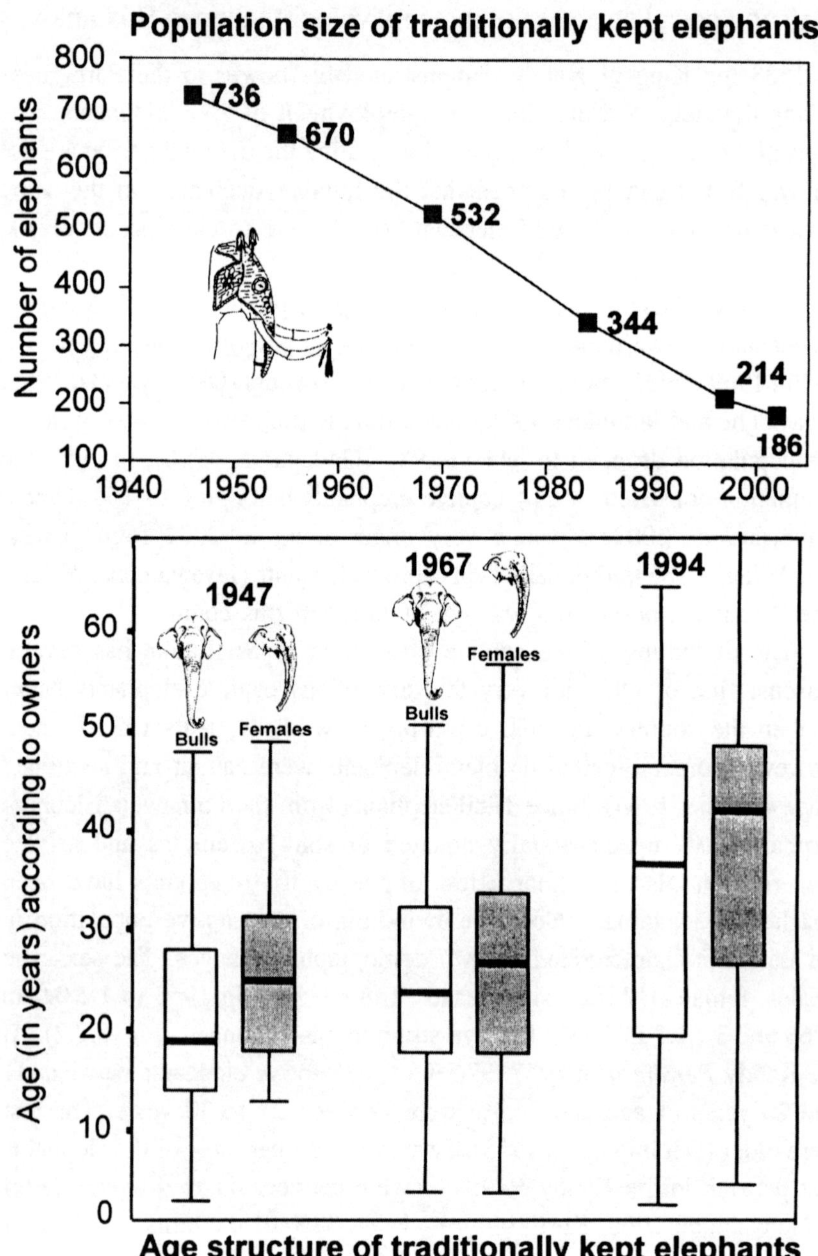



Fig. 3.2.1    Size of captive population in Sri Lanka and age of captive elephants in 1947 (Deraniyagala 1955), 1967 (Kurt & Nettasinghe, 1967) and 1994 (Godagama, 1996)

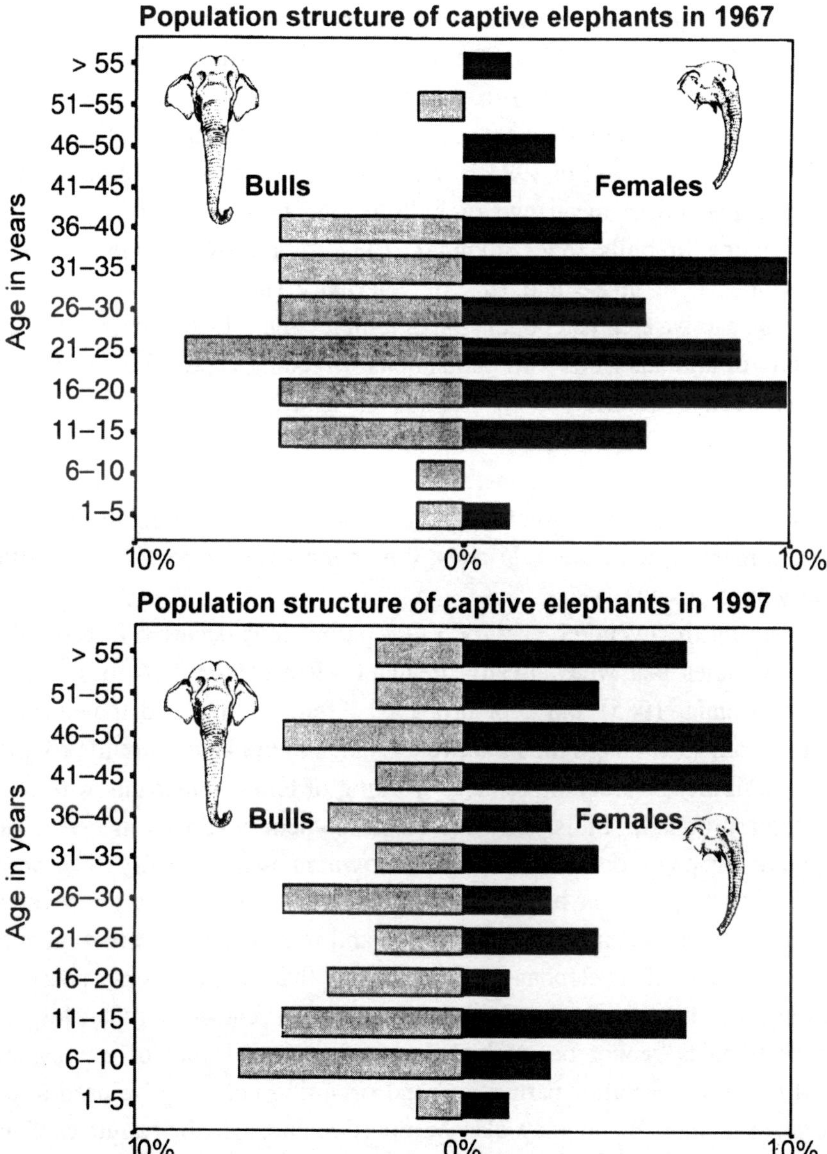



**Fig. 3.2.2** Social structure of elephants at the Kandy Perahera in 1967 and 1997 given as percentage of the total population (Sources: Kurt & Nettasinghe, 1968; Kurt, 2001)

by owners and mahouts and the mean age of their elephants was calculated. For bulls it was 28.1 ± 15.4 years (n = 145) and for females 34.1 ± 15.2 years (n = 115). The difference is significant. The mean age of bulls in 1967 was 24.9 ± 10.4 years and by 1994 it was 33.3 ± 11.7 years, i.e. significantly higher. The mean age of females in 1967 was 27.4 ± 11.7 years and in 1994 it was 39.4 ± 15.6 years, i.e. significantly higher. The lower mean age of bulls is due to a generally lower life expectancy in bulls (e.g. Sukumar, 1987) and most probably also a consequence of abuse and lack of veterinary care.

The captive elephant population of temples and private owners consists mainly of animals from wild populations. Breeding captive elephants has never been a tradition in Sri Lanka, as it was stated already by the English sailor Robert Know: "Neither will they ever breed tame ones with tame ones; but to ease themselves of the trouble to bring them meat, they will tie their two forefeet together, and put them into the woods, where meeting with the wild ones, they conceive and go one year with the young" (Knox, 1966).

During the twentieth century 5 to 7 parturitions occurred in captivity by 3 females that were already pregnant when captured (Norris, 1952; Deraniyagala, 1955) and 2 or probably 4 females that had mated with captive bulls. Although the Pinnawela Elephant Orphanage would provide the possibility of breeding females working or temple elephants with one of the reproducing bulls, only 7% of the owners would be in favour of captive propagation. However, most owners would gladly buy new elephants despite their high prizes (Godagama, 1996). There are several reasons for this retention: Some owners and mahouts believe that mating would weaken their elephants. Some believe that mating is only possible during musth of the bulls. Furthermore, owners fear economical losses, since females cannot be worked during the second year of pregnancy and the first year after parturition, and offspring can only be used after 5 years of age. But it may also be questionable, whether adult captive bulls that are permanently kept under extreme restrictions are not psychologically 'castrated' and unwilling to mate, and therefore physically unable to reproduce (e.g. Kurt, 1995). For the conservation of the species, captive elephants in the hands of temples and private owners are not necessary.

83% of the so-called 'working' elephants belong to private owners and 17% to temples and Dewalas. 15% are considered as too young or too old for work. 85% are used for specific services. Most of them haul, heave or carry logs and boards in saw mills or construction establishments. In the last few years captive elephants have gained a certain importance for tourism (Godagama, 1996).

## Owners, mahouts and accidents

The following notes stem form the study of Wasantha Kumari Godagama (1996): Of 82 elephant owners, all are Buddhists, 40 are wealthy landlords, 28 businessmen (wood merchants, construction or tourism entrepreneurs), nine government employees, two Ayurweda doctors, two mahouts and one post-master. They keep between 1 to 12 elephants (mean: 2) with the following motives: status symbol, love for elephants, family tradition, taking part in *peraheras* (processions) or as a symbol of good luck. However, the paramount motive is income. According to their size and the work they have to do elephants are hired out for Rs. 500 to 1200 per day (mean Rs. 920). An important function of elephants is parading during Buddhist processions. 6 such *peraheras* take place in August or September and 2 in January or February. A quarter of the owners loan their elephants free of charge for these festivals. A third asks for payment from certain temples. The rest charge all temples a fee, i.e. Rs. 200 to 400 per elephant a day. In general the temples are responsible for providing food for the elephants and maintenance of the mahouts during these festivals. However, the *Nawam Perahera* in Colombo makes an exception. It has become fashionable during the last years due to increasing tourism. Mahouts taking part in this spectacle with their elephants receive expensive gifts such as radios, sewing machines, watches, bicycles or cash of Rs. 1000 to 2000 per year, during which the elephants are engaged for 3 to 30 days (mean 11 days) at *peraheras*.

Out of 137 mahouts interviewed by Wasantha Kumari Godagama, 135 of them were Buddhist and 2 Hindu Tamils. They were between 20 to 75 years (mean 38 years). 25% were single and 75% married. They started their career as golaya, as helper, between 9 to 40 years (mean 18 years). According to their abilities and opportunities, they take over full responsibility of their own elephant after 3 months to 13 years. Most of the 137 mahouts changed their jobs frequently and depending on their ages, have led upto 20 different elephants. This contradicts the widely believed opinion that a mahout spends a life-long time together with one elephant. Another prejudice is that the father of a mahout was himself a mahout. In Sri Lanka this is only true in 33% of the cases. One third of mahouts are sons of farmers and one third are sons of employees or labourers. The monthly income of a mahout varies between Rs. 1000 to 5250 according to his age, the value of the elephant and the agreements between mahouts and owners of the elephants. 77% of the mahouts are paid directly by the owners. 23% have other agreements, e.g. the mahout has to pay a certain amount to the owner every month. The rest must be used for his own maintenance and salary, for the maintenance of the golaya, who normally gets no salary, and the food and occasional medicine for the elephant. (The summary of the research by Wasantha Kumari Godagama, 1996).

According to their provenience, training, motivation and character the society of Sri Lankan mahouts is extremely heterogenous, as compared to those e.g. from Tamil Nadu in South India or Myanmar, who are in charge of extensively kept timber elephants. Among the *Kurumbas* of Tamil Nadu (Kurt, 1992) and the *Karen* of Myanmar (Heidinger, 1997) the profession of mahout or *oozie* is strictly confined to certain families, where traditions of guiding even potentially dangerous elephants are kept alive. Out of 82 bulls, studied by Anouk D. Ilangakoon (1993) on the occasion of the *Navam Perahera* in Colombo and the *Esala Perahera* in Kandy, 33 (40.2%) were considered aggressive and 16 (19.5%) had killed one or more humans. Out of 99 females 24 (25.2%) were reportedly prone to aggression and eight (8.0 %) had killed humans. In most cases the working elephants had intently killed their mahouts. Most of the killed mahouts were drunk at the time of the incident. 4 out of 5 mahouts regularly drink *arrak*, *tody* (a kind of palm wine) or illegally distilled spirits. The mean daily ration is estimated to be 0.3 litres and the highest 1.5 litres (Godagama, 1996). Especially the mahouts of very large bulls, amongst which most man-killers are found, can be considered to be heavy alcoholics (Kurt, 1993). Alcohol drinking mahouts often lose their jobs. They are not free in the choice of their next elephant. Only one third of all interviewed mahouts would like to work with a killer elephant.

These cold blooded mahouts boast that leading a killer bull increases their social status and that the owner of the dangerous elephant hardly dares to interfere with their training and leading style. This gives them the freedom to punish their charges at will and demand extreme working output, which increases the income of the torturer (Godagama, 1996).

In the following lines some accidents are listed according to 'Kala Santha' (The Island, 6th August, 2003), which happened in the last years. In 1996 Raja, a young tusker kept in the Dehiwala Zoo, killed his trainer, and 18 months later killed another elephant keeper. In July 1998, the full grown tusker Nawam Raja, chased his mahout. A few days later the same elephant carried the sacred casket at the Kandy *Perahera*. The following day, when the keeper wanted to remove the chain from a hind leg, Navam Raja knocked him over, gored him and threw him in rage. The mahout succumbed to his injuries. In 1999 Singharaja, a former Pinnawela orphan of about 6 years of age, went berserk while walking in a procession and started attacking spectators and chased his mahout. One foreigner was thrown off a wall and suffered a fracture on her leg. In the same year a bull killed his mahout during the *Kaduwela Perahera*. And in 2001 Mathali, one of the breeding females at Pinnawela killed a mahout.

## Keeping systems

The earliest record concerning care of captive elephants is decidedly the *Hastyayurveda* and seems to have been written down in the fifth or fourth century BC. The *Arthashastra* of Kautiliya (300 BC–300 AD) deals with capture, keeping, care and training of elephants in two chapters. The *Matangalila* of Nilakantha, another antique source, may go back a thousand years or even to a much earlier date. Next to these 3 standard texts there are a number of other old Buddhist and Hindu sources referring to the management of captive elephants such as the Tanjore Manuscript (a compilation of several older texts), the *Gajanirupana*, the *Gajacikitsa*, the *Gajalaksana*, or the *Brhat-Samhita* of Varahamihira (c.f. Edgerton, 1931). All these rules and regulations were the basis of captive elephant management in areas under the control of Buddhist and Hindu rulers in South Asia.

In India, since antiquity 2 elephants were kept in one stable and each elephant had its own pen, which measured double the animal's height, width and length. The elephants stood on a wooden floor and either free or fettered with chains or ropes on two poles. Next to the main door the pen had a second opening where the dung was taken out. At night the elephants had to stay in another pen. Its floor was convex, i.e. it was kept lower at its centre. Straw mats or palm leaves were used to cover the gabled roof, and during the cold and rainy season the sides of the construction, were kept open during hot and dry seasons. War and riding elephants were kept within the town-walls, but elephants in musth or those under training had to stay outside. The elephant stables were always built on dry grounds relatively far away from the next water bodies. Similar constructions or at least rudiments of these still exist for state and temple elephants in Thailand (Frädrich, 1993). In the Mysore Zoo as well as the elephant centre of Jaipur, elephants have to spend the night on similar convex floors as was the case in antiquity.

The captive elephants were already fed with green grass, hay and straw and occasionally with sugar cane as well as balls of boiled rice mixed with, molasses, milk, butterfat and/or yoghurt. Occasionally alcohol, boiled meat or broth were added. Working elephants in Tamil Nadu still receive rations of boiled rice and many of the ingredients known since antiquity (Kurt, 1993; Krishnamurthy and Wemmer, 1995). In antiquity the daily activity of captive elephants was already dominated by work and relatively little time for feeding remained. Accordingly the owners had to provide them with easily digestible rations to compensate possible losses and these rations had to consist of carbohydrates and fats (Kurt, 1992). The provision for meat and broth is unknown, but working elephants in Kerala still occasionally receive such rations during the rainy seasons. During the sixteenth and seventeenth century most bulls were given alcohol to render them extremely aggressive in war. Today Sri Lankan elephants kept by temples and private owners are fed normally only with green fodder, mainly the stems of the Kitul palm (*Caryota urens*), leaves of Jak trees (*Artocarpus integra*), Cocos (*Cocos nucifera*) and Kitul palms.

Kautiliya's *Arthshastra* describes more or less the same aids for maintenance and guidance of elephants as *Deraniyagala* listed in 1955 for Sri Lanka. They include chains for fettering as well as additional chains for feet and neck, as well as a noose of leather or tendon. A ribbon or rope around the neck of the elephant is used to fix the naked feet of the mahout who sits either on the shoulder or on the neck of the animal and guides it. People riding on the back can hold onto a chest girth, which also serves to hold blankets and caparisons carried during processions. Since antiquity the mahout guided the elephants with a sharply pointed stick and the well known elephant-hook, *ankus* or *ankusa*, which carries a pointed straight end and a sharp hook. The Kurumbas in Tamil Nadu, however, never use an *ankus* (Kurt, 1993). With their stick the mahouts drives the elephant, and the riding mahout can use the stick as baton to show for example, the elephant how high a load should be lifted. The more or less sharply pointed straight end of the *ankus* is used to keep the elephant at a distance, and the hook is used to bring the elephant closer. Furthermore, the hook once served to force war elephants to keep their heads high to protect the mahout sitting on the neck (Lahiri–Choudhury, 1992).

The multiple and diverse deployment of captive Asian elephants requires various keeping systems (Kurt, 1995; Kurt & Mar, 2003) whose extremes can be defined as follows (Table 3.2.1): Working elephants in jungle villages live in extensive keeping systems. They are used for riding, carrying and towing. During their resting periods they live, with hobbled front feet, in the nearby forest, where they find food and meet their tame and wild conspecifics. In South India, depending on tradition, additional fodder is given. Examples are the timber elephants of the Myanma Timber Enterprise (MTE) of Myanmar or those under the charge of forestry authorities in Assam, Tamil Nadu, Karnataka or Kerala where captive elephants are increasingly used as riding animals for wildlife patrolling, conservation research and tourism.

Intensive keeping systems (Table 3.2.1) are those in which animals are kept by temples or private owners more or less individually, fed exclusively on prepared fodder and, shackled with longer or shorter chains by a hind foot or by both one hind and one forefoot at night time or if idle. Contact with captive conspecifics is avoided. Contact with wild conspecifics is unlikely because intensive keeping is concentrated to

**Table 3.2.1** Definitions of management systems

|  | Extensive | Intensive | Alternative |
|---|---|---|---|
| *Examples:* | Jungle based timber and wildlife camps | Temples, urban tourist centres | Modern zoo, elephant parks |
| *Mainly used for:* | Riding, transport, hauling, heaving | Parades, circus tricks, begging | Display, reproduction, research |
| *Daily activities, food:* | | | |
| – Free movements: | Forefeet hobbled but free in nearby forest | 12 – 22 h/day chained in stands or stables | Chains and hobbles exceptionally |
| – Activity centres: | ± Determined by elephants | Determined by man | ± Determined by elephants |
| – Quality of food: | Divers | Monotonous | Divers |
| – Choice of food: | ± By elephants, except prepared rations | By man | By man |
| – Food intake: | ± Permanently except when bathing & sleeping | 1–3  feeding periods/d | Several feeding periods |
| – Bath, skin care: | By man and elephants | By man | By elephants |
| – Place for sleep: | Determined by elephants | Determined by man | Determined by elephants |
| *Social behaviour:* | | | |
| – Pop structure: | Heterogeneous | ± Homogeneous | Heterogeneous |
| – Social contacts: | Permitted | Hindered | Permitted |
| – Stereotypies | Absent | Very common | Rare |

[Table 3.2.1 Contd.

Contd. Table 3.2.1]

| | Extensive | Intensive | Alternative |
|---|---|---|---|
| ***Reproduction:*** | | | |
| – Reproduction rate: | Relatively high | Very low | Relatively high |
| – Neonate mortality: | Relatively low | Very high | Relatively low |
| – Rate of raising: | Relatively high | Very low | Relatively high |
| – Role of 'aunts' | Important | Not existent | Important |
| ***Musth:*** | | | |
| – Musthbulls kept: | Permanently in chains | Permanently in chains | In bull enclosure or with herd |
| – Social contacts: | Not permitted | Not permitted | Permitted |
| – Musth period: | Relatively short | Relatively long | Relatively long |
| ***Owners & mahouts:-*** | | | |
| – Ownership: | Mainly government | Mainly private | Private |
| – Soc.status owner | High | High | High |
| – Soc.status keeper | Relatively high | Increasingly very low | Relatively high |
| – Skill of keeper | High | Increasingly low | High |
| – Fluctuations of mahouts/keepers | Low | High | Low |
| – Tool used: | Stick or blunt pointed *ankus* | Stick or rod, blunt pointed *ankus,* sharply pointed knife and *ankus,* poles, spears | Blunt pointed *ankus* |
| – Use of drugs: | Rare | Increasingly common | None |
| – Accidents: | Rare | Increasingly common | Very rare |

urban regions. Today intensively kept elephants are rarely employed for work (e.g. sawmills), but increasingly used in religious processions, political demonstrations or wedding ceremonies. Restricted movements and the practical absence of contacts to conspecifics lead to extreme aggressions towards handlers and to stereotypic behaviour (see Section 5.4).

The intensive keeping system, as it is still present in privately owned and temple elephants in Sri Lanka, has been taken over by western circuses and in former times also by zoological gardens. The keeping system in traditional zoos as well as in the Pinnawela Elephant Orphanage in Sri Lanka can be considered as intermediate between extensive and intensive. During the day the elephants are kept free in the grassy area but during the night they are shackled or otherwise kept solitary in small cages.

During our studies in 1997 Beate Ludescher (2001) described the keeping systems of 122 elephants; 56 of them lived in the Pinnawela Elephant Orphanage, seven in the Dehiwala Zoo near Colombo, and 19 in private ownership (Riverside Elephant Park; Elephant Village, Habarana; Elephant Bath, Katugasthota). Data on further 40 elephants were collected during the *Esala Perahera* in Kandy. These animals were kept on temporary stands. Beate Ludescher described the stands, the method of fixation, the tools to guide elephants as well as the objects used for work, e.g. the mouth pieces and bells.

## Stands

Stands were described according to the following criteria: are they covered or do the animals stand more or less unprotected? What does the floor consist of (e.g. concrete, sand, mud)? What is the size of the area reachable to the fettered animal and how is the elephant fettered? Generally three categories could be distinguished: animals kept in their usual stands, animals in provisional stands and animals with no stands. Similar roofed stands with concrete floors, as already described for certain elephants in the Pinnawela Elephant Orphanage, were found in the gardens of the Dalada Maligawa (the Temple of the Tooth) in Kandy. The unroofed stands in the Riverside Elephant Park were placed in immediate vicinity of a river and their concrete floor was accordingly moist. The elephants of the 'Elephant Village' had similarly equally deplorable stands. The bulls had to stand on more than 2m high mounds of dung and foul plant parts. Here, the tusker with the longest tusks in Sri Lanka, the world famous 'Milangoda Tusker' was found. Although this animal's tail and the left hind leg were heavily infected, the animal had to lie during the whole day in a small brook and pose for tourists.

The most horrible conditions were found in the 'Elephant Bath' tourist facility. Due to the nearness of the river the floor was permanently wet. The animals were in bad physical condition. The elephants of Kastugasthota were likewise in bad condition but due to alternate circumstances: the bull, who was in musth at that time, and the female, had to stay unprotected day in and day out in the sun on a sandy place. Here also, the hygienic conditions were far from optimal. Contrary to the above mentioned cases, the elephants of the Habarana establishment were kept almost exemplary, although their stands were unroofed, but at least they were on sandy dry ground, well shaded by large trees and meticulously cleaned. The stone slabs under the hind feet were not present. Accordingly, the elephants could stand in a normal position and not in an unnatural one which permanently stresses the joints of the extremities.

An extreme situation was found in the Dehiwala Zoo. The elephants had to stand under a corrugated iron sheet roof on a concrete floor, where they were fettered very tightly for about 23 hours per day. All these elephants were in bad physical condition and showed serious behavioural defects, e.g. stereotypies (see later). The provisional stands for elephants that came to Kandy during the *Esala Perahera* are not very different from normal stands. Some of the elephants had to stay on stone slabs or the lawn within the temple areas. Outside the temple walls elephants had to stay on asphalt. Their chains could be attached to permanent rings which were fixed into the temple wall. Additional elephants were kept on two terraces on a hill behind the Temple of the Tooth. Here, permanent rings were fixed into the ground. The elephants had to stay on barren ground or on wooden flooring made out of the bark of Kitul stems. Each animal had to stand with the hind legs on a slab in an unnatural position.

## Chains and other implements

Beate Ludescher (2001) examined 107 elephants which were attached to their stands by chains that were tied around their wrists or ankles. These chains had diameters of 1 to 4 cm, lengths of 4 to 8 m and were fixed to trees, pillars, poles or rings. 8 elephants were only chained by a hind ankle, 3 by one ankle and both wrists, one by all extremities and 85 by one fore and one hind extremity. A further 18 elephants were not fettered during the observations, but 16 of them carried chains on one wrist and two elephants had a chain around one wrist and one ankle. In all of these cases the fettering chains were placed over the neck, shoulder or back of the animals. 2 infants and 2 neonates in Pinnawela were never chained. Considering the arrangements of chains it became obvious that the combination 'right front' and 'left hind' was much more frequent than all other possible variations. Fixation of the left ankle was found significantly more frequently than of the right one, and the right wrist is used more often for chaining than the left one.

Obviously, elephants are almost always chained by the same extremities. This contradicts the comments of many mahouts who assert, that chains are changed daily from one to the other side. Most probably the variant 'right wrist-left ankle' is the easiest method for a right-hander, as most mahouts are, to chain an elephant. The hind extremities are normally fettered by standing parallel to the animal and facing forward, i.e. the right hand is nearer to the ankle when the mahout stands on the left side of his elephant. In case he wants to fix the chain on the right ankle he has to stand on the right side of the animal, but to shackle the chain he has to look backwards and hence, he loses close control over the animal for a moment. Fettering wrists is carried out vice versa. i.e. the mahout approaches the elephant from the front. In this case it is easier for him to use his right hand without loosing control when he fetters the right wrist.

Sri Lankan working and temple elephants often have to carry their fettering chain, which is still fixed to a wrist but placed over their shoulders or backs, when they walk. If the elephant intends to run away the chain slips down and inevitably the animal stands on it, which immediately prevents it from carrying out its intention, as elephants walk or run in such a manner that the hind foot is placed at the same spot where previously the forefoot was. Therefore, the elephant will stand on the free end of the chain and thereby abruptly and painfully block its intended escape. However, there are many known cases where clever elephants learned the trick to pull down the fettering chain from their back and hold it high up with their trunk before they run away. Such events must have been known since Indian Antiquity and were counteracted by a more sophisticated 'chain-work' as Kautiliya called it in the *Arthashastra*. In Sri Lanka three such methods are still used: (1) A second chain is fixed to a wrist and wound around the neck of the elephant. This chain is kept so tight that only small steps are possible. If the elephant tries to break out, it strangulates itself. (2) A further chain is fixed to an ankle and swung around the chest. If the mahout straightens this chain, one hind leg is blocked. (3) Fore, and in most cases hind feet, are connected with a chain allowing only small tripping steps. Since 1959, when an adult bull ran amuck during the Kandy *Perahera* killing 14 people and severely wounding 129, all elephants taller than 2 m have to be hobbled with such connecting chains between fore and hind feet to be allowed to take part in the procession (Jayewardene, 1994).

Most working elephants carry upto 4 ropes around their necks. They are used as foot rest for the mahout, and to hold bells and the 'mouthpiece', i.e. a 30 to 50 cm long, cylindrical plait of Coconut or Kitul palm fibers, which is fixed to a chain. In order to carry or pull loads the elephant has to clamp the mouth piece between its molars. This method was introduced under British rule in Sri Lanka and South India to replace the rather complicated dragging-gear, as it is still used in Assam and Southeast Asia and which can cause serious skin wounds. However, the mouth-piece-method can lead to deep abrasions of the molars if used permanently on only one side of the elephant's mouth. In one instance, it was observed how an elephant had to carry a rope, which was connected to a wooden stringer on the nape of its neck. By continuously turning the stringer the riding mahout could tighten the rope around the elephant's neck and thus exert control.

The tool used to guide and punish the elephant is in most cases the *ankus*. In Sri Lanka these hooks are fixed to a stick with a length of 120 to 250 cm and a width of 1.5 to 2 cm. The straight point measures between 2 to 5 cm and occasionally upto 10 cm for the control of certain adult bulls. Sometimes the opposite end of the stick is also pointed or protected by an iron cap. Tall and potentially dangerous elephants are guided with a longer *ankus* with a sharper point than the smaller, rather harmless one generally used. Wooden and bamboo sticks are other devices used to guide and punish the animals. However, beating with the stick is harmless compared to the pricking and cutting induced by a sharply pointed *ankus*. Furthermore, each mahout carries a knife, which is used for cutting food and as guiding stick. However, this knife is also used to painfully control the elephant.

## 3.3  Wounds, Scars, Ulcers, Abnormalities of the Limbs

The hygienic conditions under which many working elephants in Sri Lanka have to live, are deplorable. Permanent chaining of non-working animals and frequent participation in processions, during which both front feet and hind feet are more or less tightly shackled together, cause skin abrasions and wounds, which generally do not heal well in elephants. Even correctly treated injuries take two to three times as long to heal than in horses (Salzert, 1972). Freshly caught elephants and those born in captivity, which are chained for taming processes, are particularly vulnerable. False mounting of foot chains result in deep abrasion and cut injuries which causes more or less severe inflammations of the deeper tissue layers. Lack of care may result in skin necrosis, ulcers and phlegmons which severely impair the general condition, often with fatal outcome. Elephant skin is highly prone to subcutaneous pus formation (Overview in Salzert, 1972; Schmidt, 1988). Badly fitted girths can lead to skin irritations with severe consequences and even small wounds caused by the hook can lead to abscesses. Bulls in particular are exposed to harsh punishments, which are equally part of the cause for their being dangerous as well as a result thereof.

### Injuries of the feet

Scars, ulcers and wounds are easily detectable on the elephant skin, their cause however, is often difficult to diagnose. Scars appear whitish or flesh coloured throughout the life of the animal, because the skin no longer contains dark pigments after wound healing. These scars can be the result of wounds caused by chains, the *ankus*, knives, calluses induced by lying on hard ground or boils generated by filaria. Injuries to the feet are the most frequent of all medical problems in the keeping of zoo, circus and working elephants (e.g. Salzert, 1972; Schmidt, 1988). Out of 140 examined elephants by Wasantha Kumari Godagama (1998) 96 (68.6%) had ulcers on their feet and 76 (54.3%) had split hooves. The cause for these fissures, which lead more or less deeply from the outer edge of the hoof, inwards, are, according to Evans (1910) and Schmidt (1988), the result of standing on wet ground and lack of hygiene. Many mahouts state that the cause of hoof splitting is the long marches or work on hard ground. Foot problems are more frequent in older elephants than in younger ones.

In 1997 Beate Ludescher (2001) and Barbara Rill examined 43 (17 males, 26 females) elephants that were of 20 or more years for split hooves, scars, wounds and abscesses on their feet, as well as for visible grooves in the hooves. These grooves run parallel to the hoof edge, are recognisable as bulging bands and possibly originate through profound changes in nutrition or keeping conditions. We compared bulls and females, as well as animals with a normal body weight development, having similar weights as their wild peers, and animals with retarded growth whose estimated weight was at least 10% below the expected normal weight for their age group. Although not all comparisons were significantly different, the following tendencies were discernible (Fig. 3.3.1): wounds, abscesses, chain scars and split hooves appear more frequently on the hind feet than on the front feet. Grooves appear more or less equally on front and hind. These findings suggest that the hind chains execute more pressure on the skin as is confirmed by direct observations. An elephant pulling on its chains will try to escape forwards and not backwards. In addition, the hind legs stand on wet ground drenched with urine and fouled with faeces more than the front feet.

Bulls appear to have more split hooves and grooves, more scars, abscesses and wounds on their feet than females (Fig. 3.3.1). This is probably a result of them being exposed to more rigorous constraining methods and longer periods of standing chained due to musth than females are. In a comparison between animals with normal body weight and those that are underweight it is noticeable that the grooves and hoof splitting is extremely different in animals that are underweight. Being

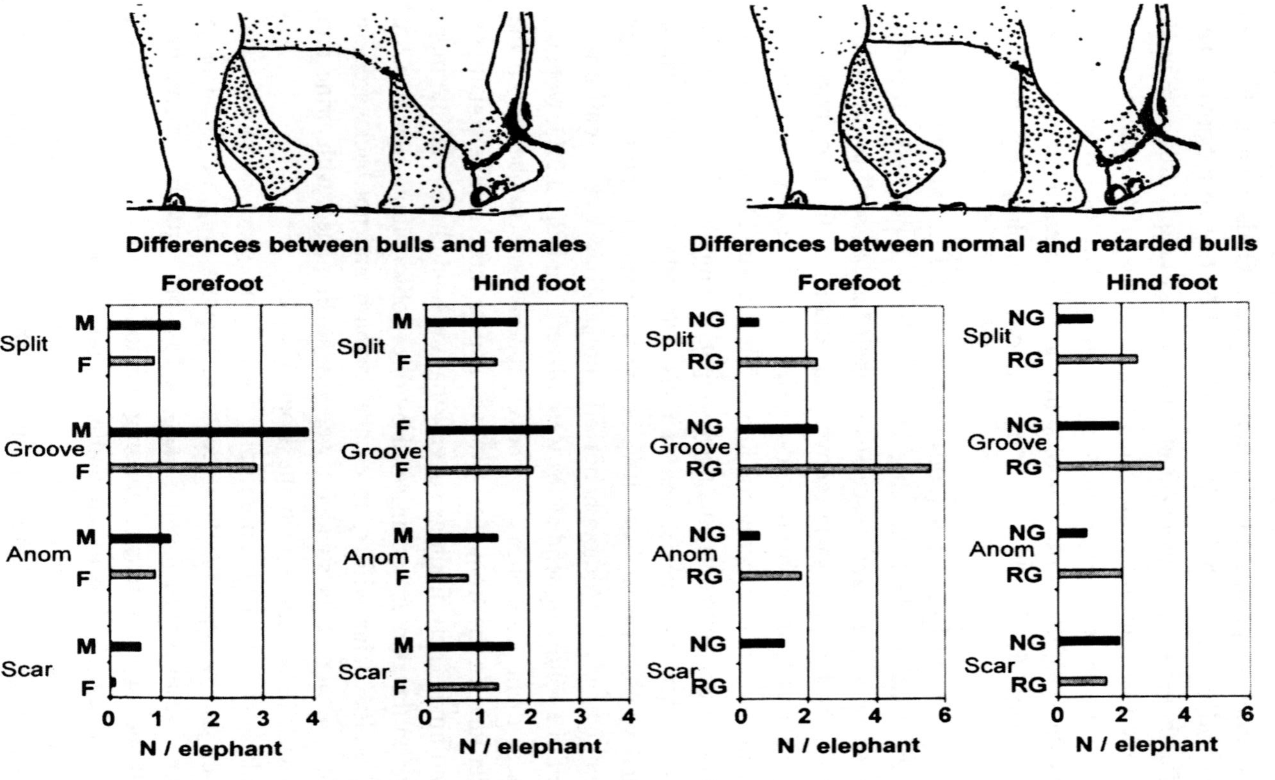


**Fig. 3.3.1**  Passively inflicted wounds on the feet of 43 captive elephants at Kandy according to sex and body growth M: bulls; F: female; NG: normal growth; RG: retarded growth; Anom: Abnormalities

underweight, among other causes, can be the result of inadequate feeding and over exertion. Wounds, abscesses and scars are however, the same in both categories.

Abnormal postures are caused by false, constantly one-sided chaining and the resulting non-physiological angles of the joints (e.g. Salzert, 1976) as a consequence of tightly chained front or hind legs in connection with uneven patches in the ground, such as stone slabs or wood-blocks, which can be extreme particularly under the hind legs (see Section 3.2). The result is 'knocked-knees' with unnaturally inverted feet, and often unnaturally formed, very long single hooves. In some extreme cases abnormal postures finally result in deformed limbs. The studies show that deformed limbs are to be found more frequently in bulls than in cows, and more frequently in animals which are underweight than in animals with normal weight.

## Injuries on the whole body

In order to describe wounds, scars and abscesses on the entire body we pooled our own data on 105 elephants with those of Godagama (1996) on 140 elephants. We defined 16 parts of the body comprising 616 skin injuries (Fig. 3.3.2). As was to be expected skin injuries were most frequently found on the hind feet and around the area of the hind ankle (24.5%). Numerous, i.e. 5% of injuries were found on the trunk, the skin on the hip joint, the tail, in the area of the knee and elbow joints, i.e. on all sensitive areas of the body and those that are important for locomotion and movement or which serve as points of support for the recumbent body. If one only compares wounds which are most likely the result of mechanical injuries caused by the elephant hook or the mahout's knife, one obtains a similar picture with 369 puncture wounds. They are most frequent on the hip joint (9.5 %), followed by the elbow (8.9 %), the heel /tarsal joint (8.4 %), tail (7.9 %), knee joint (7.6 %), trunk and wrist/ carpal joint (7.0 %), neck and shoulder (6.5 %), lower belly including penis sheath or vulva respectively (6.2%) and ears (5.4 %). Once again, it is the most sensitive areas and joints that are being stabbed.

Apart from the wounds caused by chains on neck, feet and wrists, these findings suggest that most scars and abscesses are not the result of inflamed calluses or filaria boils. Rather they were caused by mechanical

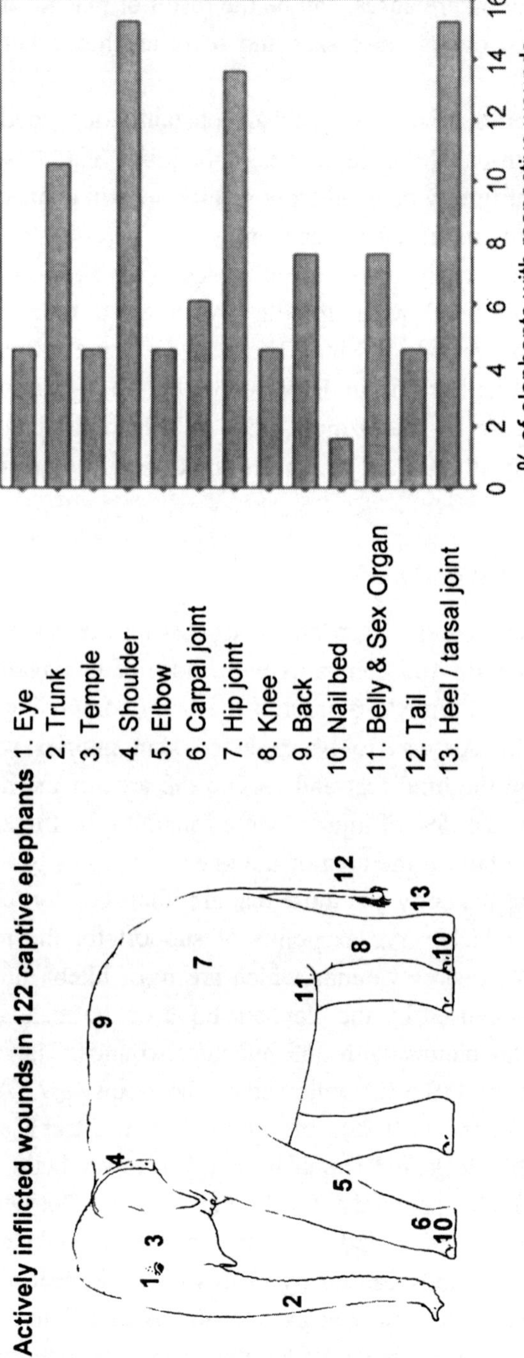


Fig. 3.3.2   Distribution of actively inflicted wounds in 122 captive elephants at Kandy. Data are given as % of all wounds (Source: Ludescher, 2001)

injuries actively administered by the mahout with knife and *ankus*. That this is in fact the case is clearly illustrated by the comparison between 20 years old and older animals. Bulls, which are considered to be more aggressive towards humans, possess nearly three times as many stab wounds caused by *ankus* and knife as do females. The same applies to normal growing, well nourished bulls (which enter musth regularly). They posses more of these wounds than do bulls with retarded growth.

The indifference which mahouts, elephant owners and other so-called elephant friends in Sri Lanka show towards these easily visible and recognisable skin injuries and illnesses, is striking. Three examples will illustrate this: The Milangoda Tusker, the bull with the longest tusks, had a vile smelling necrotic infection at the root of his tail, which was spreading to the left hind leg, and which was obviously not being treated. The animal has to lie in the water during the day, to be photographed and filmed by innumerable tourists. The Maligawa Tusker, a large bull which carries the shrine containing the holy relic during the *Esala Perahera*, has nasty foot injuries and stands mostly right next to the Mahaweli river on marshy ground. The elephants at the 'Elephant Bath' near Pinnawela fare similarly. These animals are so riddled with skin wounds, that they cannot be over-looked.

The findings of a survey by Wasantha Kumari Godagama (1996) showed that 54 out of 82 elephant owners had no idea about medical care for their animals, 67 of them used the services of an *aliwedamahattaya*, a traditional elephant-doctor, 15 of them relied on western medicine. Out of 137 mahouts 107 preferred traditional medication and 30 western medication. About half the mahouts use mostly hand written books concerned with simple illnesses such as digestion problems, small wounds and abscesses. About half the elephant owners declared that they personally collect medicinal plants for their elephants. Next to the feet, the eyes are the organs most affected in captivity.

## 3.4 Eyes – Injuries and Diseases

The eyeball of an adult Asian elephant has a diameter of about 40 mm and is therefore nearly identical to that of humans. Upper and lower eyelids, supported by cartilaginous tarsal plates, as well as very long eyelashes guard the eye. The lachrymal gland is replaced by the Harderian gland, which opens onto the nictitating membrane, or third eyelid. This membrane is moved by a cord of the orbital muscle, which is typical in elephants. The lachrymal duct is vestigial and non-functional (Duke-Elder, 1958). At the median corner of the eye one can frequently observe a certain amount of fluid. If this amount increases this develops into a modest to extreme serous exudate, which runs down a groove in the skin from the median corner of the eye and finally ends in a dry mass.

In a study conducted in 1997 by Pamela Kerschbaumer (2001) on 121 elephants in Sri Lanka, observed during the Kandy *Perahera*, at Pinnawela elephant orphanage and at Dehiwala Zoo, these were classified into 4 groups of eye colour: dark brown (54.4%), light brown (38.0%), honey colour (12.4%) and grey (9.00%). Among these four main colour groups more subtle colourings could be distinguished, e.g. bluish-grey and brownish-grey, walnut brown and chestnut brown, hazelnut brown and amber, as well as gold and yellow. 3.3% of the elephants had a light brown or golden speckled iris respectively. Furthermore, the left and right eyes were not always identical in colour. Combinations such as grey/dark brown, grey/light brown and honey/light brown occurred in 10.7% of all animals. A light golden ring around the pupil occurred in 11.6% of the elephants. This is with all probability the *Arcus scleralis*, described for the first time for both elephant species by Henning Wiesner (1992), the morphological equivalent of the *Arcus lipoides* in humans. This ring occurred in every ninth elephant between the ages of 1–15 years, and in every third elephant between 16–25 years. The ring was absent in older elephants. It occurred in 9.0% of all studied elephants in both eyes, and in 2.6% only in one eye. The ring occurred once with grey, three times with dark brown and ten times with light brown eyes. The visibility of this ring depends greatly on the angle of incoming light.

The field of vision of an elephant is impaired by the massive forehead bulges and supra orbital crests, especially with lowered head. Its vision does not differ much to that of a horse and does not decrease in dull light. However, direct sunlight very obviously reduces its vision. The round pupil is generally a mammalian feature of adaptation to shade. The eyes are relatively prominent, i.e. lie far in front in the orbital fossa, which could be a predisposition for traumas in dense bush land. The eyes of 300 culled elephants from Tsavo National Park in Kenya, in many cases showed opacity of the cornea, in 42% *conjunctivitis follicularis* and in 5.4% affected lenses from opacity to advanced cataracts or lens displacement. As in those days in Tsavo there was very little shade available to elephants as a result of tree destruction by the elephants themselves, and the ground structure generated great heat reflection, it can be assumed that at least part of the eye defects were induced by intensive sun rays. In general however, eye defects are rather rare in both African and Asian elephants. In the 300 elephants studied during 1998 and 1999 in Ude Walewa National Park, at least one female had one-sided opacity of the cornea. A similar, rare case in Sri Lanka is described by McCaughley (1965), and Sikes (1971) saw a very old cow in a large herd of African elephants, which was obviously blind with total opacity of both corneas. In captive Asian elephants, however, eye diseases are frequent and are the result of too intensive sunrays, traumas, dust, draught and infections. The first symptoms are usually swollen eyelids and eye discharge (Evans, 1910; Krishnamoorthy & Wemmer, 1995).

During the examination of 121 tame elephants, Pamela Kerschbaumer (2001) refrained from using any narcotics or sedatives, she did not palpate nor manipulate with instruments. 66 (54.6%) of the examined elephants had pathological defects on one or both eyes. The most conspicuous among these were slight to extreme serous discharge (34.7%), prolapse of the nictitating membrane (26.4%), cataracts (12.4%), conjunctivitis (3.3%) and retraction or loss of the bulbous (3.3%). Serous discharge is a side effect of inflammatory diseases and is a result of conjunctivitis and sometimes of the prolapse of the nictitating membrane.

Serous discharge was found in 34.7% of the 121 examined elephants. In 14.7% of the cases only one eye was affected and in 20.7% both eyes were affected. Minor discharge was found in 15.7% of the cases in the right eye and in 17.4% in the left eye. Medium discharge was found in 8.3% in the right eye and in 9.9% in the left eye. Extreme discharge occurred in 1.7% on the right and in 2.5% on the left. Irrespective of the amount of discharge in 73 elephants with normal growth the mean value for eyes with discharge was 0.44 ± 0.78, in 48 elephants with retarded growth the value was 0.73 ± 0.84, which was significantly higher. This result suggests that the method of keeping captive elephants directly or indirectly influences the amount of discharge from the eyes. This is corroborated by following comparisons. In 60 working elephants the mean value for eye discharge was 0.35 ± 0.66 cases per animal. The respective value for seven zoo elephants was 0.57 ± 0.98 and for 54 elephants in Pinnawela Elephant Orphanage 0.78 ± 0.98. Eye discharge occurs significantly less in working elephants than in the Pinnawela elephants. The elephants in Pinnawela stand at the same place, which is hardly cleaned, every night. In 18 Pinnawela elephants, which stood on natural ground, a mean value of 0.39 ± 0.78 cases of discharge per animal was obtained, in 36 elephants standing on concrete during the night the value was 0.97 ± 0.91, which was significantly higher.

The nictitating membrane of the elephant is usually visible as a thin strip, pigmented on the edge, in the nasal corner of the eye. If there is a prolapse, this membrane covers more or less of the eye-ball. This can occur on one or both sides. The causes are mechanical (reduction or retraction of the bulbous, retrocession through loss of water), induced by enlargement of the membrane itself (tumours) or as sign of a reduced sympathetic tonus following mechanical or toxic defect (Schmid, 1988). Of the 121 examined elephants 17 (14.0%) had prolapse of the nictitating membrane on both eyes and 15 (12.4%) on one eye, of which 7 right and 8 left. In 39 elephants with normal body growth the mean for prolapse per animal was 0.15 ± 0.39 (n = 39), in elephants with retarded body growth the respective value was 0.50 ± 0.75 (n = 28), which was significantly higher.

Cataracts can be divided into two types, the true cataract (*Cataracta vera*), which appears whitish-grey and the false cataract (*Cataracta falsa*), which appears brownish-black. The cataracts observed in the following context were of the first type. We observed both total (*Cataracta totalis*) as well as partial (*Cataracta partialis*) cataracts. Cataractogenic blindness manifests itself in reduced to total loss of sight but retention of awareness of light and dark with opaque lens. Some possible causes of opaqueness of the lens are pathological defects of the lens epithelium, destruction of the epithelium, swelling of the fibrous tissue, liquefaction of the fibrous tissue and pathological deposits. During the embryonic stages lens opacity can develop as a result of severe metabolic disturbance, lack of vitamins and intoxication.

In contrast to this *Catarata senilis* or *Cataracta traumatica* are diseases acquired at a later stage in life or a sign of old age. Furthermore, opacity of the lens can be caused by infections, allergies or parasites. In 121 examined elephants, 23 (19.0%) had one-sided and one had (0.8%) both-sided opacity of the lens. Godagama (1996) received a similar value (22.1%) in 140 elephants. In 35 zoo and working elephants between 1 to 15 years no opacity of the lens could be found; in 32 elephants over 30 years the mean value for lens opacity was 0.31 ± 0.47 (n = 32). In Pinnawela, the respective value for animals between the ages of 1 to 15 years was 0.25 ± 0.49 (n = 44) and for those between 16 to 30 years 0.57 ± 0.53 (n = 7), which was significantly higher than for zoo and working elephants. In 73 elephants, with normal body growth the mean value for lens opacity per animal was 0.14 ± 0.35, and in 48 elephants with retarded body growth the value was 0.31 ± 0.51. The difference is significant. This last result and the fact that in Pinnawela even young animals show lens opacity, could be an indication of nutritional deficiency or infections due to the never clean concrete and stone floors or the long hours of standing in the glaring sunlight in the river. In working elephants, lens opacity occurred in only the older elephants. According to the owners and mahouts none of these animals had cataracts when they were young. The blood serum-glucose level was within the normal range. This indicates that cataracts in older working elephants are the result of trauma, for instance through a severe blow on the eye (Godagama, 1996), as has also been reported for working elephants in southern India or Myanmar(e.g. Evans, 1910; Krishnamoorthy & Wemmer, 1995).

The most frequent causes of non-infectious conjunctivitis (*C. simplex*) are exogenous, such as a foreign object (sand or dust), mechanical irritation through abnormal positioning of the eyelid or parasites and allergies. Pathogenic conjunctivitis is rather rare and is caused by pathogens such as *Staphilococcus aurea* or *Streptococcus pygogenes*. Loss of bulbous through trauma occurred in four cases, three of which were in Pinnawela. A 42-year-old bull had had both eyes shot by poachers before being captured and brought to Pinnawela. This appears to be a normal procedure among the Buddhist rural community. Killing of an elephant is taboo to them. The blinding of the animal, which causes damage to crops, is not (Kurt, 1992).

In the Elephant Orphanage at Pinnawela 41 (73.2%) of 56 examined elephants had pathological defects of the eyes. In the remaining 65 examined elephants in the surroundings of Kandy and at Dehiwela zoo this was the case in only 25 (38.5%) animals. In Pinnawela 68.3% of the affected elephants stood chained next to each other at night on the bare concrete or stone floors. Those elephants standing on grass or soil far apart under the coconut palms constituted 31.7% of the affected group. Although these standing places were seldom to never cleaned during our stay, the dust and dirt in the buildings with concrete and stone floors was even worse. Most of the working and zoo elephants with eye diseases (44.0%) stood on sandy soil covered with excrements.

A comparison of frequencies of the described eye diseases in relation to age class and living conditions clarifies the above said. In 17 working and zoo elephants between 1 to 15 years eye diseases per animal occurred in 0.65 ±0.0 cases, in 18 elephants between 16 to 30 years the value was 0.33 ± 0.84, and in 32 elephants older than 30 years the value reads 1.25 ± 1.44. In Pinnawela eye diseases occurred in 44 elephants between the ages of 1 to 15 years with a frequency of 1.52 ± 1.34 per animal, in seven 16 to 30 year olds with a frequency of 2.00 ±1.41. These are significant differences. Differences were also to be found in the comparison of occurrences of the above mentioned eye diseases, expressed as general health state by body weight development. In 73 elephants showing normal body weight measures, per animal 0.96 ± 1.27 cases occurred, in 48 with retarded body growth, the value was 1.54 ± 1.34. Again this is a significant difference.

It appears that eye diseases are irrelevant to elephant owners in Sri Lanka, although in the oldest classic Hindu and Buddhist texts the frequent mentioning of eye disease is striking. In the oldest text from the *Hastyayurveda* from the 5th to 4th century BC the *hastipa* (person in charge of the elephant) is responsible for eye diseases. In a paragraph on nutrition the relevance of water and milk for eye sight and the relevance of purified butter for the treatment of the eyes is emphasised. In later times the texts appear to have been wrongly rendered. In any case in former times mahouts applied purified butter and various oils externally on their elephants. Not only in large areas of southern and south-eastern Asia, but also in many western zoos and circuses, it long ago became the practice to paint a ring of oil around the elephants' eyes. This dark ring has no healing powers, although many mahouts and keepers maintain that it is an important prophylaxis against colds. Henning Wiesner (1992) very convincingly points out the apotropaic character of the marking: the black ring around the eye, and therefore its emphasis, serves as protection against the 'evil eye'. This custom was later continued without knowledge of the original connection.

The classical as well as the modern texts on the art of healing in elephants indeed correctly point out that elephants are extremely sensitive to strong sunlight. Therefore, working elephants have to stand in the shade during midday. During the British colonial era they even carried sunshades over their eyes during the day (Krishnamoorthy & Wemmer, 1995). In well run enterprises animals with eye diseases are first kept in shady places before any treatment is applied (e.g. Gerbet, 1994). In contrast to this in Pinnawela, where eye diseases are very frequent also in young animals, the elephants are made to stand precisely between 10.00h to 12.00h and between 14.00h to 16.00h, during the brightest time of the day, in the shallow glaring waters of the Maha Oya river to pose for tourists. Shade is also absent on the grass area where they stand during the remaining time.

Keeping of tame elephants on dirty ground, brutal blows on the eyes, false or inadequate nutrition as well as the absence of shade during the most sun intensive hours of the day are predisposing and usually also triggering factors for eye diseases, which sooner or later can lead to total blindness. It is about time that elephant managers and owners dealt with these problems seriously.

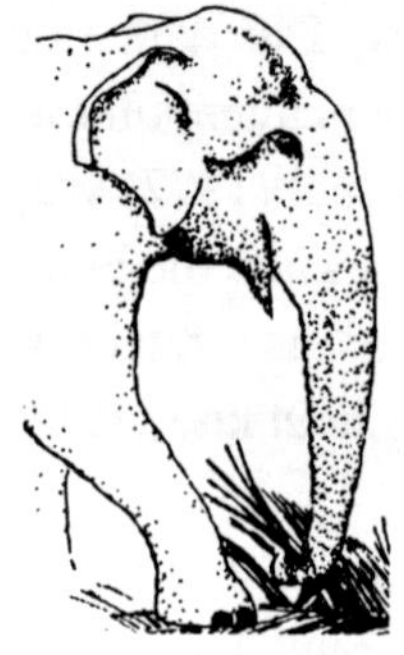

# 4

# Food, Tool Use and Sleep

## 4.1  Body Growth and Daily Food Intake

In Asian elephants large quantities of food stimulate rapid body growth and reproductive success in either sex. Well-nourished females seem to invest in body growth and in their own energy and maintenance resources, but well-nourished bulls invest in slim but tall and impressive body size. In males shoulder height, next to musth is the most prominent attribute for female choice (see Section 5.3), and high quantities of food trigger the phenomenon of musth, when the testosterone level is raised, and behaviours associated with reproduction, olfactory marking and aggression are increased. In captive Asian elephants, body growth varies considerably between individuals and populations. Many females kept in western zoos are obese, with body weight upto 80% more than the weight of wild-living ones, but certain captive elephants in Asian establishments are characterized by retarded body growth, which has been attributed so far to stress during taming, overwork and malnutrition (e.g. Sukumar *et al.*, 1988). As the remaining wild populations of *Elephas maximus* are threatened with extinction, the  importance of captive propagation of the species increases. This applies mainly to populations of jungle based working elephants, where captive reproduction occurs regularly, but so far not on a self-sustainable basis (e.g. Mar *et al.*, 1997; Sukumar *et al.*, 1997). In zoos, reproduction has until recently in Europe been unimportant for the conservation of the species. Nevertheless, studies on food intake and body growth are of considerable importance for the welfare of Asian zoo elephants with a strong tendency towards obesity.

Many captive elephants are orphans which have been separated from their natal family group at a young age, and it can be expected that their body growth is retarded; but so far no studies have been carried out on comparing patterns of body growth in orphans and captive born elephants, which grow up with their mothers at least until weaning. Such a comparison should include animals living under the same management conditions, e.g. the same feeding regime, and they should be of an age upto 20 years, the time when in wild elephants at least 90% of the shoulder height and 75% of the body weight are reached (Sukumar *et al.*, 1988). In the Sri Lankan Pinnawela Elephant Orphanage, which harbours the largest captive herd of Asian elephants, both criteria prevail and comparative studies on feeding ecology and ethology were carried out. Therefore, the present chapter discusses three questions: (1) Is body growth (shoulder height and body weight) of captive elephants similar to body growth of wild ones or is it retarded? (2) What is the origin of animals with retarded body growth? (3) What are the possible reasons for retarded body growth?

## Body growth of captive born and orphaned elephants

In Pinnawela shoulder height and thorax circumference of 24 males and 29 females were measured during July 1997 and (whenever possible) during August 1998 (Table 4.1.1 and Table 4.1.2). Body weight (W; in kg) was calculated by the formula given in Section 2.2. for the same 24 males and 29 females. Measurements of shoulder height and estimates of body weight were compared to values from wild Asian elephants (Sukumar, 1992). Shoulder height and body weight of Pinnawela elephants were considered to be normal when the respective values were within a range of $\pm$ 10% of the values given for wild elephants, and they were considered to be retarded, when they were below minus 10%. Animals which had normal shoulder height, but body weight below the normal value, or vice versa, were considered to have a slightly retarded body growth.

The shoulder heights of all 6 captive born males upto 18 years were within or above the range of $\pm$ 10 % of the appropriate values of wild-living males (Fig. 4.1.1). This was also applied to 8 orphans, but in 5 orphans shoulder height was below minus 10% of the shoulder height of

**Table 4.1.1**  Shoulder height (S, in cm), chest girth (T, in cm) and estimated weight (W, in kg) of male elephants in the Pinnawela Elephant Orphanage. Measurements were taken in July and August 1997 and 1998: captive-born. Abr.: Abbreviation of name; nd: no data; Mo: mother

| Name | Abr | Data of birth (year of birth) | S (cm) | | T (cm) | | W (kg) | | Remarks |
|---|---|---|---|---|---|---|---|---|---|
| | | | 97 | 98 | 97 | 98 | 97 | 98 | |
| *Neonates: 0–2* | | | | | | | | | |
| Arjuna* | aj | 7.2.97 | 110 | 118 | 148 | 156 | 241 | 287 | Mo: Sukumali |
| Isuru* | is | 27.10.96 | nd | nd | 168 | nd | nd | nd | Mo: Anusha |
| Mihindu | mi | (1996) | nd | 108 | nd | 201 | nd | 436 | |
| Nandimithra* | na | 27.9.95 | 140 | 151 | 190 | 169 | 505 | 610 | Mo: Rejina |
| *Infants: 2–5* | | | | | | | | | |
| Anura | an | (1995) | 150 | nd | 220 | nd | 726 | nd | died, Jan. 1998 |
| Aruna* | ar | 10.4.94 | 160 | 185 | 225 | 231 | 810 | 987 | Mo: Komali |
| Sumana | sm | (1994) | 153 | 153 | 236 | 249 | 852 | 949 | |
| Esala* | es | 1.7.93 | 168 | 201 | 241 | 238 | 976 | 1139 | Mo: Kumari |
| Singharaja | sg | (1993) | 170 | nd | 244 | nd | 941 | nd | |
| Jatila | jt | (1993) | 177 | 178 | 257 | 261 | 1097 | 1128 | |
| Tharaka | tk | (1992) | 177 | nd | 247 | nd | 1004 | nd | |

[Table 4.1.1 Contd.

Contd. Table 4.1.1]

*Juveniles: 6–10*

| | | | | | | | | | |
|---|---|---|---|---|---|---|---|---|---|
| Jayathu* | jy | 5.5.91 | 172 | 180 | 245 | 245 | 1012 | 1586 | Mo: Indi (Dehiw.) |
| Kapila | kp | ( 1991) | 164 | 173 | 225 | 227 | 739 | 874 | |
| Chula | ch | (1990) | 172 | 186 | 247 | 255 | 1028 | 1185 | |
| Nilagiri | ng | (1989) | 194 | nd | 273 | nd | 1417 | nd | |
| Rangiri* | rg | 16.8.89 | 193 | 202 | 285 | 283 | 1536 | 1586 | Mo: Anusha |
| Sanka I | sk | (1989) | 179 | 186 | 225 | 221 | 887 | 890 | |
| Ravana | rv | (1988) | 182 | 180 | 245 | 250 | 1071 | 1102 | |
| Suranimala | sr | (1988) | 184 | 200 | 265 | 278 | 1267 | 1514 | |

*Adults: > 15*

| | | | | | | | | |
|---|---|---|---|---|---|---|---|---|
| Kandula | ka | (1980) | 230 | 233 | 363 | 363 | 2819 | 2855 |
| Jandura | ja | (1972) | 269 | 264 | 370 | 373 | 3425 | 3416 |
| Kadira | kd | (1970) | 252 | nd | 382 | nd | 3420 | nd |
| Neela | ne | (1970) | 270 | nd | 402 | nd | 4058 | nd |
| Vijaya | vj | (1966) | 262 | 262 | 430 | 434 | 4505 | 4589 |
| Raja | rj | (1954) | 273 | nd | 455 | nd | 5246 | nd |
| Sanka II | sa | (1944) | 265 | nd | 410 | nd | 4143 | nd |

**Table 4.1.2** Shoulder height (S, in cm), chest girth (T, in cm) and estimated weight (W, in kg) of female elephants in the Pinnawela Elephant Orphanage. Measurements were taken in July and August 1997 and 1998; *captive – born. Abr.: Abbreviation of name; nd: no data; Mo: mother

| Name | Abr | Data of birth (year of birth) | S (cm) | | T (cm) | | W (kg) | | Remarks |
|---|---|---|---|---|---|---|---|---|---|
| | | | 97 | 98 | 97 | 98 | 97 | 98 | |
| *Neonates: 0–2* | | | | | | | | | |
| Ashokamala* | aa | 7.8.98 | nd | 80 | nd | 113 | nd | 65 | Mo: Komali |
| Kumari II | k2 | (1996) | nd | 115 | nd | 170 | nd | 332 | |
| Sapumali | sp | (1996) | nd | 112 | nd | 175 | nd | 343 | |
| Surangani | sn | (1996) | nd | 112 | nd | 166 | nd | 309 | |
| Meena | ma | (1996) | nd | 110 | nd | 170 | nd | 318 | |
| *Infants: 2–5* | | | | | | | | | |
| Amali* | arn | 16. 4.94 | 165 | 176 | 232 | 231 | 888 | 939 | Mo: Mathali |
| Anuradhika* | ak | 15.11.93 | 166 | 168 | 232 | 238 | 893 | 952 | Mo: Anuradha |
| Sandali | sd | (1993) | 147 | 154 | 233 | 240 | 798 | 887 | |
| *Juveuiles: 6–10* | | | | | | | | | |
| Kanthi | kt | (1991) | 175 | 190 | 245 | 246 | 1029 | 1127 | |
| Lasanda | la | (1990) | 180 | nd | 253 | nd | 1129 | nd | |
| Sarna II | s2 | (1990) | 152 | 159 | 225 | 238 | 754 | 883 | |
| Sharmi | sy | (1990) | 147 | 152 | 227 | 234 | 742 | 816 | |
| Shanti | si | (1990) | 147 | 151 | 230 | 246 | 762 | 896 | |
| Thilaka | ti | (1990) | 155 | nd | 219 | nd | 729 | nd | |

[Table 4.1.2 Contd.

Contd. Table 4.1.2]

| | | | | | | | | | |
|---|---|---|---|---|---|---|---|---|---|
| Kirimenika | ki | (1989) | 162 | 174 | 230 | 243 | 840 | 1007 | |
| Kamani | km | (1989) | 180 | nd | 260 | nd | 1192 | nd | |
| Menika* | me | 16. 8.89 | 181 | 184 | 264 | 270 | 1236 | 1315 | Mo: Komali |
| Ramnya | ry | (1989) | 192 | *204* | 279 | 281 | 1465 | 1579 | |
| Noni | no | (1988) | 173 | 173 | 240 | 248 | 977 | 1043 | |
| Pun chi | pc | (1988) | 195 | nd | 296 | nd | 1674 | nd | |
| Sarna I* | sl | 18.12.87 | 200 | nd | 285 | nd | 1592 | nd | Mo: Mathali |
| Tharnmenna | tm | (1988) | 184 | 204 | 250 | 261 | 1127 | 1362 | |
| Soma | so | (1987) | 177 | 180 | 284 | 292 | 1399 | 1504 | |
| *Sub-adult: 11–12* | | | | | | | | | |
| Ninja | nj | (1985) | 192 | nd | 260 | nd | 1272 | nd | |
| Nikini | ni | (1984) | 175 | nd | 253 | nd | 1098 | nd | |
| Sukumali* | su | 19. 9.83 | 207 | 203 | 310 | 310 | 1949 | 1912 | Mo: Kumari |
| Mauri | mu | (1983) | 188 | 186 | 257 | 272 | 1217 | 1349 | |
| Anuradha | ad | (1982) | 222 | 217 | 315 | 330 | 2159 | 2316 | |
| Rejina | re | (1982) | 210 | nd | 315 | nd | 2042 | nd | |
| Ranmali | | (1982) | 184 | 180 | 286 | 280 | 1505 | 1383 | |
| *Adults: >15* | | | | | | | | | |
| Mahaweli | mw | (1981) | 216 | nd | 292 | nd | 1805 | nd | |
| Komali | ko | (1971) | 217 | 218 | 330 | 338 | 2316 | 2441 | |
| Mathali | mt | (1970) | 225 | 225 | 343 | 346 | 2594 | 2640 | |
| Kumari | ku | (1965) | 214 | 216 | 313 | 304 | 2055 | 1938 | |
| Anusha | as | (1949) | 208 | 206 | 322 | 320 | 2113 | 2067 | |

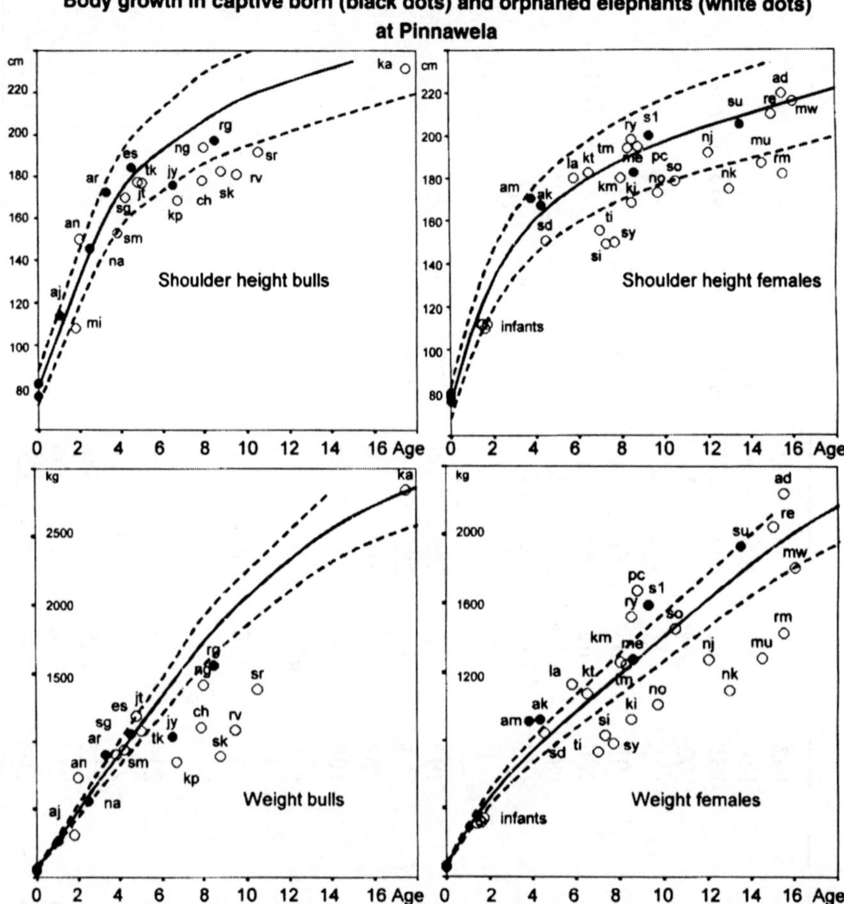



Fig. 4.1.1   Shoulder height (in cm), estimated weight (in kg) and age (in years) of males and fremales in the Pinnawela Elephant Orphanage compared with the mean shoulder height and weight of wild males or wild females, respectively (Sukumar *et al.*, 1997). Animals measured are marked with the abbreviations given in Table 4.1.1

●     : Animals born in Pinnawela

O     : Orphans

_____ : Mean.

----- : Deviation of 10% from mean

wild males. Estimated body weight of 4 captive born males and 6 orphaned males were within or above the 10% limit. The neonate orphan Mihindu and all males between the age of 6 to 10 years had weights below the 10% limit, but the captive born Jayathu and Rangiri, which could be considered as having a slightly retarded body growth, were heavier than orphans of the same age. The shoulder heights and estimated body weights of all 5 captive born females were within or above the ± 10% limit (Fig. 4.1.1). The same was applied to 9 orphaned females. But in 13 orphaned females shoulder height as well as body weight were below the 10% limit. The juvenile female Soma showed retarded growth in shoulder height, but normal weight, and the female Mahaweli showed retarded weight but normal shoulder height, i.e. both were considered as slightly retarded.

In 6 males older than 18 years neither retardation in shoulder height nor weight were observed. The same was applied to the 2 adult females Komali and Mathali, but in the 2 adult females, Kumari and Anusha, shoulder height and body weight were below the given 10% limit. The results confirmed that captive born young elephants grew in the same manner as wild conspecifics, except the Pinnawela-born Rangiri and the Dehiwala-born Jayathu, which showed slight retardations in body weight. 17 orphans (8 males, 9 females) had normal body growth and 20 orphans (5 males and 15 females) had retarded body growth.

Retardations in body growth might be triggered by several factors such as illness or stress due to capture or separation from their families. The most obvious hypothesis, however, would consider the daily food intake of each individual as the main reason affecting body growth.

## The daily food intake

In Pinnawela the diet consisted mainly of four food types: (a) 2 to 3 m long stems of the Kitul palm (*Caryota urens*), cut in half, with a diameter of 20 to 30 cm, (b) leaves of Coconut palms (*Cocos nucifera*) with a length of 2 to 3 m, as well as (c) leaves of Kitul palms with a length of 1 to 2m and (d) branches of the Jack tree (*Artocarpus integra*). The water content of the four food types was measured at the laboratories of the University in Colombo: In *Caryota* stems it was 90%, 50% in *Cocos* leaves, 60% in *Caryota* leaves and *Artocarpus* branches. Daily food intake was measured by direct method as the difference between the food offered in the late afternoon and the amount left over next morning at 8 a.m. The food was weighed on large spring-scales with an accuracy of about 2 kg. As most elephants in Pinnawela were fettered while feeding in such a way that neighbours could snatch food from each other, direct measurements could be collected only from a restricted number of animals kept relatively isolated. In 1997 such measurements were possible with seven animals and in 1998 with five animals. Weight measurement of food eaten by a particular elephant was carried out twice on two consecutive days and then the average was calculated. According to direct measurements daily food intake of juvenile and adult elephants ranged between 44 to150 kg (wet weight) or 11.3 to 41.7 kg (dry weight).

Next to direct measurements daily food intake was estimated by the following indirect method: Focus animals were observed at a distance of 4 to 6 m by two team members. They measured the time (t) necessary to prepare a trunkful of food in seconds (s) and estimated the amount of food (m) prepared and brought to the mouth in grams (g) by simulating the preparation process with samples of the same stems, leaves and branches which the elephant was feeding from. The samples representing a mouth-filling were bundled and weighed on two spring-scales with an accuracy of $\pm 5$ g. Altogether, 1477 data pairs of m and t were established, taken from nine males and eleven females. The maximum intake of a certain food type ($M_{a\,to\,d}$) was given in kg/h and calculated as follows:

$$M_{a\,to\,d} = 3.6m_{a\,to\,d} / t_{a\,to\,d}.$$

In Pinnawela the elephants were fed mainly during the night. Feeding time per 24 h (T) could be estimated with an accuracy of 0.5 h from data collected during observations on sleeping behaviour. For estimates of the daily food intake of particular elephants we had to take into consideration that each elephant fed on four different food types with different maximum intake, and particular individual preferences were not obvious. Hence, the data from direct observations was generalized and it was assumed that each of the four food types was consumed in the same percentages ($p_{a\,to\,d}$) as they were offered; i.e. 57.2% of the daily food intake (wet weight) consisted of *Caryota* stems, 24.2% of *Cocos* leaves, 8.4% of *Artocarpus* branches and 10.2% of *Caryota* leaves. Taking the water contents of each food type into consideration, the daily dry weight intake consisted of 22.6% *Caryota* stems, 47.9% *Cocos* leaves, 13.3% *Artocarpus* branches and 16.2% *Caryota* leaves. This daily eaten mixture of 4 different food types (F) was estimated first as a maximum intake per h, according to the formula: $F = 100 [(m_a /p_a) + (m_b/p_b) + (m_c/ p_c) + (m_d/p_d)]$. The daily food intake was $D = TF$.

The reliability of this rather complicated indirect method was tested in animals where both methods were adopted. For Sanka I the calculated wet weight intake was 101.4% of the actual value, for Chula the respective value read 95.5%, for Rangiri 103.1%, for Neela 107.9%, for Sukumali 98.6% and for Soma 102.3%. The mean value for all 6 tested elephants read $101.4 \pm 4.2\%$; hence calculated values were corrected by a factor of 0.99. The calculated values for daily dry weight intake compared to the  measured ones read 104.3% for Sanka I, 102.2% for Chula, 107.6% for Rangiri, 98.0% for Neela, 108.9% for Sukumali and 91.1% for Soma. The mean value for all 6 tested animals was $102.00 \pm 6.6\%$; hence calculated values were corrected by a factor of 00.98.

Daily wet weight intake ranged from 30.5 kg in the orphaned juvenile female Shanti to 160.2 kg in the adult male Neela. The daily dry weight intake had values between 7.6 kg for Shanti and 40.1 kg for Neela. Among 5 juvenile males between 7 to 9 years daily food intake was larger for the Pinnawela-born Rangiri than for orphaned males. The daily food intake (dry weight) of the juvenile female Shanti with extremely retarded body growth was 7.6 kg and less than that of the Pinnawela-born Menika (18.1 kg) or orphaned juvenile females without retarded body growth. The same applied to the slightly growth-retarded female Mahaweli with 9.5 kg food intake per day and  the growth-retarded adult female Anusha with 12.6 kg food intake per day, as compared to the lactating females Rejina (25.8 kg) and Anuradha (17.5 kg) with normal growth.

Juveniles with normal or slightly retarded body growth daily ate $13.2 \pm 9.2$ kg of *Caryota* stems, $7.7 \pm 4.5$ kg of *Cocos* leaves, $11.1 \pm 5.3$ kg of *Caryota leaves* and $7.3 \pm 4.0$ kg of *Artocarpus* branches. For juveniles with retarded body growth the respective values read $10.3 \pm 6.9$ kg, $7.2 \pm 4.3$ kg, $7.2 \pm 4.3$ kg and $6.7 \pm 6.7$ kg. With the exception of *Cocos* leaves the differences are significant. Concluding from these findings, it can be said that elephants with retarded body growth had a smaller daily food intake than Pinnawela-born elephants and orphans with normal or only slightly retarded body growth (Fig. 4.1.2).

Daily food intake (dry weight) expressed as percentage of body weight ranged from 0.6% in the old adult female Anusha to 2.1% in the

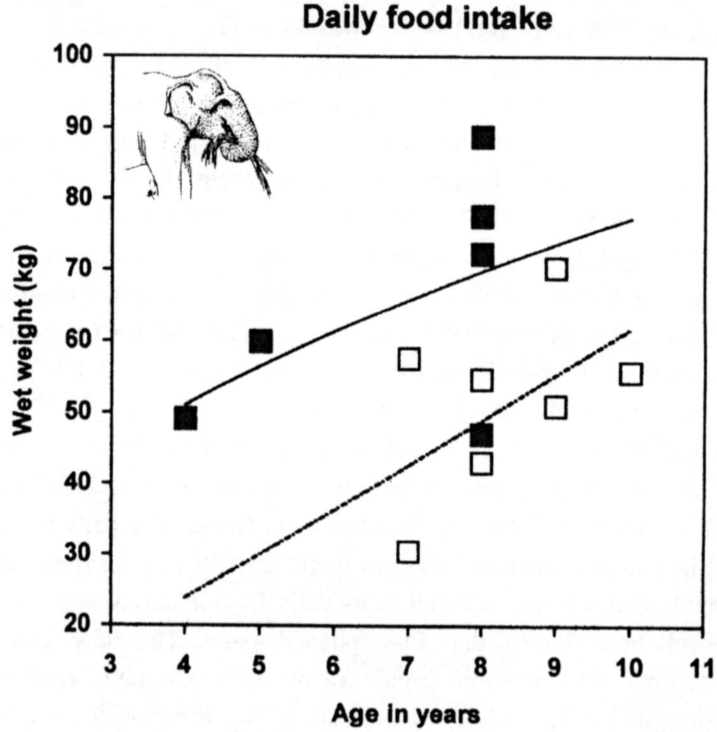



**Fig. 4.1.2.** Mean of daily food intake according to age and social status (Data from Kurt, 2001)
■ Pinnawela-born
□ Orphans

captive born infant Anuradhika. In general, younger animals had higher values than older ones. Some orphans with retarded body growth had relatively small values, e.g. Nilagiri (0.8%) and Shanti (0.9%), but the juvenile male Ravana had a rather high value (1.6%). Similar results were found when the daily food intake (dry weight) was expressed in g per kg of physiological body weight ($W^{0.73}$). In infants, juveniles and sub-adults, i.e. growing animals, the values ranged between 56 g/kg in Nilagiri and Shanti, respectively, and 135 g/kg in Anuradhika. The corresponding range for adults read 52 g/kg in Mahaweli and 94 g/kg in Jandura.

Before concentrating on questions concerning different parameters of feeding, such as feeding time, frequencies of mouth fillings and amount of food collected per mouth filling, a comparison is given of our data with results on daily food intake found by other researchers on elephants living under conditions different from those of Pinnawela.

## Comparison with wild, working and zoo elephants

Wild Asian elephants spend between 9 to 19 hours daily in searching and preparing food (McKay, 1973; Vancuylenberg, 1977; Kurt, 1992), and African elephants are believed to consume 1.5% (dry weight food) of their body weight daily (Laws *et al.*, 1970). But sub-adult and adult wild Asian elephant bulls can consume upto 1.9 % of their body weight in just 12 hours of feeding indicating much higher daily quantities; and crop-raiding bulls are capable of consuming upto 12 kg (wet weight) per hour within 6 to 7 hours, i.e. an amount with a dry weight of upto 1.5% of their body weight (Sukumar, 1992). In a clan of wild elephants in the Sri Lankan Ruhuna National Park juveniles, sub-adult and adult females prepared food and ate during 10 to 11 hours daily and consumed 3.8%, 2.7%, and 2.2% of their body weight in the form of short grass (dry weight). Sub-adult and adult bulls prepared food and ate in 10 to 11 hours daily and consumed 1.9% and 1.3% of their body weight (Kurt, unpublished).

The Pinnawela elephants spent between 6 to 8.5 hours daily in feeding, and were fed a mixture of *Caryota* stems, *Cocos* and *Caryota* leaves and *Artocarpus* branches totalling a wet weight between 4.7 kg (Shanti) to 21.6 kg (Neela) per hour. Per day they consumed a dry food weight between 0.6% (Anusha) to 2.1% (Anuradhika) of their body weight, averaging less than the amount  indicated by studies on wild Asian elephants. Furthermore, they ate less than working elephants in Kerala (southern India). This applied even to animals with normal body growth: For example, the adult bull Neela with an estimated body weight of 4058 kg consumed 40.1 kg (dry weight) per day, the sub-adult female Sukumali (W: 1931 kg) consumed 21.2 kg, and  the infant male Esala (W: 1060 kg) 15.0 kg. The daily nutritional requirement for Kerala elephants, which depended mainly on *Caryota* leaves, was estimated to be 46.0 kg for a bull of 4000 kg, 35 kg for a sub-adult female of 1900 kg, 24 kg for a young bull of 1100 kg, following the formula 142 g (dry matter) per kg physiological body weight ($W^{0.73}$) for growing elephants and 108 g per kg physiological body weight for full-grown elephants (Ananthasubramaniam, 1992). Although the data concerning water content of food were the same as in the Indian study, results were below the recommendations given above and read 93 to 135 g/kg in infants, 56 to 107 g/kg in juveniles, 85 to 99 g/kg in sub-adult females, and 47 to 93 g/kg in adults. In a clan of wild elephants feeding mainly on short grass in the Ruhuna National Park, the daily dry weight intake per kg physiological body weight was 241 g in juveniles, 205 g in sub-adult and 178 g in adult females, as well as 154 g in sub-adult and 125 g in adult males (Kurt, unpublished).

The relatively low daily food intake in the Pinnawela population is not due to portions being too small for the daily quantitative requirement, as none of the animals could eat up all the food offered. The possible reasons to explain the conspicuous difference between the food intake of Pinnawela elephants and working elephants in Kerala or wild ones in the Ruhuna National Park, as well as the fact that certain elephants had retarded body growth, can be discussed under four aspects: (1) The health of elephants, (2) the nutritional and daily metabolic energy cost, (3) the digestive abilities of elephants, the preparability and digestibility of food plants, as well as (4) behavioural abilities and constraints.

(1) During the studies the veterinary care for Pinnawela elephants was practically absent and it was obvious that certain elephants suffered under digestive problems, e.g. Ravana, whose body weight had to be considered as retarded, although he had the comparatively highest relative food consumption of all juveniles with a dry matter intake of 1.6% as compared to body weight and 107 g per kg physiological body weight. Ninja, Nikini, Kirimenika and Sanka I could also be considered as sick and apathetic; all of them had extremely retarded body growth and were not integrated in the social set of the Pinnawela herd (see Chapter 5).

(2) The daily metabolic energetic cost of Pinnawela elephants could be considered as comparatively low. Juvenile bulls were occasionally engaged in play-fighting and adult bulls had to carry food to the night

quarters – an easy task compared to the daily work of timber elephants hauling and heaving heavy stems. The main activities of Pinnawela elephants comprised hardly more than preparation of food and walking between night quarters, grassy area and river, totalling some 2 to 3 km daily, which was less than the distances covered by wild and working elephants (Kurt, 2001). With working elephants the daily food requirement fluctuate enormously according to the distance covered and the load carried. The phenomenon has been well-known since the times of the British administration in South Asia, and since then, hard-working timber elephants in India have received, according to their body size and work, additional daily food concentrates of boiled rice, wheat, sugar etc., which were offered next to natural food found by the animals themselves during the night in nearby forests (e.g. Evans 1910; Toke Gale 1974). Man-made food concentrates make preparation of food unnecessary and daily food intake may take place in a relatively short time, as the apparent digestibility of a certain type of food correlates negatively with its crude fibre contents (Löhlein, 1999). Asian elephants living in western zoos and tending to obesity ate daily rations of hay or food mixtures with a dry weight of 1.03% to 1.30% of their body weight (Foose, 1982; Hackenberger & Atkinson, 1982; Hackenberger, 1987; Fujikara *et al.*, 1989), which is about as much as the daily rations for the Pinnawela elephants. Neither zoo nor Pinnawela elephants live under physical stress. One might ask, however, whether it is feasible to compare relative rations of extremely different food types such as hay, carrots and pellets on one side and tough leaves and palm trunks on the other.

(3) Wolfgang Löhlein (1999) used chrome oxide as a marker to measure the rate of passage during digestion in 4 elephants from Pinnawela and 7 from the Munich Zoo. In the juveniles Sandali and Esala from Pinnawela the marker first appeared after 14.5 hours and the last evacuation containing chrome oxide was voided after 45.0 hours. In 2 zoo elephants of similar ages the respective values read 14.0 hours and 43.8 hours. In 2 adult Pinnawela bulls the marker first appeared after 21.8 hours and lasted up to 57.0 hours. In four adult zoo elephants the respective values read 21.0 hours and 58.5 hours. Significant differences between the elephants from Pinnawela and those from Munich were not found, although their food was completely different.

In zoo and circus elephants the apparent digestibility for hay was found to be between 29.0% to 43.8% (Benedict, 1936; Reichard *et al.*, 1982; Fujikara *et al.*, 1989). South Indian working elephants, which were fed only with *Caryota* leaves, had an apparent digestibility of 64.5% to 82.5%. The digestibility of energy contents of hay was found to be between 40.0% to 54.9% and from *Caryota* leaves between 54.9% to 75.0%. The reason for this unexpected difference is not clearly known. But it is known that green fodder which seems to be tough, such as *Caryota* leaves, has a higher relative water content than dry hay which seems to be tender. It could be assumed that a relatively high water content enhances the efficiency of digestion of food nutrients in general and crude fibres in particular, and a relatively low food intake is digested better than a larger one (c.f. Ananthasubramaniam, 1992). Furthermore, it must be considered in the present context that the microfauna in the digestive tract of zoo elephants is, due to medicaments, less diverse than in wild or working elephants (Smith *et al.*, 1982).

(4) The behavioural abilities and the constraints prevailing under the rather restricted conditions during feeding periods in Pinnawela are discussed in the next section.

## Feeding time and food portions

To a certain extent, the food of Pinnawela elephants was prepared by man; nevertheless, it still demanded some preparation by the elephants. Our study led to the opinion that Pinnawela elephants received enough food, but some of them had not enough time to prepare and eat enough of it. The daily time spent on feeding ranged from 6 hours in the orphaned juvenile male Sanka I to 8.5 hours in the adult males Jandura and Kandula. In juveniles with retarded body growth the daily time spent on feeding had a mean of 6.5 ± 0.3 hours, i.e. significantly shorter than for captive born animals and juveniles with normal or slightly retarded body growth with a mean of 7.7 ± 0.7 hours. Infant and juvenile males spent a mean time of 6.5 ± 0.3 hours on feeding and infant as well as juvenile females spent significantly more time on feeding with 7.5 ± 0.6 hours per day. Juvenile animals with retarded body growth spent less time on feeding than all other elephants observed. They spent considerable time on weaving, i.e. stereotypical behaviour (see Section 5.4), during which food may be eaten, but in considerably smaller amounts. But juvenile males also had comparatively short feeding times. They were fettered next to each other and often engaged in aggressive behaviour, which must have hindered them from feeding and hence caused indirectly slightly or fully-retarded body growth in all juvenile males.

But differences in daily food intake were also due to differences in the amount of food per trunkful (Fig. 4.1.3) and the time spent in preparing it. For *Caryota* stems juveniles with retarded growth had mean amount per trunkful of 79.2 $\pm$ 59.3 g and used a mean time of 33.2 $\pm$ 19.5 s to prepare it. For juveniles with normal or slightly retarded growth the according values read 110.3 $\pm$ 51.2 g and 38.4 $\pm$ 18.1 s. The differences were significant for the amount of food per trunkful and preparation time. For *Cocos* leaves, juveniles with normal or slightly retarded body growth the mean trunkful was 72.2 $\pm$ 38.9 g and the mean preparation time 37.3 $\pm$ 12.4 s. For juveniles with retarded growth the value for trunkful was significantly less with 59.5 $\pm$ 28.5 g, but the time was not significantly different with 35.1$\pm$ 14.8 s. Feeding on *Caryota* leaves, the juveniles with normal or slightly retarded growth had the following mean values: 115.1 $\pm$ 46.4 g and 39.5 $\pm$ 11.5 s. For juveniles, with retarded body growth the according values read: 62.1 $\pm$ 370.2 g and 35.0 $\pm$ 16.5 s. The differences were significant. For *Artocarpus* branches the comparison between juveniles with normal or slightly retarded body growth and juveniles with retarded growth resulted in the following values 63.1 $\pm$ 34.0 g and 43.0 $\pm$ 34.1 g and 33.5 $\pm$ 11.1 and 31.1 $\pm$ 22.6 s respectively.

Such comparisons are of special interest when considering animals of the same age but with different body growth. The Pinnawela-born Rangiri, for e.g., collected more *Caryota* stems per trunkful than Nilagiri and Chula as well as more *Cocos* leaves and *Artocarpus* leaves than Nilagiri, but used more time to prepare them. Nevertheless, the maximum intake per hour of *Caryota* stems, *Cocos* leaves and *Artocarpus* leaves were significantly higher for Rangiri than for Nilagiri. Rangiri prepared more *Caryota* stems per h than Chula, Ravana, and Sanka I, more *Cocos* leaves than Sanka I, and more *Artocarpus* leaves than Chula and Sanka I. Ravana collected more food per trunkful than most of the other orphaned juvenile males, but in most cases required significantly more time to prepare it. However, Ravana quite often had a larger

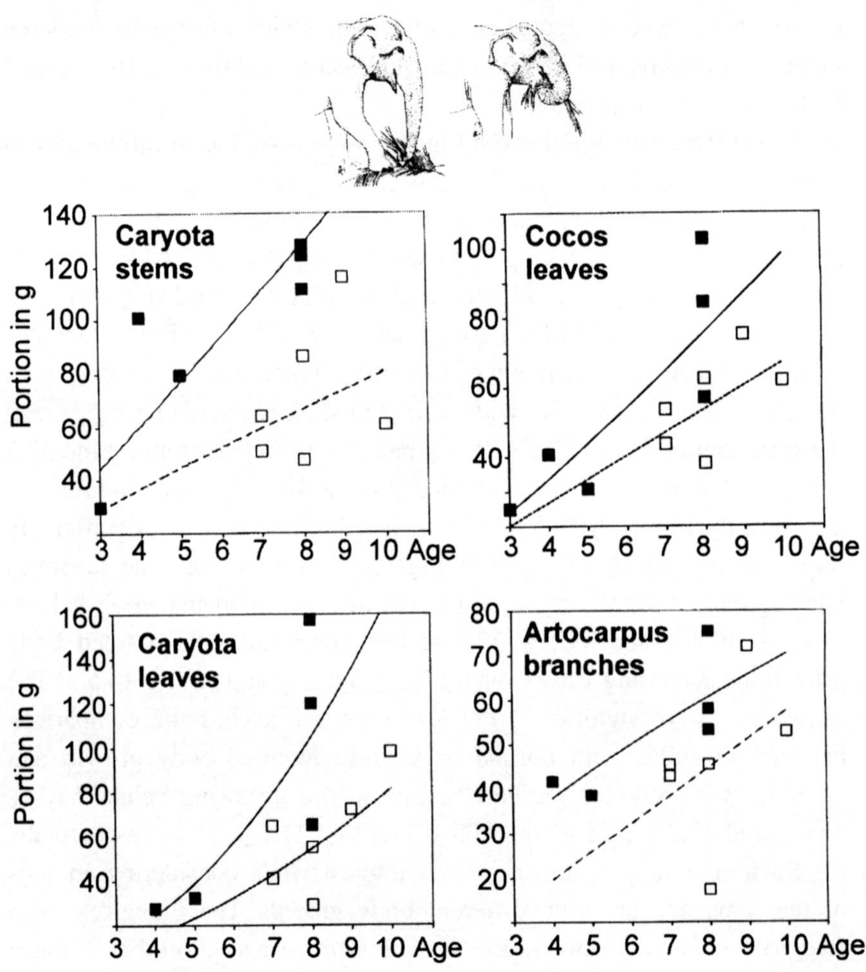



**Fig. 4.1.3**   Mean wet weight of food portions prepared from *Caryota* stems, *Cocos* leaves, *Caryota* leaves and *Artocarpus* branches according to age and body growth.

■  Normal body growth

□  Retarded body growth

maximum food intake per h than other orphaned juvenile males. Only the adult bulls required more time than Ravana to prepare a portion of *Cocos* leaves as well as *Caryota* stems and leaves, but ate twice the amount of Ravana. The extremely growth-retarded juvenile female Shanti, e.g., needed about the same time to prepare a trunkful of *Cocos* leaves as the slightly growth-retarded juvenile female Soma, the normal-growing juvenile female Punchi and the lactating female Rejina, but collected less food per trunkful. In all four food types Shanti collected significantly less food per trunkful than Menika and Punchi. To prepare *Cocos* leaves, *Caryota* stems and leaves, the slightly growth-retarded adult female Mahaweli needed insignificantly less time than the sub-adult females Rejina and Anuradha, but collected significantly less food per trunkful.

Maximum food intake per h depends first of all on the virtuosity of the trunk tip to prepare a relatively large amount of food and less on the frequency of bringing the food to the mouth. Daily food intake is also a function of the daily feeding time. Some of the juvenile males were fettered next to each other and hence often engaged in play-fighting. Accordingly, feeding time was shortened. However, adult bulls were kept in such a manner that they had no direct contacts with other elephants. Their feeding time was considerably longer.

## 4.2  Methods and Traditions of Food Preparation

Asian elephants are occupied with feeding whenever possible. They use a number of feeding behaviour patterns in affiliative as well as aggressive social interactions, for example by preparing food for offspring and feeding partners or alternatively by stealing prepared food from a lower ranking partner. All these behaviours were also observed in the Pinnawela herd. Certain elements of feeding behaviour are included in certain forms of skin care  (e.g. beating branches against the body) and are only absent during bathing and sleeping. Since the digestive system of elephants is not very efficient, an optimal feeding strategy would be required to prepare as much food as possible and bring it to the mouth as fast as necessary. Elephants have rationalised the appetitive behaviour of feeding to a high degree. They can simultaneously walk and chew one food portion, while the next one is prepared and brought to the mouth by the trunk, and when a food portion has already been prepared before the former portion is thoroughly chewed, it is stored between ventral trunk base and lower lip or tusk when this is present

The foregoing chapter stated that certain orphaned elephants in Pinnawela had retarded body growth. Furthermore, it became clear that some orphans were smaller and lighter and ate less than the captive born elephants. Sub-adult and adult reproducing females were taller and heavier and had a higher daily food intake than non-reproducing females of the same age. Furthermore, we postulated that in orphans with retarded body growth the process of socialisation was disturbed. The corresponding hypothesis reads: females who reproduced know more effective methods in food preparation than childless females and transmit their experiences to their offsprings. Accordingly, their offsprings grow faster and reach reproductive status earlier than socially isolated orphans.

As the methods of food preparation can differ between wild mother-offspring groups living in different environments, it has been suggested earlier that they are transmitted from older to younger animals, and could therefore be considered as matrilinear traditions (Kurt, 1992). In Asian elephants matrilinear traditions have been described or suspected in several respects, ranging from annual migrations to vocalizations, (Wemmer & Mishra, 1982) as logical consequences of extraordinary capacities of learning and discovering new methods to cope with an array of problems as well as an astonishing power of recollection (eg. Rensch, 1957; Shoshani, 1998).

In the following pages, three questions will be dealt with: (1) How do captive elephants prepare different food types? (2) Are methods of food preparation different between captive born elephants with normal growth and orphaned elephants with retarded growth? (3) Is the similarity of methods in mother-offspring-pairs  more pronounced than between other members of the study-herd?

A team of 20 researchers collected data during the *Esala Perahera* in Kandy in July and August 1997 as well as during a five month period in 1997 and 1998 in the Pinnawela Elephant Orphanage. In Kandy, we observed each of 26 animals (8 tuskers, 6 maknas and 12 females) during a period of 2 h while they prepared *Caryota* stems. In Pinnawela, we monitored food preparation in 30 animals of different ages and origins

(14 males and 16 females), when eating and preparing the four main food types (stems and leaves of *Caryota*, leaves of *Cocos*, branches of *Artocarpus*). While observing elephants we concentrated on single acts of preparation. An act of preparation started, when the focus animal touched a part of a food plant and ended when particles from this plant were brought to the mouth. In Pinnawela, 5825 acts were described; 1620 stemmed from 20 individuals chopping up *Caryota* stems; 1707 acts from 22 elephants preparing *Cocos* leaves and 1364 acts were observed in 17 elephants preparing *Caryota* leaves; 1134 acts of food preparation were observed in 16 animals, while they were separating leaves and twigs from *Artocarpus* branches.

The food offered to the Pinnawela elephants was rather monotonous as compared to that of wild conspecifics (c.f. McKay, 1973). Many grass species are preferred parts of the diet of wild elephants, but the Pinnawela animals never obtained grass. They tried, however, to eat small portions of different grass species on their daily route to the river. The four main food types stemmed from crop plants. They were parts of full-grown trees which are hardly used by wild elephants, e.g. stems of palms with a height of several meters hardly belong to the diet of wild elephants. But to a certain extent palm leaves or branches of the Jack tree were morphologically similar to young trees and palms, the main food of elephants living in rain forests.

## Methods of food preparation

Ethologists observing elephants in the open grasslands in the Ruhuna National Park would hardly face problems of describing methods of food preparation. The short grass is kicked with the forefeet, bundled with the trunk and then rubbed with the 'trunk hand' (i.e. the distal part of the trunk) against a forefoot to separate substrate and insects from edible parts. Certain adult bulls reach the same goal by lifting up bundles of the scalped grass portions with the trunk and dropping them in such a manner that the omnipresent strong Kachan wind cleans the roots. In Pinnawela, things were not as easy. Each elephant received an impressive bundle of tree trunks, leaves and branches in the evening. After checking the food, they started to prepare the food by comminuting it, and separating edible parts from unwanted ones.

Parts of plants to be prepared for feeding could be fixed by one forefoot, the trunk, the tusks or the mouth before and while manipulating another part with one forefoot, the trunk hand, one tusk or the mouth and the molars. The following patterns of food manipulation were distinguished: (1) Parts of food were bitten off with the molars, or (2) torn off with the trunk or with trunk and forefoot. (3) Forefoot and trunk broke off food parts. (4) Bark was pulled off. (5) Branches were bitten until they split. (6) One forefoot or the trunk turned off food. (7) Stems were slit with one tusk. They could be (8) kicked, (9) crushed or (11) planed until some parts broke off, or they were (10) levered, i.e. one stem was used as abutment to break a second one with a forefoot. (12) Broken-off parts of stems and small leaves were swept to heaps before being brought to the mouth; (13) small food particles were plucked off. (14) Bundles of leaflets were collected in the trunk hand and then stripped of from the leaf vein. (15) Large branches were held in the trunk and beaten to the ground until they broke. Palm stems were repeatedly lifted with the trunk or the tusks and then (16) dropped or (17) thrown to the ground until they broke into pieces.

Since food plants were fixed and manipulated with forefeet (1), trunk (2), tusks (3), as well as mouth and molars (4) the number of preparation methods was higher than the forms of manipulations distinguished above. Altogether we defined 33 methods of food preparation (Table 4.2.1) eight of them were found only in the elephants studied in Kandy. The methods found in Pinnawela are presented in Fig. 4.2.1. Each of the 33 methods was marked with a four-figure number; the first figure stood for the organ which fixed the food plant, the second figure stood for the manipulative organ. If one or both forefeet were used for fixation or manipulation, respectively, the number used was 1.2 that stood for fixation or manipulation with one or both tusks, 3 for fixation or manipulation with the trunk and 4 was used when food was fixed or manipulated with the mouth and/or the molars. The last two figures in our four-figure code were those given to each of the 20 manipulative patterns; e.g. 1307 means 'snap off (twigs or leaves) with the trunk, while the part of the plant concerned is fixed with one forefoot', and 4317 means 'snap off twigs and leaves with the trunk, while the plant is fixed with the mouth'.

**Table 4.2.1** Description of food preparation methods. Numbers given in brackets mark methods which were only found in Kandy

| Number | Element | Description |
| --- | --- | --- |
| 3401 | 'bite off' | Trunk brings food to mouth, and parts of it are bitten off. |
| (1202) | 'tear off' | A stem is brought ± in vertical position. With a tusk bark is pulled off. |
| 1302 | 'tear off' | Food is fixed with a forefoot. The trunk tears off parts. |
| 4302 | 'tear off' | Food is fixed with the mouth. The trunk tears off parts. |
| 1103 | 'break off' | Food is fixed with a forefoot, the other forefoot breaks off parts. |
| 1303 | 'break off' | Food is fixed with a forefoot. The trunk breaks off parts. |
| 3103 | 'break off' | Food is fixed with the trunk. One forefoot breaks off parts. |
| 4303 | 'break off' | Food is fixed with the mouth. The trunk breaks off parts. |
| (1204) | 'pull off' | A stem is brought ± in vertical position. With a tusk parts are pulled off. |
| 1304 | 'pull off' | Food is fixed with a forefoot. The trunk pulls off parts. |
| 4304 | 'pull off' | Food is fixed with the mouth. The trunk pulls off parts. |
| 3405 | 'bite' | Wooden parts are taken in the mouth. Molars bite until wood splits. |
| 1106 | 'turn off' | Food is fixed with a forefoot. The other forefoot turns off parts. |
| 1306 | 'turn off' | Food is fixed with a forefoot. The trunk turns off parts. |
| 4306 | 'turn off' | Food is fixed with the mouth. The trunk turns off parts. |
| (3207) | 'slit' | A stem is brought ± in vertical position. The tusk slits it. |

[Table 4.2.1 Contd.

Contd. Table 4.2.1]

| | | |
|---|---|---|
| 1108 | 'kick off' | Food is fixed with a forefoot, the other forefoot kicks off parts. |
| 3108 | 'kick off' | Food is fixed with the trunk. A forefoot kicks off parts. |
| 1109 | 'crush' | A stem is fixed with a forefoot, the other forefoot crushes it. |
| (4109) | 'crush' | A stem is brought in vertical position and kept with the mouth. One forefoot crushes it. |
| (2309) | 'crush' | A stem is lifted with both tusks and then thrown to the ground. |
| 1110 | 'lever' | A stem is put over another stem, fixed with a forefoot, the other forefoot presses until the stem brakes. |
| 3110 | 'lever' | As 1110, but stem is fixed with the trunk. |
| 1111 | 'plane' | A stem is fixed with a forefoot, the other forefoot planes off parts. |
| 0312 | 'sweep' | Small food particles are swept together with the trunk and brought to the mouth. |
| 1314 | 'pick off' | Food is fixed with a forefoot. The trunk fingers picks small particles and collect them in the trunk hand. |
| 4314 | 'pick off' | As 1314, but food is fixed with the mouth. |
| 1315 | 'strip off' | Food is fixed with a forefoot. The trunk strips off leaves or other parts. |
| 4315 | 'strip off' | As 1315, but food is fixed with the mouth. |
| 0316 | 'beat' | Food is collected in the trunk hand and beaten against the ground, a forefoot or against the body. |
| (0317) | 'drop' | A stem is lifted with the trunk and then dropped to the ground. |
| (0318) | 'throw' | A stem is lifted with the trunk and then thrown to the ground. |
| (2318) | 'throw' | A stem is lifted with both tusks and then thrown to the ground. |

| 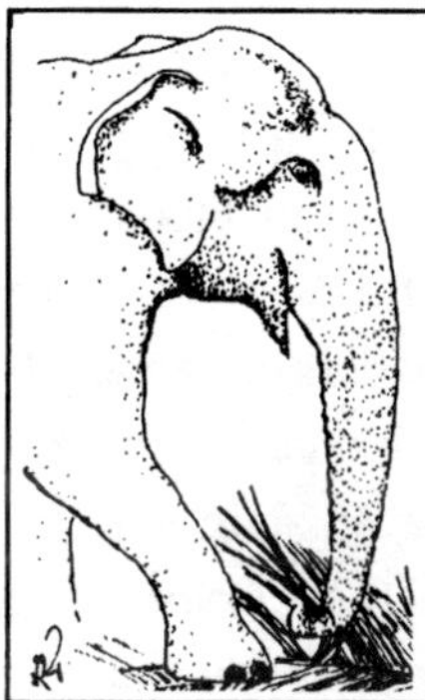 | | Manipulation (preparing of food portion) | | |
|---|---|---|---|---|
| | | **(Other) forefoot**<br>**1**  | **Trunk hand**<br>**3**  | **Mouth (molars)**<br>**4** 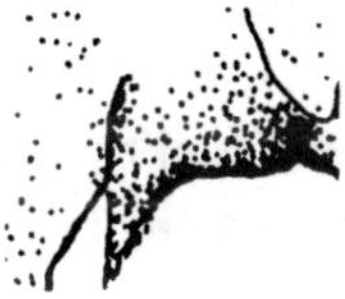 |
| **Fixation (hold the food plant)** | **1** <br>Forefoot | 1103 break off<br>1106 turn off<br>1108 kick off<br>1109 crush<br>1111 plane<br><br>(inefficient methods) | 1306 turn off<br>1302 tear off<br>1303 break off<br>1304 pull off<br>1314 pick off<br>1315 strip off | |
| | **3** <br>Trunk hand | 3103 break off<br>3108 kick off<br>3110 lever | | 3405 bite<br>3401 bite off<br><br>(inefficient methods) |
| | **4** 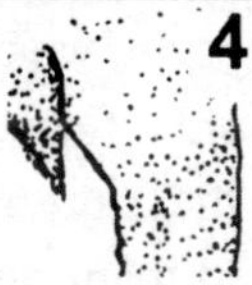<br>Mouth (lip &<br>trunk base) | | 4303 break off<br>4306 turn off<br>4314 pick off<br>4302 tear off<br>4304 pull off<br>4315 strip off<br><br>(inefficient methods) | |

**Fig. 4.2.1**  Methods of food preparation in the Pinnawela herd according to fixation and manipulation. Methods were distinguished as efficient and inefficient.

From *Caryota* stems only the sago – like 'marrow' is eaten. It looks and tastes (also for humans) like candyfloss. This marrow is protected by a tough wood and bark rich in siliceous crystals. *Caryota* bark is traditionally used to cover the ground of the standing places where captive elephants in Sri Lanka are chained. 26 elephants observed in Kandy used 23 methods to crush and chop *Caryota* stems, 8 of these methods were never observed in Pinnawela. In many of these 8 methods tusks played an important role (in one case also tushes) to split, debark or break the stems. In Kandy 5 tuskers and 5 maknas used 9 methods, 6 of which were used by both. But tuskers never levered stems on an abutment, and only one makna used one of his short tushes to slit open a *Caryota* stem. This method was normally adopted by tuskers. But tuskers always used only one of their tusks. It hardly took them one minute to comminute a large stem. Full-grown adult bulls could smash stems alone by their weight and power (e.g. elements 1109, 2309, 4109). Females used at least 4 to 5 minutes to reduce a *Caryota* stem to smaller pieces, before they could prepare it to edible size. But 2 young bulls and 1 adult female which we observed in Kandy, used 12 to 16 minutes to smash a *Caryota* stem. They lifted it and let it fall or threw it with full trunk power to the ground. These two methods were developed most probably during the process of taming, when the forefeet were fettered close to each other and could hardly be used for food preparation.

The 24 elephants we observed in Pinnawela were fed with split *Caryota* stems, hence the preparation of edible parts was easier. Altogether they adopted 19 different methods. But only two of them were used regularly. In 43.5% of 1720 acts observed they tore off parts of the marrow with the trunk, and in 18.8% they kicked them off. Three methods occurred occasionally (e.g. in 5.5% to 5.7%) 14 methods could be considered as rare, since they occurred only in 2.9% or less.

The 22 monitored Pinnawela elephants ate only the lanciform leaflets of the *Cocos* leaves which were 40 to 60 cm long and distained the veins which were 90 to 180 cm long. To prepare *Cocos* leaves they adopted 16 methods. The three most prominent were 1315 'strip off' (33.4%), 1302 'tear off' (18.1%) and 4315 'strip off' (with mouth fixation; 8.8%). Two of the methods occurred occasionally (4302 and 3401). But the rest was used in less than 5% of all acts observed.

From the *Caryota* leaves the Pinnawela elephants ate the fish-tail-like leaflets which were 10 to 18 cm long, but not the unwieldy veins. The five most common methods were 1302 'tear off' (with forefoot fixation, 27.0%), 1314 'pick off' (22.6%), 1315 'strip off' (17.9%) 4302 'tear off' (with mouth fixation 10.4%) and 4315 'strip off' (with mouth fixation; 7.6%). The other 12 methods occurred 3.7% or less. From *Artocarpus* branches only the leaves and thin twigs were eaten. The 16 elephants observed used 15 methods to prepare the branches. Eight of them occurred in more than 6% of all acts observed, the other 7 occured rarely, i.e. in 4.1% or less often.

The four food types demanded different preparation methods or at least different frequencies of the methods were distinguished. *Caryota* stems and *Cocos* leaves were prepared more easily than *Caryota* leaves and *Artocarpus* branches, as can be seen from a relatively low number of methods which were adopted regularly. The number of methods to prepare food depended on experience and power, i.e. from social classes and social status, as discussed below:

## Methods and social classes

Three adult bulls in Pinnawela used altogether 9 different methods for preparing all 4 food types. The sub-adult female Ninja used 7 methods, but the 7 mothers altogether used 20.5 captive born neonates adopted altogether 22 methods, but 3 orphaned neonates only 10. 3 captive born juveniles knew at least 19 different methods and 8 orphaned juveniles at least 18. When comparing the number of methods used by individuals to prepare specific food types, the following values were found for each age class: Captive born neonates adopted in the mean 9.9 ± 2.8 methods and orphaned neonates 5.0 ± 2.9 methods, i.e. significantly less. The 3 mothers with the smallest offspring used a mean of 9.5 ± 1.9 methods, and the 3 mothers with the oldest offspring a mean of 10.5 ± 1.9 methods. The values of all mothers were significantly higher than the values of captive born juveniles or the values of the non-reproducing sub-adult Ninja with a mean of 6.5 ± 0.7 methods. From these results it can be concluded that mothers and their neonates used more different methods to prepare their food than bulls or orphaned young animals. But the difference did not concern juveniles. These findings may give the first indication that offsprings learn a wider array of food preparation methods in contact with their mothers. Furthermore, one could assume that juvenile found out efficient methods after they acquired the necessary knowledge.

Differences between social classes were also found when methods were categorized as 'efficient' or 'inefficient'. A method was considered as being efficient, when the mouth and/or the molars were neither used for fixation nor manipulation of the food, which means that the process of mastication was not interrupted while an efficient method was applied. In the Pinnawela herd 11 efficient methods were noted, and they were adopted in 68.3% of all the 5825 observed acts of food preparation, i.e. they were used much more frequently than inefficient methods. Food preparation methods could also be categorized as 'direct' or 'indirect'. Methods in which the trunk hand is used for fixation and/or manipulation of food plants were considered as direct, since the trunk hand could bring food portions directly to the mouth. In methods where a forefoot, the mouth, the tusks or tushes fixed or manipulated the food plants, preparation was more complicated, i.e. inefficient because the trunk hand had to bundle the single food portions first, before it could bring them to the mouth. In the Pinnawela herd only 6 indirect methods could be observed, and they occurred only in 7.7% of all acts. Indirect methods could also be considered as inefficient. To prepare *Caryota* stems inefficient methods like 1106/4306 'turn off' occurred only in orphaned juveniles, and the orphaned neonate Anura was the only member of the herd to adopt the extremely inefficient method 1111 'plane', with a very small amount eaten per time.

To prepare *Cocos* leaves, captive born juveniles adopted efficient methods in 77.8 ± 16.0% of all acts observed; orphaned juveniles, however, used them only in 66.5 ± 21.9%. For younger elephants the method 1314 'pick off' seemed to be very efficient. To eat *Caryota* leaves and *Artocarpus* branches, captive born neonates used this method in 7.6 ± 7.7% of all acts. But captive born juveniles used this method in 51.8 ± 25.6%, i.e. significantly more often. Orphaned juveniles adopted 'pick off' in only 3.0 ± 4.5%, i.e. significantly less. Another efficient method for young elephants was 0312 'sweep'. With the exception of Jatila, this method was only found in captive born-neonates (7.4 ± 4.5%). In the 3 mothers with the youngest offsprings 'sweep' method was found in 2.8 ± 4.1%. The 3 females with the oldest offsprings used 'sweep' significantly more often (5.6% ± 6.0%).

The 'sweep' method could only be used, after reducing parts of food plants to small pieces by other methods and by indirect and inefficient ones. To these categories belonged for e.g. 4302 'tear off', 4304 'pull off' or 4315 'strip off'. To prepare *Artocarpus* branches, *Cocos* and *Caryota* leaves, the three mothers with the youngest offspring used these 3 methods in 11.6 ± 10.0% of all acts they performed. This was significantly more often than the 3 mothers with the oldest offspring (5.6 ± 6.0%). Furthermore, it was observed that the mothers with the youngest calves used direct and efficient methods relatively seldom. The tendency became clear that with increasing age of the neonate or infant direct and efficient methods are adopted more and more often, but 50 to 60 months after the last parturition, which means shortly before the next one, direct and efficient methods become relatively rare again. With the adoption of inefficient and indirect methods, mothers had the possibility of 'feeding' their small neonates and providing them with food reduced to small pieces. A similar behaviour was also observed in wild-living herds. Furthermore, in the Ruhuna National Park matriarchs fed younger members of their family units with grass bundles in moments of supposed danger (Kurt, 1991).

## Similarities of methods within mother-offspring groups

For each of the elephants we observed in Pinnawela we calculated the frequencies of methods to prepare the four food types, i.e. the number of acts of the same method expressed as percentage of all acts used to prepare one of the four food types. Similarities could be expressed quantitatively as well as qualitatively. The quantitative similarity of food-preparing methods between two particular elephants was expressed as the sum of percentages of methods common to both. The qualitative similarity was calculated according to the formula: 200c / (a + b); 'a' stood for the number of methods used by one animal, and 'b' for the number of methods used by the other animal; 'c' represented the number of methods used by both of them. Values were given separately for each of the four food types (Fig. 4.2.2).

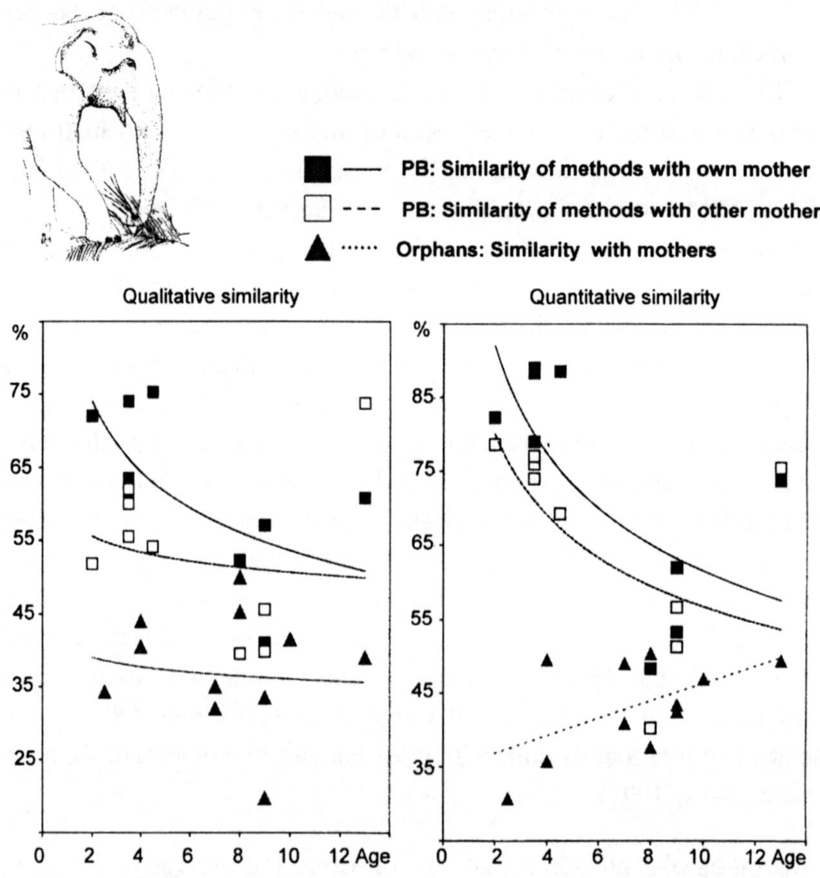



**Fig. 4.2.2**   Mean values of quantitative and qualitative similarities in food preparation methods between young elephants and their mothers and other mothers

Similarity values of all captive born neonates and their respective mothers were pooled. The mean value for quantitative similarity reads 69.2 ± 11.5% and for qualitative similarity 85.4 ± 9.2%. The respective mean values for similarities between all captive born neonates and all other mothers except for their own mothers were 55.2 ± 15.0% and 74.1 ± 12.2%, respectively, i.e., they were significantly smaller. All orphaned neonates had quantitative and qualitative similarity values with all mothers of 38.1 ± 9.0% and 36.4 ± 11.3% respectively. Both values were significantly smaller than the respective values of captive born neonates. However, on comparing the values of single mother-neonate pairs with each other, significantly higher quantitative values with their own mothers in contrast to other mothers were only found in Nandimithra and Esala, and only Esala had a significantly higher value of quantitative similarity with his own mother as compared to other mothers.

In captive born juveniles similar tendencies were obvious, but neither quantitative nor qualitative similarity values were significantly different between an offspring and its own mother, and the values of an offspring with other reproducing females. The respective values read for quantitative similarity 48.1 ± 17.8% and 41.4 ± 16.9 %, for qualitative similarity 52.4 ± 5.7 % and 49.2 ± 12.7%. Orphaned juveniles with normal body growth had similar values with mothers as captive born juveniles, i.e. 43.8 ± 14.7% for quantitative similarity, and 49.0 ± 10.4% for qualitative similarity. For orphaned juveniles with retarded body growth, however, the respective values were significantly smaller than for captive born juveniles or orphaned juveniles with normal body growth and read 33.7 ± 16.0%, and 42.7 ± 11.2%, respectively.

The captive born sub-adult female Sukumali and her mother Kumari had mean similarity values of 60.8 ± 15.2% or 75.2 ± 9.6%, respectively. These values were not significantly different from similarity values Sukumali had with other mothers. But the orphaned sub-adult female Ninja, with retarded body growth, had significantly smaller similarity values with mothers (39.1 ± 13.9% or 49.4 ± 12.7% ). Furthermore, values of quantitative and qualitative similarities between orphaned juveniles were compared, but no differences were found between elephants fettered next to each other and animals feeding in larger distances from each other. Obviously methods of food preparation were not adopted from one another.

Captive born neonates living together with mothers and allomothers, had several advantages when compared to orphaned neonates. They were not only suckled regularly, they knew more different methods to prepare their food (Fig. 4.2.3) and – as Franziska Martin (1998) had shown, their motoric abilities in trunk use were more advanced than in orphaned neonates. Furthermore, it must be considered that the oral cavity of neonates is small and the grinding surface of the first three

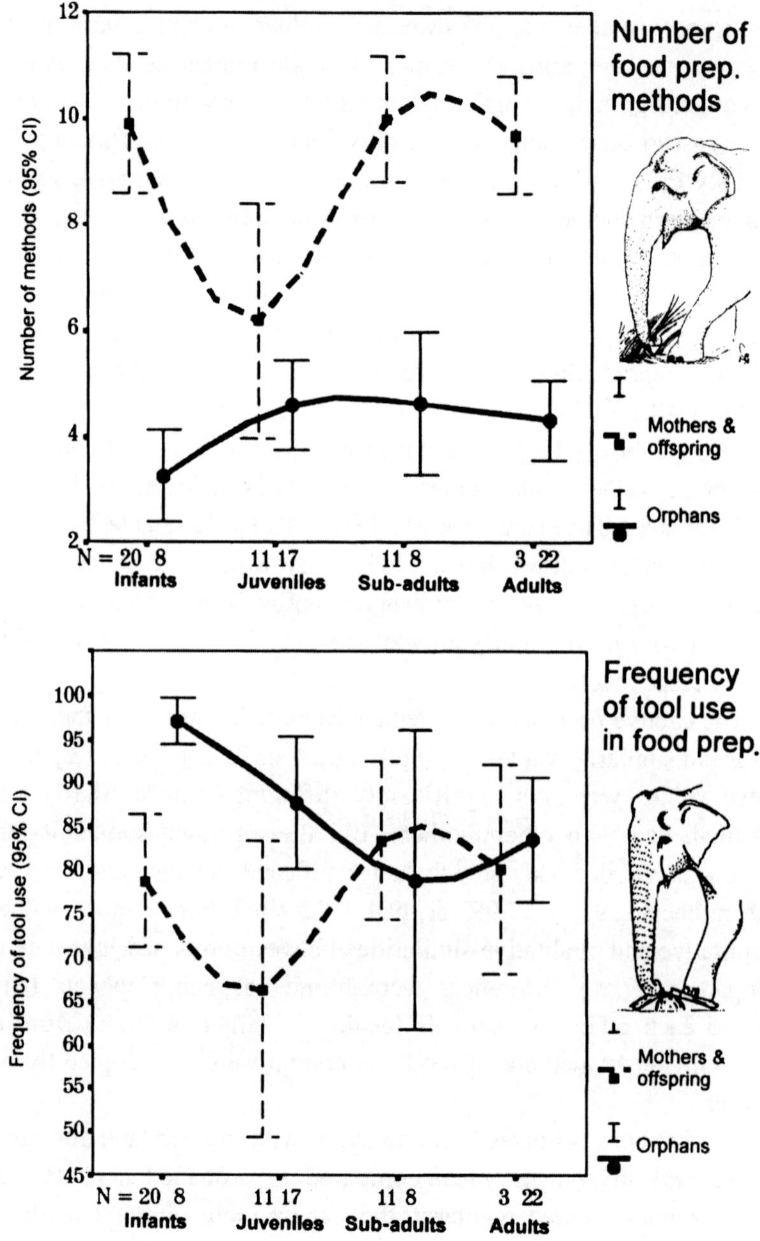


**Fig. 4.2.3** Methods of food preparation and elements with tool use (all and in narrow sense) in Pinnawela-born elephants and their mothers and orphans. Data presented with an interval of confidence of 95%. N: number of scans

molars is hardly useful in tackling larger food portions. In the sensitive period, when relative body growth is highest, the ability to prepare food has to evolve before weaning has taken place. The ability to prepare a certain food type, was acquired rapidly: in August 1998, 17-month-old Arjuna developed the technique to use one of the forefeet to comminute *Caryota* stems. On the 1st of August, he was successful in only one of 86 attempts. On the 7th of August, one of the five attempts was successful. But one week later the method had been completely mastered.

Methods seemed to be developed specifically for each food type, as demonstrated in the following example: The captive born Amali managed to prepare a mean weight of 30 g of *Caryota* marrow in about 27s. Anuradhika, five months older and also captive born, was able to separate 104 g of marrow from wood and bark within the same time. However, in preparing *Cocos* leaves, Amali was already as skilled as Anuradhika and yielded a maximum ration of 3.7 kg/h as compared to 3.6 kg/h. In Pinnawela the captive born elephants could move around freely and select certain partners upto the age of 2 years. In 1998, 20 months old Arjuna spent less time in close vicinity to his mother Sukumali than close to his grandmother Kumari or close to Komali. Both functioned as allomothers and suckled Arjuna. At the same time the 8-month-old Saliya spent as much time close to his mother Mathali as to his allomother Rejina. Different methods to prepare food could be adopted from partners, which were often visited but not necessarily closely related. This could have been the reason for the extreme similarities of methods between offsprings, mothers and allomothers. Neonates growing up with mothers and allomothers manipulated food particles or ate often shortly before or shortly after being suckled and spent, when feeding, most time in close vicinity to adult and sub-adult reproducing females. Furthermore, they concentrated on food which had already been prepared by them with inefficient and indirect methods. It can also be assumed that these food particles contained some saliva of mothers or allomothers, which must have been known to them with regard to taste. In any case, data presented here advocate the opinion that neonates adopt methods of food preparation from mothers and allomothers. Whether they do so by imitation or whether they are actively assisted by older females in this learning process is not known.

## Life histories and methods of food preparation

Our theory seems to be contradicted by simple observations anyone can make who is interested in Asian elephants. Wild elephants, as well as elephants living in temples, timber camps, zoos or circuses prepare short grasses by kicking them from the ground, bundling them and cleaning them on the front leg or forefoot. All healthy elephants break, bite, kick and bundle their food before they eat it, and even rather complicated methods such as 'lever' are developed sooner or later and are definitely not a product of tradition. However, seen from a standpoint of evolutionary biology, the fact that elephants learn a certain efficient method some time or other during their lives is not the only one that counts. What counts is the fact that they learn direct and efficient methods as early as possible. In a long-lived species with an extremely long period of body growth early learning is vital for fast growth at a young age in order to reach reproductive status as soon as possible. Captive neonates growing up within a family group apparently find a social environment which promotes fast adoption of efficient methods. This could be shown by the data presented above.

The comparisons of methods between captive born and orphaned juveniles revealed that both categories adopt more or less the same number of methods to prepare their food plants. But in orphans the cognitive processes obviously develop more slowly, hence also less efficient methods can evolve (Fig. 4.2.3 and Fig. 4.2.4). Of course this could mean that captive born offsprings learn the more efficient methods from their mothers, but, as already mentioned, more concrete data are necessary. Accordingly, the daily food intake of orphaned juveniles is smaller than in captive born animals. The Pinnawela-born Menika for example, ate more of all the four food types per hour than the same aged orphaned Shanti with extremely retarded body growth, but more or less as much as the same aged orphaned Soma with normal body growth (see Section 4.1). Between captive born juveniles and normal-growing orphans differences of methods are hardly recognizable. Orphaned juveniles with normal body growth and captive born juveniles had high similarity values with reproductive females. It was assumed that these females knew the most successful methods to prepare food. But how could certain orphans find out efficient methods without having had close contacts with such females in

Pinnawela? Perhaps, orphans with normal body growth learnt efficient methods just accidentally. It could also have been that they learnt to prepare food efficiently before they became orphaned and were brought to Pinnawela. In case the second assumption is correct, orphans with normal body growth should have come to Pinnawela relatively late. This hypothesis is proved best with sub-adult and adult females, since reaching reproductive status depends on body condition, and hence the ability to maximise food intake. We found a certain tendency backing this theory. Nevertheless, it should not be generalized, as the following examples show:

The adult female Kumari came to Pinnawela as a neonate. Her body growth was retarded. She gave birth to her first calf relatively late, i.e. at the age of 19 years. The 3 sub-adult females Ninja, Nikini and Mauri, as well as the adult Mahaweli also came to Pinnawela as neonates. They had retarded or at least slightly retarded body growth. Until 1999 they had never reproduced. The Pinnawela-born Sukumali grew up with her mother and had her first offspring at the age of 13 years. So far our assumption was correct. But 4 of the 10 sub-adult or adult females behaved differently: Anuradha came to Pinnawela as a neonate, grew in a normal way and parturiated the first time at the age of 12. Rejina also grew in a normal way and gave birth for the first time when she was 13, although she had already come to Pinnawela at the age of 2. However, Mathali and Komali came when they were relatively old, i.e. as juveniles, to Pinnawela. Nevertheless, they only had their first calves at the age of 18. It must be stressed, however, that little was known about the social structure of the Pinnawela herd in former times; but it is known that certain orphans were nursed by allomothers, however, the first study on Pinnawela elephants and their social structure had been carried out in a very general manner (Poole *et al.*, 1997).

The findings of this chapter can be summarised as follows: The knowledge of efficient food preparation is best acquired in the presence of mothers and allomothers. However, it can also be acquired in older age by accident or by trial and error. Early learning of the efficient methods is more likely in close contact with the family group, furthermore, it increases personal fitness, since it stimulates relatively fast body growth and early reproduction. Reproductive fitness is a function of relative body weight and hence a function of efficient feeding. Therefore, it can be assumed that mothers know optimal food preparation methods.

## 4.3 Tool Use

It is generally agreed that the trunk in *Proboscidea* evolved in relation to feeding on grasses into an arm-like part of the body and its tip into a skilful prehensile organ which allows manipulation of objects. In many feeding behaviours of elephants plants or parts of them are treated as tools, as shown in the previous chapter; and tool use in general occurs in elephants more frequently and more diversely than in other non-primate mammals (e.g. Beck, 1980). Asian elephants have been reported to perform at least 27 types of tool use which occur in six functional contexts (Chevalier – Skolnikoff & Liska, 1993; Kurt & Hartl, 1995).

In a narrow sense, tool use can be defined as the use of a physical implement which is manipulated to attain a particular goal. A broader definition qualifies tool use "as the external employment of an unattached environmental object to alter more efficiently the form, position, or condition of another object, another organism, or the user itself when the user holds or carries the tool during or just prior to use and is responsible for the proper and effective orientation of the tool" (Beck, 1980).

According to these two definitions, scratching the temporal gland with a small stick would be an example of the narrow concept of tool use. Elephants prepare such sticks from selected branches which they shorten and sharpen with their trunk hand, forefeet and molar teeth. Elephant bulls in musth also smear substrate and water over the temporal gland, temples and trunk base and rub the mixture against special marking trees to discourage non-musth bulls from entering into the ranges of family-units controlled by musth bulls (Kurt, 1974 & 1992). This marking behaviour can be considered as tool use in the broader sense, and the mixture of substrate and water as tool. According to prevailing data, the diversity of tool use in elephants is similar to that of tool-using primates. In the following pages, first an overview of tool use in elephants living in different environments is given. Later, tool use will be discussed under the environmental and evolutionary aspects.

Forms of behaviour regarding tool use were studied in 1967–68 for about 50 weeks in 2 wild-living populations of the Ruhuna and Lahugala National Parks (Kurt unpublished) and for about 13 weeks in Mudumalai and Periyar (South India). During the *Khedda* of Kakanakote (Karnataka) in 1968, 16 newly captured elephants were observed for 5 weeks. In

Pinnawela, 63 elephants were studied in 1997–98. Between 1978 to 1993, 61 timber elephants were monitored in South India, Northeast India, Thailand and Myanmar. In 1997, for 5 weeks 52 intensively-kept elephants were studied in Kandy. A one week study on 19 temple elephants in Kerala was carried out in 1993. Furthermore, 44 zoo and circus elephants were regularly observed in the zoos of Berlin, Munich and Zurich as well as in the Swiss National circus.

Behaviour was observed during day and night. Events regarding tool use and the context in which they occurred were recorded. Five behavioural contexts were distinguished: (1) 'Feeding and drinking', including preparation and storing of food, procuring and protecting of water sources. (2) 'Skin care' (body care), including patterns of cooling and cleaning the body surface. (3) 'Social behaviour', (4) inter-and intraspecific aggressive interactions. (5) 'Rest' and 'sleep', including patterns before and between periods of doze and sleep in standing or recumbent position.

The rich collection of examples of tool use in a total of 885 Asian elephants observed during a total period of 106 weeks allowed us to give an overview of tool use behaviour in a narrow and broader sense (Beck, 1980) in elephants living in a natural environment as well as under more or less restricted conditions. Accordingly, it can be discussed whether tool use is maintained by tradition or whether it develops independently under certain environmental conditions. In the first step, different forms of tool use will be listed according to the social and ecological contexts they occurred in. In the second step, some examples will be given of the frequencies of tool use in different social classes living in different environments.

Forty types of tool use were distinguished, 23 could be considered as tool use in a narrow sense (Table 4.3.1). In the five functional contexts tool use behaviour was found as follows: Body care (13), feeding and drinking (10), social behaviour (8), interspecific interactions (6), rest and sleep (3). Neonate and infant elephants performed 18 types of tool use, juveniles – 26 types, and sub-adults and adults – 37 types. 31 types of tool use were found in wild living elephants, 17 in newly captured ones, 22 in Pinnawela Orphanage, 20 in jungle camps, 26 in temples and 23 in zoos and circuses.

**Table 4.3.1** Occurrence of tool use in the narrow sense (*) and in a wider sense in different environments. wi = wild; kh: during and immediately after a khedda operation in Kakanakota; pi = Pinnawela; ex = timber elephants; in = temple elephants; zo = zoo elephants. N = Neonates; I = infants; J = juveniles; Sa = sub-adults, A = adults. A particular tool behaviour can be adopted occasionally (+) by a few animals of a social class, it can occur regularly in all members of a social class (++) or it can be absent (–).

| Description | Environment | | | | | | Social classes | | |
|---|---|---|---|---|---|---|---|---|---|
| | wi | kh | pi | ex | in | zo | N / I | J | Sa / A |
| **Skin care** | | | | | | | | | |
| 1. Air blowing | ++ | ++ | ++ | ++ | ++ | ++ | ++ | ++ | ++ |
| 2. Sand blowing | ++ | ++ | ++ | ++ | ++ | ++ | ++ | ++ | ++ |
| 3. Water spraying | ++ | ++ | ++ | ++ | ++ | ++ | ++ | ++ | ++ |
| 4. Throwing with substrate or plant parts | ++ | ++ | ++ | ++ | ++ | ++ | ++ | ++ | ++ |
| 5. Spraying with regurated fluid | ++ | ++ | ++ | + | + | – | + | ++ | ++ |
| 6. Beating with plants* | ++ | ++ | ++ | ++ | ++ | ++ | ++ | ++ | ++ |
| 7. Rubbing with plants in trunk* | ++ | ++ | ++ | ++ | ++ | ++ | ++ | ++ | ++ |
| 8. Rubbing with plants (tail)*o | – | – | – | – | – | + | – | – | + |
| 9. Ear scratching with stick* | ++ | ++ | ++ | ++ | ++ | ++ | ++ | ++ | ++ |
| 10. Temple scratching with stick* | ++ | ++ | ++ | ++ | ++ | ++ | ++ | ++ | ++ |
| 11. Body scratching with stick* | ++ | ++ | ++ | ++ | ++ | ++ | ++ | ++ | ++ |
| 12. Keeping stick between head and ear* | + | – | – | – | – | – | – | – | + |
| 13. Scratching on a vertical stem* | – | – | – | – | + | – | – | + | – |
| *Total* | *11* | *10* | *10* | *10* | *10* | *10* | *10* | *11* | *12* |

[Table 4.3.1 Contd.

**Feeding and drinking**

| | | | | | | | | | |
|---|---|---|---|---|---|---|---|---|---|
| 14. Beat on nails of forefeet | ++ | ++ | + | ++ | ++ | ++ | ++ | ++ | ++ |
| 15. Rubbing on forefoot | ++ | ++ | + | ++ | ++ | ++ | ++ | ++ | ++ |
| 16. Washing | + | – | – | – | – | – | + | + | + |
| 17. Soaking | – | – | – | – | – | + | – | – | + |
| 18. Shacking stem | + | – | – | – | – | – | – | – | + |
| 19. Lever on a abutment* | – | – | ++ | – | + | – | + | + | + |
| 20. Breaking on a wall* | – | – | + | – | + | + | – | + | + |
| 21. Throw on the ground* | – | – | – | – | + | – | – | – | + |
| 22. Cover water–hole* | + | – | – | – | – | – | – | – | + |
| 23. Use tube | – | – | – | – | + | + | – | + | + |
| *Total* | *5* | *2* | *4* | *2* | *6* | *5* | *4* | *6* | *10* |

**Aggression**

| | | | | | | | | | |
|---|---|---|---|---|---|---|---|---|---|
| 24. Sand blowing | ++ | ++ | ++ | ++ | ++ | ++ | + | ++ | ++ |
| 25. Beat small animals with stick* | + | ++ | + | + | ++ | + | + | ++ | ++ |
| 26. Roll object towards small animals* | + | ++ | + | + | ++ | + | + | ++ | ++ |
| 27. Sand throwing | + | ++ | + | + | ++ | + | + | ++ | ++ |
| 28. Throwing with sticks or stones* | + | + | + | + | + | + | – | ++ | ++ |
| 29. Beat with chain* | – | – | – | – | + | – | – | – | + |
| *Total* | *6* | *5* | *5* | *5* | *6* | *5* | *4* | *5* | *6* |

[Table 4.3.1 Contd.

Contd. Table 4.3.1]

| | | | | | | | | | |
|---|---|---|---|---|---|---|---|---|---|
| **Sleep** | | | | | | | | | |
| 30. Clean soil | + | – | – | + | + | + | – | + | ++ |
| 31. Prepare pillow | + | – | – | – | – | + | – | + | ++ |
| 32. Rest head on stem* | – | – | + | – | + | – | – | + | + |
| *Total* | *2* | *0* | *1* | *1* | *2* | *2* | *0* | *3* | *3* |
| **Social behaviour** | | | | | | | | | |
| 33. Offer food to conspecifics* | + | – | – | – | – | – | – | – | + |
| 34. Feeding conspecifics* | + | – | – | – | – | – | – | – | + |
| 35. Kicking substrate towards conspecifics* | + | – | – | – | – | – | – | – | + |
| 36. Throwing substrate towards conspecifics* | + | – | – | – | – | – | – | – | + |
| 37. Mix musth–secretion with water | ++ | – | + | + | + | + | – | – | + |
| 38. Rub musth–secretion on tree | ++ | – | – | – | – | – | – | – | + |
| 39. Masturbate on a stem* | – | – | – | – | + | – | – | + | – |
| 40. Cleaning of offspring with plants* | + | – | + | + | – | – | – | – | – |
| *Total* | *7* | *0* | *2* | *2* | *2* | *1* | *0* | *1* | *6* |
| **Total** | **31** | **17** | **22** | **20** | **26** | **23** | **18** | **26** | **37** |

## Tool use in food preparation and drinking

During the first days of life trunk movements are hardly coordinated and the neonates occasionally step on the tips of their trunks. They drink with the mouth – not only the milk from their mothers and occasionally aunts (which is always done by young elephants), but also water. In older elephants the trunk functions among other things – also as a siphon, it can suck water or mud or very rarely, fine-grained sand or earth and blow it unerringly towards certain parts of their bodies, conspecifics or putative foes. But long before this is possible, neonates exercise while they play – how to blow and spray. The same applies to manipulation. This pattern is as hereditary in newborn elephants as the manipulative tendencies in neonate chimpanzees. For example: the chimpanzees grip sticks with their hands and swing them, but before they can use these sticks as instruments, they have to become familiar with their use. In young elephants this means, for example: first they must master the manipulation of grass, before they can use it as a tool and are able to clean it from sand or insects.

Grass can be cleaned by several methods. For example: it is bundled with the trunk hand in such a way that single blades are parallel, and all the roots are on one side and all the tips on the other side of the bundle. The side with the roots is then beaten against the nails of one forefoot or rubbed against one carpal joint. Sometimes elephants drop these bundles so masterly that strong winds separate sand and insects from the grass. In such cases the grass bundle eaten later can be considered as a tool. Although all healthy elephants develop similar methods of preparing their food sooner or later, tool use in context with feeding differs according to the environment and, hence, the food offered. Captive elephants can throw *Caryota* stems with such force to the ground that they break. The same is achieved when they use a second tree trunk as abutment on which the food is broken (Fig. 4.2.4). Such behaviours have never been observed in wild elephants. The same applies to the following method: captive elephants fettered closely to a wall sooner or later learn that strong and large branches can easily be broken by kicks of a forefoot after having propped up the branch against the wall. In such cases floor and wall could be considered, so to speak, as tool.

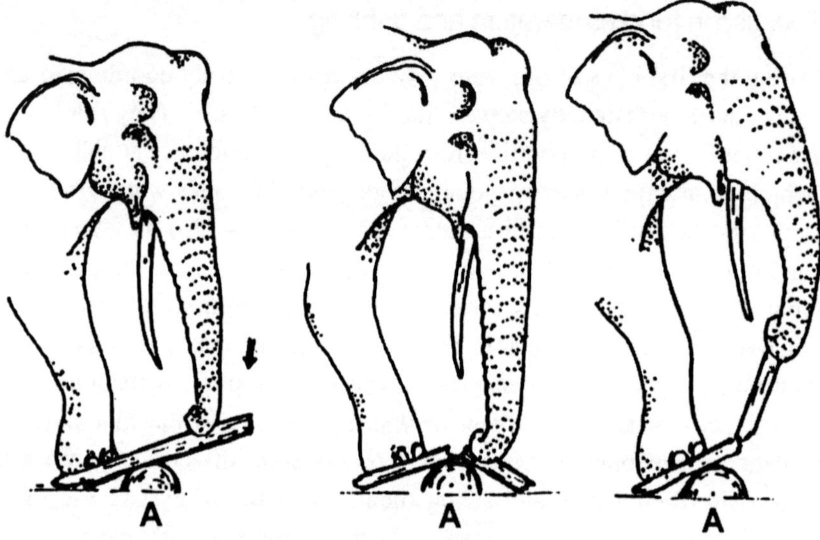


**Fig. 4.2.4**    Tool use in food preparation. On an abutment a thick branch or stem is held with one forefoot and broken with the other forefoot or the trunk.

Some of the elephants observed in zoos and circuses stored portions of hay and straw on their heads, necks or shoulders to safeguard them from their neighbours. But wild elephants were never observed to collect such food deposits, although they sometimes used plants to protect their bodies against sun and ectoparasites. Wild elephants occasionally shook trees with such force that the ripe fruit dropped to the ground. Captive elephants never had a chance to behave in such a way, but one female in the Berlin zoo in Friedrichfelde soaked dry bread in water before she ate it. In the Periyar Tiger reserve of Kerala wild elephants regularly wash grass bundles in the nearby lake before eating them.

Tool use rarely occurred in the context of drinking: Wild elephants are capable of digging waterholes, for e.g. during the dry season in the sandy bed of the Menik Ganga in the Ruhuna National Park. To protect these important sources against smaller competitors, such as wild boars, deer and bears, they covered them with large branches. Former hunting elephants kept in a glass house in Bangkok, used water tubes as tools, i.e. they held them so masterly with their trunk hands that water could be sprayed into the mouth for drinking or over their head, back and flanks for cooling. Some of them had even learnt how to turn on the tap, but they never turned it off. The same observations were made by Ruedi Tanner in the zoo of Zurich.

## Tool use in skin care behaviour

The greatest variety of forms of tool use was found in the context of skin care behaviour. They are used to cool the body, to protect it against ectoparasites, and to treat wounds. Although the dermis, the actual skin, has a thickness of about 2 cm, which is covered by a thin epidermis of 1 to 2 mm, elephants should not be considered as insensible pachyderms. On the contrary: skin as well as hair can be taken as very good indicators for their welfare (e.g. Hediger, 1990; Kurt, 1999). Furthermore, the dermis harbours papillary protuberances rich in papillary vessels which give the skin the typical rough, uneven form. The papillary pattern is most prominent in the distal part of the trunk, on the legs and on the tip of the tail. Ectoparasites, such as elephant lice (*Haematomyzus*), find easy access to fresh blood on the protuberances (Hart & Hart, 1994). In their range countries Asian elephants suffer under a heavy load of ectoparasites, such as leeches, mites, ticks, elephant lice, gnats and horseflies. In one single Nepalese elephant camp 14 species of horseflies, drinking blood from elephants, were found (Hart & Hart, 1994).

Wild elephants regularly take a bath, wallow and rub their bodies against trees or rocks and keep ectoparasites at a distance by trunk swaying, ear flapping movements or well-directed lashings of the tail. Such forms of behaviour and similar ones are the most common. Tool use occurs mainly in a broader sense. Air, water or mud are blown towards certain regions of the body. A full-grown Asian elephant sucks about 10 l of fluid up to a height of about 40 cm into its trunk. The water is either used for drinking or skin care, and accordingly it is sprayed into the mouth or over certain parts of the body.

But Asian elephants can also take water from their mouths – even when they have not drunk before. In a conspicuous way they repeatedly contract the cheek muscles in pumping movements. The fluid seems to be a mix of saliva and water stemming from the *Bursa pharyngialis*, a muscular pocket, lying just behind the root of the tongue and formed by muscles of the larynx (Tennent, 1867; Shoshani, 1998). Elephants also 'blow' sand towards certain parts of their bodies. Actually they do not 'blow', but throw the sand, i.e. they collect a portion of sand in their trunk hand, but blow heavily at the moment of throwing, which leads to a finer dispersion of the substrate. Finally, wild-living elephants throw substrate and plants on their bodies as protection against sun and ectoparasites. This behaviour, as well as the blowing of air or substrate, also appears in the context of aggressive behaviour.

Seven out of 13 tool behaviours in the context of skin care could be considered as tool use in the narrow sense. The use of branches and palm leaves as fly-swatters is well-known in the West, since Charles Robert Darwin discussed tool use in his work *The Descent of Man* (1871) and also mentioned the Asian elephant. When kept in chains, captive elephants occasionally defoliated certain branches or palm leaves which were considered as suitable for fly swatting. They never ate such instruments, but put them handily next to the forefeet, as observed by Benjamin and Lynette Hart (1994) in Nepal and by our team in Pinnawela. Elephants scratched certain parts of their bodies, mainly ears, cheeks and temples with small sticks, which could be as small as a match or as long as a pencil. These small sticks were prepared with trunk hand, forefeet and molars. They could also be used to remove leeches from their skin. A female zoo elephant became known for having got rid of an abscess on the inside of her trunk by pulling the trunk over a stick which was 60 cm long. The tip of the instrument which had been sharpened by her before penetrated the abscess so that the pus could flow off (Beck, 1987).

Only wild elephants were observed to store prepared sticks and twigs behind the ear. In rather rare cases zoo elephants rubbed their skin with parts of plants which were pinched under the tail (Chevalier-Skolnikoff & Liska, 1993). This behaviour could not be observed during our studies. But in Kandy we knew a young bull who scratched his flanks with the help of a stem which he had first put in a more or less vertical position with his trunk (Fig. 4.3.1). With the exception of 3, all tool behaviours in the context of skin care behaviour were observed in all 6 environments distinguished (Table 4.3.1). Out of 13 elements, 10 were seen in neonates and infants, 11 in juveniles and 12 in sub-adults and adults. A more differentiated picture is given by a quantitative analysis of the data.

**Fig. 4.3.1** Tool use in the context with body care. A young captive bull scratches head and body flanks with the help of a log held in his trunk.

## Frequencies of tool use in skin care behaviour

In wild elephants of the Ruhuna National Park the frequency of all skin care behaviours was significantly higher at water holes than in cases when the elephants were feeding or walking (Fig. 4.3.2). For neonates, the respective values read 2.31 + 0.71 acts / 10 min (n = 22) or 0.53 + 0.23 acts / 10 min (n = 43), and for adults 1.41 + 0.53 acts / 10 min (n = 147) or 0.20 + 0.15 acts / 10 min (n = 227) respectively. But the frequency of all skin care behaviours was 2-fold to 7-fold in newly-captured elephants in the *Khedda* of Kakanakote. These animals were fettered by all four legs, they were led two to three times daily to the nearby river for drinking and bathing. The respective values were 6.01 + 5.20 acts / 10 min (n = 89) for neonates and 6.49 + 5.37 acts / 10 min (n = 54) for adults. The striking difference between wild and newly captured animals could be related to a certain degree to the fact that

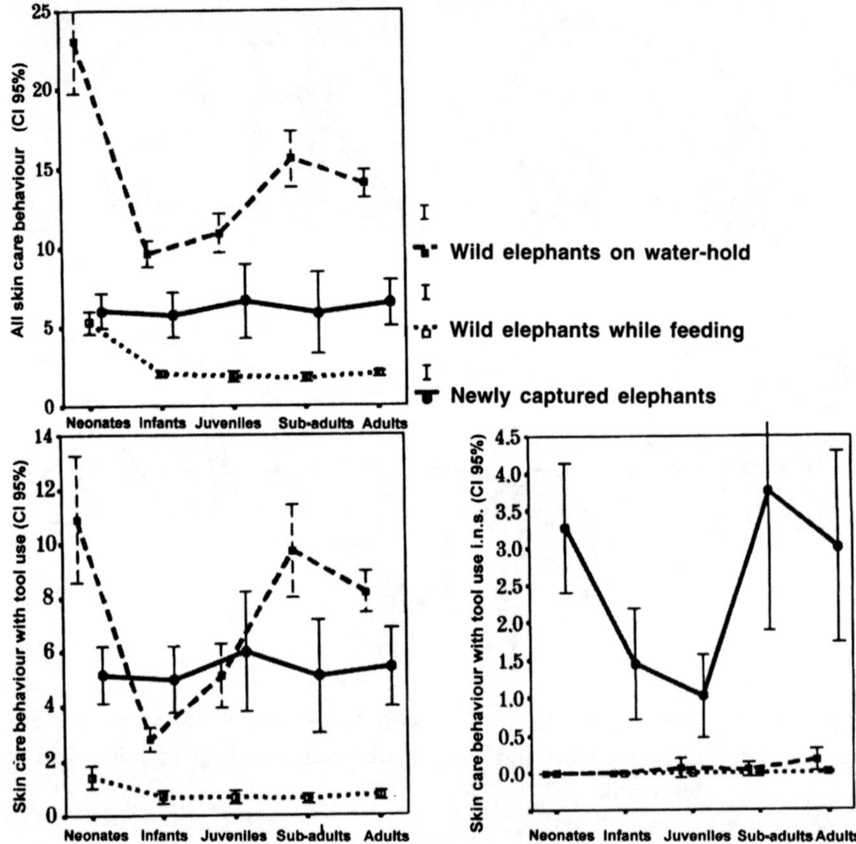



**Fig. 4.3.2**   Frequency of all skin care behaviour, skin care behaviour with tool use and tool use in the narrow sense in wild elephants on water-holes and while feeding, as well as in newly captured elephants after the Kakanakote-*Khedda*.

newly captured elephants were extremely aggressive, and elements of skin care behaviour appeared regularly also in the context of interspecific aggression. But the lower frequency of skin care behaviour in wild elephants relied mostly on the fact that they avoided direct radiation by a clever selection of shady places during the hottest hours of the day. Furthermore, they avoided, whenever possible, damp areas with a high density of ectoparasites (c.f. Evans, 1910).

During *khedda* operations, it was usual to weaken the elephants in the stockades by withdrawal of water and by exposing them to direct sunlight often for several days. The same methods have been customary until now in the process of taming (Kurt, 1992 & 1993). Permanent direct sun radiation and lack of water, which could be used to cool the body, weaken elephants to such a degree that, in the long-run, individual fitness is seriously threatened (Kurt, 2001a.). The same applies to ectoparasites. They sting and bite painfully, transmit pathogenic agents such as trypanosomes or plasmodia, and, under certain circumstances, they can draw considerable amounts of blood. Asian elephants seem to be resistant to endemic diseases, such as malaria or *surra*, which are transmitted by ectoparasites. But loss of blood could become crucial to the maintenance of fitness (Hart & Hart, 1994). It is unknown how high it might be in elephants, but one single horse can be bitten 4000 times in one day by horseflies and suffer from a loss of four litres of blood (Tashiro & Schwadt, 1953). The loss of the body's own resources such as proteins and haemoglobin could also reach a critical level in elephants too, especially in intraspecific competition for reproductive success, as lactating females and newborns are mostly attacked by ectoparasites (Hart, 1990).

Often the strategies to protect the body from radiation or ectoparsites could not be distinguished. A thick layer of moist soil or plant parts, for example, protects from both. But tool use is not the only possibility. In wild elephants the percentage of all tool behaviours in the context of skin care behaviour was higher at waterholes (mean: 32.6%–61.1%) than during walking or feeding (mean: 22.9%–31.4%), and in newly captured elephants (mean: 74.3%–81.9%) it was much more pronounced than in wild ones (Fig. 4.3.2). In wild elephants the frequency of tool use in the narrow sense was practically nil, but in the newly captured animals from Kakanakote it was between 1 to nearly 4 acts / 10 min. From the data presented here it can be concluded that tool use in the context of  skin care behaviour occurred in all the populations studied, but it was used only rarely as long as other behaviours could be employed to avoid strong radiation and ectoparasites, e.g. avoiding time and place with strong radiation and ectoparasites.

## Tool use in aggressive intraspecific and interspecific encounters

When threatening an opponent, elephants shift the weight of their body forward, raise their head and spread their ears, thus seemingly enlarging their appearance. In cases of uncertainty, the threatening animal throws sand, as described earlier, as displacement activity. But in aggressive encounters sand is always thrown upwards, never to the flanks. The use of elements from skin care behaviours in situations of conflict is usual in many species of mammals and birds. While attacking, elephants run towards their opponents in the threatening posture with the trunk slightly rolled inwards. The opponents could be humans, vehicles or different animal species. If a direct contact is unavoidable, the attacking elephant kicks with the forefoot and hits with the trunk in an upward movement. Several tool use behaviours could be derived from kicking and trunk beating: sand, soil, stones or parts of plants are kicked or rolled towards the opponent. Furthermore, they can be taken into the trunk hand and thrown with a quick forward thrust of the trunk. The initial velocity of the missile is accelerated by a sudden step forward and stretching the head forward. Adult bulls can fling larger objects unerringly upto distances of 20 m, and these can be so heavy that they can break arm bones in a man (Kurt, 1992). However, wild elephants throw objects not only to hit an opponent, they also use this pattern, when they cannot see the opponent, but suspect where he could be hiding.

Our data revealed that young elephants attack comparatively small animals, such as different species of snakes, monitor lizard (*Varanus bengalensis*), blacknaped hare (*Lepus nigricollis*), peacock (*Pavo pavo*) and Indian wild boar (*Sus scrofa*). However, crocodiles (*Crocodilus pallustris*, *C. porosus*), jackal (*Canis aureus*) and leopard (*Panthera pardus*) were only attacked by sub-adult and adult animals. Considering their immense size one may ask why elephants reacted so aggressively towards relatively small animals, as soon as they were close to them. Young elephants most probably scared away reptiles, hares, peacocks or boars just for fun. Nevertheless, even adult elephants avoided too close contact with other animals. Normally they would scare away even the small herons and other smaller birds, when these insectivores tried to rest on their heads and backs, as was regularly seen in buffaloes, deer and boars. However, in the swampy feeding grounds of Lahugala, where ectoparasites, mainly leeches, were extremely abundant, adult bulls made an exception and allowed these birds not only to rest on their backs, but they also allowed mynahs (*Acridotheris tristis*) and flycatchers (*Muscicapidae*) to clean their skin by picking away the leeches.

Amongst the larger south Asian carnivores, only the tiger preys occasionally on young elephants (e.g. Tu Yin, 1967), which led to the development of special strategies of defence used by family groups (Kurt, 1992). The tiger does not exist in Sri Lanka, but crocodiles can become harmful by biting in self-defence. Amongst about 80 individually – known elephants in the Ruhuna National Park in 1967–68, at least 2 had easily visible holes in their trunks, a fact which seriously complicated drinking. In his remarkable book, Jayantha Jayewardene (1994) listed more and more serious cases of crocodile bites. In any case, elephants were always very careful when entering a waterhole, and family groups never allowed younger members to go into the water first (Kurt, 2001a). But the elephants of Ruhuna met self-defending crocodiles not only at muddy waterholes, but also far away from the next watering place, mainly at or close to carcasses of starved buffaloes or deer killed by leopards. The carcasses were often found in or close to waterholes or on elephant paths, since all other larger animals used them. Therefore, close contacts with carcasses were inevitable for the elephants.

Carcasses did not only attract scavengers like jackal, wild boar, monitor lizard or crocodiles, but also an immense variety of insects as well as many smaller rodents, which gnawed on the bones. Their predators followed immediately: e.g. insectivorous birds and snakes. The bite of a cobra (*Naja naja*) or a Russel's viper (*Daboia ruselli*), deadly for humans, leads to immediate death only in young elephants, in older ones it is only supposed to be painful (c.f. Jayewardene, 1994). Elephants will do everything to protect their neonates. But they do not seem to distinguish between a harmless monitor lizard and a crocodile, or between harmless and poisonous snakes. They were always careful when close to a carcass, and some of them always attacked all larger animals, whether they were snakes, boars, jackals, leopards or crocodiles. When adult elephants would grab a crocodile by their back or tail, they treated it like a tool, which means that they threw it away or beat the reptile's head on stones or stems.

Wild adult elephants of the Ruhuna National Park had a frequency of aggressive interspecific behaviour of 0 to 3.8 acts / h. Juveniles of either sex never showed these patterns. However, in the Uda Walawe National Park, juveniles regularly mock-charged vehicles. In Ruhuna interspecific aggression was not found in infant females, but infant bulls performed 1.2 aggressive acts / h, which was the second highest frequency next to adult bulls in musth (3.8 acts / h). The frequency of tool use in the context of interspecific aggression, i.e. sand throwing, kicking, rolling and throwing of objects was between 0.2 acts / h in neonates and 0.9 acts / h in adult females. In musth bulls, who are generally more aggressive than other elephants, the frequency read 0.5 acts / h and in infant bulls 0.6 acts / h. Sand-throwing already occurred in neonates, kicking and rolling of objects already in infants, but throwing of stones and pieces of wood only in sub-adults and adults.

The newly captured and tethered elephants in the *khedda* of Kakanakote tried to attack humans and dogs with all possible means, and they threw, kicked and rolled objects towards them; but the mahouts and their assistants always repaid such behaviours with thrashing and beating with sharpened sticks, inflicting wounds on head and trunk. However, such corporal punishments cannot completely tame elephants. In Kandy, certain captive adult bulls waited for the right moment to grab one front chain, lever it up rapidly in such a way that the chain immediately started to undulate and finally hit and wound persons standing close to the place, where the front chain was fixed.

Captive adult bulls often throw objects. At the Berlin Zoo, for e.g., the paddock of a bull could only be covered with fine sand because the animal notoriously threw stones towards visitors. Three adult non-musth bulls, observed during 24 h in Pinnawela and Kandy, had a mean frequency of 0.05 throws / h and a mean frequency of 0.21 trunk thrusts / h towards their opponents. But 3 adult bulls in musth, in total observed for 42 h, had a frequency of 1.3 throws / h, that was 26 times more than in non-musth bulls (see Section 5.1). Wild-living sub-adult and adult bulls threw objects twice as often as captive bulls, which were not in musth, but they used this behaviour 13 times less often than captive musth bulls. Finally, it must be mentioned that captive males occasionally prepared the wooden missiles and kept them handy, waiting only for the first opportunity to use them.

## Tool use in social behaviour and sleep

Differences of tool use between wild and captured Asian elephants were most pronounced in the context of social behaviour, which could be related to the fact that captive elephants often lacked adequate social partners. It was observed that only in wild elephants particular group members fed certain conspecifics by putting food portions into their mouths. Only wild elephants kicked soil towards each other and threw soil and plants towards other group members. Although wild and captive musth bulls smeared their temporal glands regularly with water and mud and mixed these substances with the excretion of the temporal gland, it was observed only in wild elephants that the head was rubbed later against the tree where the mixture was deposited. Female elephants that deliver neonates without the help of man wiped the neonates dry with bundles of plants.

In captive elephants only one tool use behaviour in the context of social behaviour was found, which was never observed in wild ones, and it was observed only in one juvenile bull kept in Kandy. First, he got up on his hind legs and skilfully propt up a tree stem in vertical position between breastbone and ground. Then he masturbated against the stem (Fig. 4.3.3). Masturbation regularly occurred in wild bulls, they beat their erected penises towards the sternum, but they never used tools.

Wild elephants slept on specific places with dry soil. Before lying down they cleaned the soil with forefeet, trunk and occasionally with the branch kept in the trunk hand. Furthermore, they used forefeet and trunk to make pillows with the substrate. In the circus they used straw to make a pillow. During colder periods they covered head, ears and the slightly rolled up trunk with the substrate. In Kandy, the bull with the longest tusks, the so-called Milangoda tusker, hardly slept in recumbent position, but rested and slept in standing position after he had put a *Caryota* stem under the base of one tusk. Similar behaviour was seen in Sama II, the juvenile female in Pinnawela who had lost a forefoot in a land mine accident, often buttressed her head on a *Caryota* stem for sleep in standing position.

## Comparisons with African elephants and Primates

The following data, and also that of others authors (e.g. Beck, 1980; Chevalier-Solnikoff & Liska, 1993), will show that tool use behaviours in elephants are learned and sometimes complex behaviour patterns, and not merely stereotype genetically fixed behaviours. Accordingly, elephants

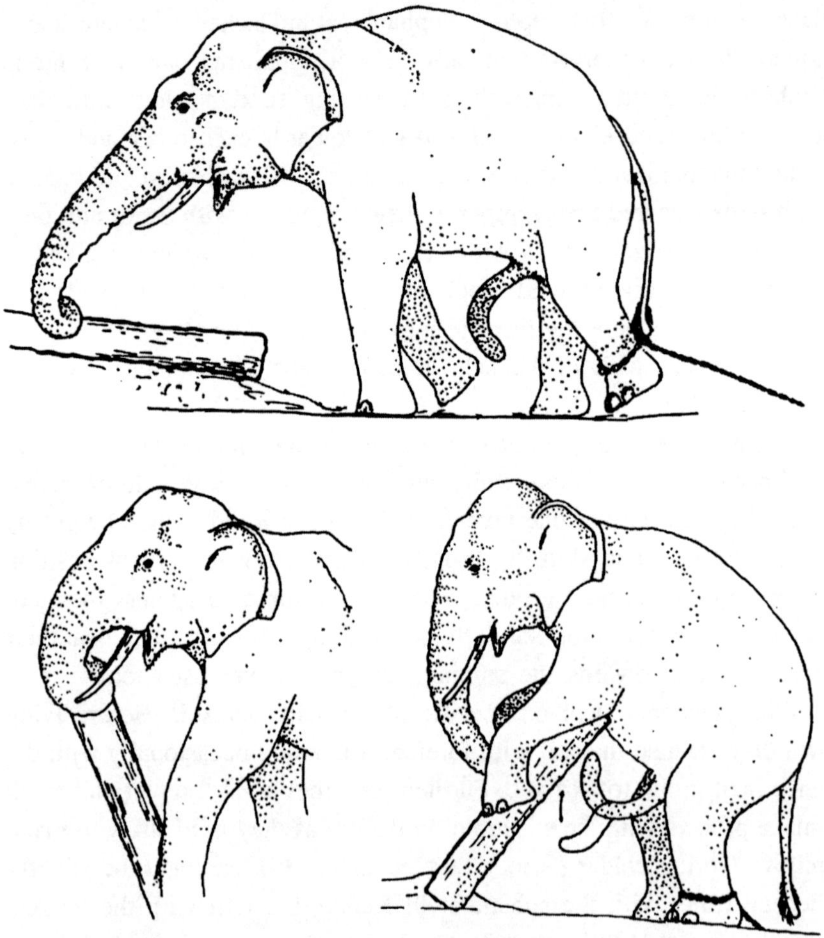

**Fig. 4.3.3**  Tool use in the context of sexual behaviour. A juvenile bull holds up a trunk and masturbates on it.

are able to apply various forms of tool use with varying frequency under different living conditions. The fact alone that elephants which have never been in contact with each other develop the same forms of tool use under similar conditions. For example, cleaning of grass bundles in a similar manner, breaking off long branches against walls or using a vertically propped-up branch as head rest during standing-sleep. Therefore, it says a lot for their highly developed cognitive capabilities, which are partly possible – due to their proverbial great intelligence, a comparatively large, complex, slowly developing brain (Laws, 1071; Sikes, 1971; Eltringham, 1982), a long time lapse between birth and physiological as well as social maturity (Section 3.4), and the gradual acquisition of sensory-neuromuscular abilities. Experimental studies on their learning abilities show that elephants possess a high degree of understanding, even under extreme conditions, of the relationship between their own needs and events as well as objects in their environment. For example, in choice experiments they learn very quickly and possess a long-lasting memory (e.g. Rensch, 1956, 1957; Markowitz *et al.*,1975; Stevens, 1978; Kurt & Knie, 1980; Dehnhardt, 1996).

Asian and African elephants utilise tools in a very similar manner, though tool use in connection with dead conspecifics in Asian elephants was never found, as has been described repeatedly for African elephants (e.g. Douglas–Hamilton & Douglas–Hamilton, 1975; Chevalier-Skolnikoff & Liska, 1993). Despite observing Asian elephants on repeated occasions in the vicinity of carcasses one never saw them try to feed the carcass or attempt to close wounds with plant parts, nor try to cover the carcass with substrate or branches. It has been reported though, that Asian elephants occasionally cover dead humans with branches and earth (Deraniyagala, 1955). Wild African elephants utilise tools with a frequency of 1.45 acts/h, those living in zoos with a frequency of 22.8 acts/h (Chevalier–Skolnikoff & Liska, 1993). Apart from the varying methods of cleaning grass, the Asian elephants show similar differences in the frequency of tool use between wild and captive ones. Whilst quantitative tool use by Asian elephants living in captivity is higher than under natural conditions, there are also qualitative differences between wild and tame elephants.

Wild Asian elephants used 31 different tool use methods out of 40 identified ones, and freshly caught elephants used 17. The animals at Pinnawela Elephant Orphanage showed 22, working elephants kept under intensive conditions 26, and zoo and circus elephants 23 different methods of tool use. For working elephants kept under extensive conditions in logging camps the figure reads 20. These 'timber-elephants' most probably display further tool use behaviours in connection with their work, however, respective studies are missing to date.

Although interspecific comparisons can be debated, variety and frequency of tool use in elephants appear to be similar to tool use behaviour in primates (Beck, 1980; Chevalier–Solnikoff, 1989). In primates tool use in wild populations also occurs less often than in captive primates. Mountain gorillas (*Gorilla gorilla berengei*) during 1,000 hours of observation used tools only on a single occasion (Schaller, 1963). Wild living orangutans (*Pongo pygmaeus*), observed during 12,000 hours, had a frequency of 0.0002 acts of tool use per hour, if one excludes that they relatively frequently threw objects towards the observer (Chevalier–Skolnikoff *et al.*, 1982; Galdikas, 1982). In contrast, wild living capuchin monkeys (*Cebus capucinus*), observed during 300 hours, had a tool use frequency between 0.2 to 0.4 acts per h (Chevalier–Skolnikoff, 1989; 1990). However, the tool use frequency in wild living chimpanzees (*Pan troglodytes*) appears to be high and its diversity is unachieved by any other species. Chimpanzees utilise more methods of modifying objects and preparing tools than do elephants (Beck, 1980; Goodall, 1986; McGrew, 1990). For example, captive chimpanzees spontaneously used sticks to reach and pull in some bananas, which had been placed outside the cage bars. This well-known tool use behaviour could only be observed in elephants in a single case, an African zoo elephant (Zedtwitz cit. in Rensch & Altevogt, 1954).

## Speculations on the genesis of tool use

Numerous hypotheses have speculated on the adaptive significance of learned tool use behaviour and the evolutionary biological decisive factors that lead to its development. A fundamental precondition for tool use behaviour is, according to Beck (1980), the ability to manipulate objects. Indeed, learned tool use behaviour is found most frequently among those mammals that possess a prehensile organ, the primates and elephants. These latter giants are occupied daily for hours in preparation of food and in feeding, that is, with manipulation of grass, branches, logs and other plant parts which they can utilise as a tool. Viewed in this light, the trunk and the feeding niche may be predisposing to tool use (Chevalier–Skolnikoff & Liska, 1993). Tool use though, is rare compared to all elements used in the context of feeding behaviour.

However, the frequency of single acts of tool use is obviously high as well as efficient, especially in the preparation of short grass. Tool use in connection with preparing logs is inefficient and rare. The trunk is not only the decisive organ for food preparation but also a siphon. Water, mud and – although rather rarely – fine-grained material can be sucked up by nasal passages and purposefully blown out. This method is used in five of the elements of tool use behaviour (Table 4.3.1). It belongs to the functional categories of skin care and aggression. In 30% of all tool use acts by African elephants in the wild they utilise their trunk, a unique organ in mammals (Chevalier–Skolnikoff, 1993).

A highly developed central nervous system and a prehensile organ promote tool use behaviour. The absence of other specific, biological attributes enforce it, according to J. Alcock (1972). He postulates that tool use is more frequent in those animals that occupy for their group atypical ecological niches. This could certainly apply to elephants: their closest relatives are the sea cows (*Sirenia*), large herbivorous sea mammals which live in shallow waters, and the first real *proboscideans*, the *Moeritheres*, are similar in body size and way of life to the now living hippopotamus (e.g. Kurt, 1987; Shoshani, 1991). For aquatic warm blooded animals a large volume and a relative small body surface are adaptive characteristics in order to maintain body temperature. For terrestrial inhabitants of sub-tropical Savannahs and forests however, they pose the problem of overheating. Sparse hair growth and therefore lack of a dense fur is regarded as thermo-regulatory adaptation to tropical climate (e.g. Shoshani, 1991). However, the passive protection against intensive rays of the sun and ectoparasites, at least partially provided by a coat, is lost. In this light, the tool use behaviour in elephants could also have developed as adaptation to the living conditions through skin care and cooling of the body. In Asian elephants in the Ruhuna National Park one quarter of the known elements of tool use behaviour occur in this context (Table 4.3.1) and the frequency of single acts is higher than the frequency of acts in other tool use behaviours. In wild African elephants 83% of all tool use acts occur in connection with skin care in combination with cooling of the body, 24% occur in connection with cooling only 6% are triggered by molestation through ectoparasites (Chevalier–Skolnikoff & Liska, 1993).

Months before neonate elephants regularly feed on solid foods and are capable of drinking with the use of their trunk, they frequently manipulate objects. Often these behaviours cannot be assigned to any specific functional category. If there would be an assignment of a specific functional category at all to these behaviours then it would be the one that infants employ. The one which can be listed as methods of food preparation, and utilisation of tools more frequently during bathing time (Fig 4.3.2), than the older youngsters. Neonates are undoubtedly those members of the herd which are not dependent on preparing their own food and do not have a social role to play, and therefore possess free time and, as the milk is supplied by their mothers, they also have a relatively large amount of energy. They utilise both these factors to explore their surroundings and to make innovations, for example to improve intraspecific communication or ecological techniques. Most probably the same applies for primates as well as for elephants: "Time and energy which are used for exploring behaviour, change into lightweight capabilities, which cannot be stolen, which help to save energy and time in a harsh future" says primatologist Hans Kummer (Kummer & Goodall, 1985). From chimpanzees it is known that they make innovations when they are young, which they might only use many years later (Goodall, 1986).

Respective life histories of Asian elephants are not there till date. Nevertheless, elephant owners have experienced that animals which grow up in a captive group, and therefore have time and sufficient food at their disposal, can make innovations and become more willing working elephants later in life than singly raised and artificially fed orphans, which are not only retarded in their physical development, but also in their sensory motor development (see Section 4.2). A striking fact in this context, is that innovations only seem to occur in habitats like the south Indian Periyar Tiger Reserve or the Sri Lankan Uda Walawe Reserve, where an extraordinary high food availability and relatively short periods of feeding time provide long periods of 'leisure time'. Among these innovations are extraordinary tool use behaviours such as washing of plant parts or complex social behaviour patterns such as 'tree-breaking', whereby single bulls may provide younger bulls with sweet food supplements (Kurt, 2001).

## Tool use in tame elephants

Primates and elephants kept in species-specific holding facilities in zoos, and therefore inevitably fed by humans, provide an opportunity to test the hypothesis that free time and energy promote the development of innovations (Kummer & Goodall, 1985). This indeed applies to Hamadryas baboons (*Papio hamadryas*) (Kummer & Kurt, 1965). Asian elephants also develop new behaviours under favourable keeping conditions. With regard to tool use, the storing of food on the head or shoulders, the bundling and pushing away of food, the soaking of hard bread in water or the breaking of long branches against the wall are all part of tool use behaviour. These tool use behaviours do not occur in the wild areas; innovative zoo elephants obviously derive advantages from them. In restrictively kept elephants, meaning chained for long periods, tool use behaviours in the context of intraspecific aggression as well as skin irritations through ectoparasites are especially frequent. Riding elephants in Nepal, which are chained in the camp during the day, swatted at horseflies 15–20 times per minute and this frequency correlated positively with temperature and number of horseflies present (Hart & Hart, 1994). The use of 'fly swats' and missiles also occurs in wild elephants, although not with high frequency. Furthermore, chained elephants prepare these tools and keep them handy (see also Hart & Hart, 1994). These qualitatively and quantitatively highly developed forms of tool use possibly developed corresponding to the hypothesis of Hall (1963). According to this, fear and willingness to attack are conflicting with each other and promote the willingness to manipulate objects for later use as a tool. In this case, some tool behaviour could evolve as redirected activities under extreme keeping conditions such as permanent chaining. Comparison between elephants that grow up with their mothers and orphans, with respect to food preparation methods (see Section 4.2), it is noticeable that young orphans know fewer methods than their same aged conspecifics with mothers, but the tool use behaviour is more frequent (Fig. 4.2.3).

Without doubt it is not only the size and strength that makes the elephant a desired working animal, but also its ability to manipulate objects. This ability is practised from the beginning of training. The command *deri* teaches the elephants in training to lift their chains or the mouthpiece, which later will serve to lift or pull heavy loads. Timber elephants are trained to use tools, such as a lever, as was already applied to heave large rocks during the Middle Ages (Kurt, 1992). How far their cognitive abilities are put to use is not clear. Working elephants that have to haul heavy logs in steep terrain, immediately underlay branches or stones as wedges, if the logs are in danger of rolling downhill and injuring the elephant. U Toke Gale reports from Myanmar (1974) that logs, fell in the highlands and rafted down rivers, often get jammed between rocks along the shore during high water. Free-swimming bulls release these logs, while their mahouts wait on the shore. This suggests high cognitive and manipulative abilities. In the beginning of the last century also, circus elephants had to pull the caravans and carry the tent masts. In modern circus programmes, tricks whereby, elephants have to manipulate objects or use them as tools (Kurt & Knie, 1980) are usually absent, apart from a few exceptions. Zoo elephants kept by conventional methods usually utilise tools only in context with skin care, food preparation and resting behaviours. New keeping concepts, geared to bring more variety into the otherwise monotonous daily life of the captive elephants, should also take into consideration the cognitive and manipulative abilities, and therefore tool use abilities of elephants.

## 4.4  Sleep

A number of observations on captive elephants in western zoos and circuses had eradicated the widely held belief that elephants rarely or never lie down to sleep, and that they sleep only while standing (e.g. Benedict, 1936; Kurt, 1960; Tobler, 1992). Most records on sleep of both the African and the Asian elephants in their natural habitats are anecdotal with one notable exception (Wyatt & Eltringham, 1974).

In this chapter, the sleeping behaviour of 51 captive Asian elephants (22 bulls, 29 females) of the Pinnawela Elephant Orphanage is discussed. The study was aimed at collecting data from Asian elephants living in their native range. In addition, the study offered the opportunity to record the development of sleep behaviour and compare sleep in three different types of young captive elephants, namely (1) captive born animals that grow up with their mothers during the first few years of their lives, (2) socially integrated orphans with normal body growth and (3) socially more or less isolated orphans with retarded body growth. Orphans were considered as socially integrated, when they had regular social contacts with other elephants of the establishment and were hardly seen standing alone when the herd was standing in the paddock or in the Maha Oya.

When their social contacts were rare or absent and when they were often alone they were considered as socially isolated (see Section 5.3.)

In summer, 1997, for four nights observations were carried out between 10pm–6am (29/30.7; 3/4.8; 30/31.8; 6/7.9). In the Pinnawela Elephant Orphanage, elephants were fettered in open stables or under *Coconut* palms between 6 p.m. to 6 a.m. The animals were accustomed to the presence of people at night, and the establishment was provided with some artificial light during the night. A team of 2 researchers sat quietly and observed a group of 4 to 6 elephants at a distance of 4 to 6 m and monitored activities such as 'feeding', 'social behaviour', 'weaving' as well as 'sleep'. Sleep was distinguished in standing (SS) and recumbent position (RS). The number and duration of single recumbent sleep episodes, the duration of recumbent sleep per night, the duration of standing sleep per night and the total sleep time per night were compared. Furthermore, sleeping behaviour between animals with normal body growth and those with delayed body growth was compared. 179 complete protocols were obtained of 51 animals observed: 16 of neonates, 38 of infants, 73 of juveniles, 24 of sub-adults and 30 of adults.

It is most likely that one underestimated the time the elephants spent sleeping in standing position, as the available light did not always allowed detailed observations, i.e. one could not always distinguish, whether an elephant was only standing or actually sleeping while standing. Furthermore, it must be pointed out that the total length of daily recumbent sleep in younger animals was underestimated, as they also lay down during the day, mainly in the early morning and in the late afternoon.

## Duration of sleep

Elephants sleeping in the standing position remained motionless with their eyes closed, the trunk hanging down with its tip touching the ground and their breathing rate slowed down. While standing, sleeping elephants distributed their body weight equally on all four legs and, with the obtuse angle of the joints of shoulder, elbow, hip, and knee, they were able to stand with minimum physical effort. Slowing down of body and trunk movements usually preceded sleep in a standing posture. In the recumbent position the elephants lay on their sides, their eyes closed, their legs stretched or slightly at an angle. The trunk was slightly rolled inward and snoring was frequent. They were also observed while preparing a 'pillow' out of food plants. Apart from very few exceptions elephants were motionless in the recumbent position. When they stood up, they swung the top hind leg (on which they were not lying), manoeuvred into a sitting position and rose immediately. Occasionally, they lay down again from the sitting position and onto the other side. In 65% of all observed cases the elephants changed from one side to the other between two sleeping episodes. In general, the elephants slept as frequently on their left side as on their right side. The adult male Jandura slept only on his left side, and the sub-adult female Mauri only on her right side. Whether this was due to the way they were chained or some other reason, could not be discerned. The adult bull Raja was never observed in the recumbent sleeping position, although he did lie down during the daily bath. Raja with 5246 kg was by far the heaviest bull at the orphanage. Lying down might be very difficult for a bull of this size. He was also totally blind and maybe this made him too insecure to lie down.

The mean duration of standing sleep per night varied between 3.5 $\pm$ 0.7 min in the 2 neonate males Arjuna and Isuru and  201.3 $\pm$ 9.1 min in Raja, the blind adult bull, who never slept in recumbent position. Except for this bull, the mean duration of recumbent sleep varied between 61.5 $\pm$ 33.23 min in Mahaweli (an adult female) and 347.8 $\pm$ 41.9 min in Anura, an infant male. The mean duration of a single recumbent sleep episode was between 25.4 $\pm$ 36.4 min in the infant female Anuradhika and 108.5 $\pm$ 33.2 min in the sub-adult female Mauri. The mean total sleeping time per night varied between 108.0 $\pm$ 56.3 min in Anuradhika (who often slept during day time), and 347.8 $\pm$ 41.9 min in the infant Anura (Table 4.4.1 and Table 4.4.2).

Generalising the data one found a number of significant tendencies: The average duration of total sleep and recumbent sleep and the number of recumbent sleep episodes correlated negatively with the age of the elephants, in other words became shorter. Alternatively with increasing age the duration of standing sleep as well as the duration of single recumbent sleep episodes became longer. Adult females and bulls slept fot 169 to 267 min per night, neonates and infants for 219 to 348 min The average of recumbent sleep duration decreased from about 260 min per night in neonates to 75 min per night in old adult females and 0 in the

**Table 4.4.1** Duration (in min) of recumbent sleep of males in Pinnawela during four nights

| Name | Date (1997) | | | | Mean ± SD |
|---|---|---|---|---|---|
| | 29./30.7. | 3./4.8. | 30./31.8 | 6./7.9. | |
| *Neonates* | | | | | |
| Arjuna | 238 | 125 | 258 | 215 | 256.75 ± 59.01 |
| Isuru | 229 | 202 | 213 | 226 | 217.50 ± 12.45 |
| Nandimithra | 248 | 227 | 294 | 281 | 262.50 ± 30.58 |
| Anura | 285 | 372 | 197 | 369 | 305.75 ± 82.96 |
| *Infants* | | | | | |
| Aruna | 279 | 227 | 271 | 374 | 287.75 ± 61.88 |
| Sumana | 240 | 245 | 228 | 252 | 241.25 ± 10.11 |
| Esala | 209 | 257 | 223 | 269 | 239.50 ± 28.16 |
| Singharaja | 224 | 246 | 241 | 224 | 233.75 ± 11.44 |
| Jatila | 337 | 287 | 262 | 289 | 293.75 ± 31.34 |
| Tharaka | 321 | 313 | 209 | 218 | 265.25 ± 59.96 |
| *Juveniles* | | | | | |
| Kapila | 211 | 236 | 152 | 205 | 201.00 ± 35.32 |
| Chula | 233 | 227 | – | 262 | 240.67 ± 18.72 |
| Nilagiri | 272 | 235 | – | – | 253.50 ± 26.16 |
| Rangiri | 272 | 235 | 216 | 269 | 248.00 ± 27.14 |
| Sanka I | 200 | 257 | 350 | 326 | 231.00 ± 23.45 |
| Ravana | 238 | 245 | 227 | 246 | 239.00 ± 8.76 |
| Suranimala | 172 | 209 | 248 | 185 | 203.50 ± 33.39 |
| *Adults* | | | | | |
| Kandula | 149 | – | – | – | 149.00 |
| Jandura | 225 | 200 | 195 | 247 | 216.75 ± 24.06 |
| Kadira | – | – | 247 | 296 | 271.50 ± 34.65 |
| Neela | 149 | – | – | 309 | 229.00 ± 113.14 |
| Raja | 0 | – | 0 | 0 | 0.00 ± 0.00 |

old adult bull Raja. On comparing the total sleep time per night, the difference related to age was only about 60 min (Fig. 4.4.1). Obviously, older animals compensated less recumbent sleep with more sleep in the standing position.

In herbivore mammals the duration of daily recumbent sleep correlates negatively with  body weight (Zeppelin & Rechtschaffen, 1974) and our

**Table 4.4.2** Duration (in min) of recumbent sleep of females in Pinnawela during four nights

| Name | Date (1997) | | | | Mean ± SD |
|---|---|---|---|---|---|
| | 29./30.7. | 3./4.8. | 30./31.8 | 6./7.9. | |
| *Infants* | | | | | |
| Amali | 246 | 278 | 245 | 336 | 276.25 ± 42.68 |
| Anuradhika | 41 | 173 | 89 | 129 | 108.00 ± 56.32 |
| Sandali | 225 | 255 | 234 | 334 | 262.00 ± 49.62 |
| *Juveniles* | | | | | |
| Kanthi | 126 | 188 | 217 | 205 | 184.00 ± 40.64 |
| Lasanda | 218 | 244 | 213 | 208 | 220.75 ± 16.03 |
| Sarna II | 227 | 353 | 331 | 249 | 290.00 ± 61.37 |
| Sharmy | – | 206 | 208 | 244 | 219.33 ± 21.39 |
| Shanti | 244 | 246 | 350 | 326 | 282.25 ± 66.67 |
| Thilaka | 258 | 411 | 349 | 306 | 331.00 ± 65.01 |
| Kirimenika | 187 | 204 | – | 0 | 130.33 ± 113.19 |
| Kumani | 197 | 216 | – | – | 206.50 ± 13.44 |
| Menika | – | 231 | 244 | 227 | 234.00 ± 8.89 |
| Ramnya | 202 | – | 256 | 188 | 215.33 ± 35.91 |
| Noni | 172 | 258 | 141 | 268 | 215.75 ± 67.66 |
| Punchi | 251 | 320 | 214 | 286 | 272.00 ± 48.55 |
| Sarna I | 160 | 142 | – | – | 151.00 ± 12.73 |
| Soma | 144 | 171 | 206 | 207 | 182.00 ± 30.36 |
| *Sub-adults* | | | | | |
| Ninja | 242 | 188 | 228 | 130 | 197.00 ± 50.19 |
| Nikini | 258 | 214 | – | 196 | 222.67 ± 31.90 |
| Sukumali | 133 | 156 | 110 | 142 | 135.25 ± 19.31 |
| Mauri | – | 217 | – | – | 217.00 |
| Anuradha | 141 | 125 | 125 | 215 | 151.50 ± 43.00 |
| Rejina | 260 | 278 | 238 | 326 | 275.50 ± 37.43 |
| Ranmali | – | 232 | 204 | 247 | 227.67 ± 21.83 |
| *Adults* | | | | | |
| Mahaweli | 38 | 85 | – | – | 61.50 ± 33.23 |
| Komali | 157 | 170 | 161 | 146 | 158.50 ± 9.95 |
| Mathali | 112 | 60 | 126 | 170 | 117.00 ± 45.33 |
| Kumari | 59 | 186 | 175 | 327 | 186.75 ± 109.74 |
| Anusha | 90 | 195 | 23 | 103 | 102.75 ± 70.79 |

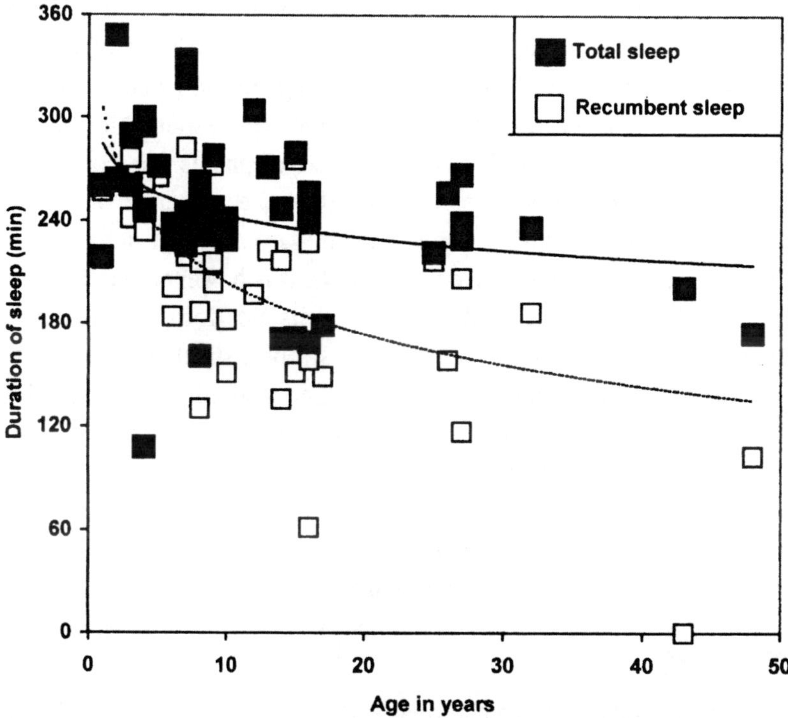


**Fig. 4.4.1** Mean duration of total sleep and sleep in the recumbent position in the Pinnawela herd

data on Sri Lankan elephants, as well as the data of other authors, who studied sleep in Asian elephants, fit well into the equation $Y = 9*W^{-0.14}$ (Y: duration of sleep in the recumbent position; W: body weight in kg).

The meaning and importance of sleep in the standing position have not been completely understood. It is possible that elephants may sleep while standing, when they are in situations where the recumbent position is not advisable (e.g. mothers with neonates). It could also be possible that standing sleep functions as a substitute when animals do not know what to do otherwise, or when (such as in some captive facilities) there is no space to lie down. The supposition that standing sleep is a lighter form of sleep (c.f. Tobler, 1992) has not been proven yet, even though there are observations which tell us that circus elephants with more physical activity during the day sleep more frequently in the recumbent position than for example zoo elephants of similar age (Kurt, 1960; Tobler, 1992). However, it is not clear, why idly-living zoo elephants often have extremely long periods of sleep in the standing position before they lie down. It can be that the researchers themselves are responsible for long rest periods in standing positions, since zoo elephants (in contrast to circus elephants) are often not accustomed to the presence of strange humans in their stables during night time and hence do not dare to sleep in the recumbent position (c.f. Hediger, 1969). It is however also feasible, that the weight of an adult elephant lying on its side for too long can be detrimental to its organs and metabolism, and sleeping standing compensates for the lost time. The elephant's heart rate interestingly increases when the animal lies on its side. This rather unusual fact probably relates to the reduced ventilation which occurs in elephants in reccumbent position and is a direct result of the fibrous attachment of the lungs to the chest wall: elephants have no pleural cavity. Therefore, it makes sense that the older and heavier an animal the less it lies down.

## Sleep profiles

Wild elephants in the Ruhuna National Park as well as in the Uda Walawe National Park rested during the hot time of the day, i.e. between 9.00 a.m. to 3.00 p.m., in the shade of large trees. During this time neonates, infants and juveniles often slept in recumbent position. Sub-adult and adult animals rested or even slept in standing position. From a few night observations in the Ruhuna National Park it is known that all the observed wild elephants also slept at night time, mainly during the second half of the night (Kurt, 1992). Hence, it could be assumed that wild elephants had at least two peaks of rest or sleep during the 24 h day. But there was no doubt that older adult bulls made an exception from this rule. They quite often slept during day time, mainly in wallows or shallow waterholes. They had sleep profiles with several peaks. The same also applied to neonates and infants, too. They regularly slept during day time, when the rest of the group was feeding, more or less under or in the shade of a sub-adult or adult female functioning as a watcher, and occasionally offspring of several mothers lay in a heap next to or over each other.

Observations in the Pinnawela Orphanage synchronised in such a way that scans were made of all animals every 5 min by all observers.

During these scans, activities were noted, e.g. whether the monitored elephants were feeding, weaving, standing, or sleeping in standing or recumbent position. According to these data, sleeping profiles were designed and given as the frequency of animals sleeping in standing or recumbent position. Such profiles were calculated for all social classes (Fig. 4.4.2).

Between 22:00 h to 06:00 h captive born infants slept occasionally (not more than 50% of all cases) between 22:15 h to 2:30 h. All the neonates always slept between 02:45 h to 4:30 h, and most of them until 06:00 h. The sleep profile of mothers followed more or less closely that of the infants: Between 22:30 h to 2:45 h they hardly slept, but between 03:00 h to 06:00 h they slept relatively often. It must be stressed, however, that it never occurred that all mothers simultaneously slept in recumbent position. At least one of them played the role of the watcher, as it is known from socially intact groups (Kurt, 1960).

In contrast to mothers and their neonates and infants, orphaned infants, juveniles and non-reproducing sub-adult and adult females had a two-peaked sleep profile between 22:00 h to 6:00 h, with a first peak between 24:00 h to 1:00 h and a second one between 3:00 h to 5:00 h. The same was applied also to sub-adult and adult bulls. The animals were fed between two sleeping peaks. Circus elephants also had two-peaked sleep profiles during the night, regardless of season and place (Kurt, 1960; Tobler, 1992).

Most probably the one-peak sleep profile of mothers in Pinnawela cannot only be related to the fact that they have suckling offspring and/ or are pregnant; they also often eat very slowly because they feed their offspring with prepared food and teach neonates and infants how to prepare food. Accordingly, their food intake must have been below the required food intake that the adult females should have. Especially for Anusha and Sukumali physiological demands must have been extraordinary. Both of them were suckling neonates from 1996 or 1997 respectively. Shortly before parturition Mathali gave birth to a calf in January 1998. She, too, must have been under extreme physiological stress. Anusha, Sukumali and Mathali had a pooled mean period of total sleep (standing and recumbent) of 204.1 ± 52.8 min per night, and this period was not significantly different from the period of total sleep of 4 other mothers

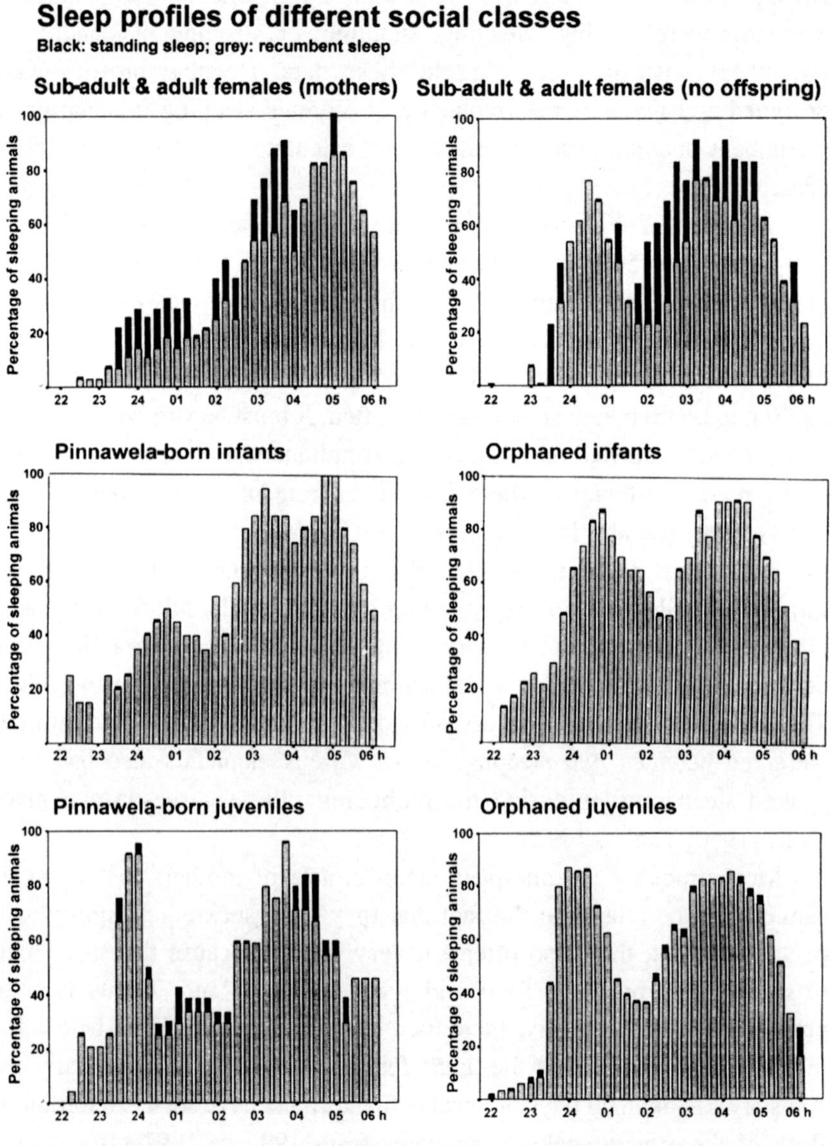

**Fig. 4.4.2** Sleep profiles of social classes: Sub-adult and adult females (mothers and childless ones) infants (Pinnawela-born and orphans) and juveniles (Pinnawela-born and orphans) between 10 p.m. (22:00h) and 6 a.m. (06:00h). Data are given as percentage of all animals belonging to the social class concerned.

(Kumari, Rejina, Anuradha, Komali) living without obvious physical stress at the time of observation. For them the pooled mean value of total sleep read 235.9 ± 69.0 min. In recumbent position Anusha, Sukumali and Mathali slept in the mean for 118.3 ± 47.1 min, the other 4 reproducing females for 193.1 ± 75.4 min. This difference was significant. Accordingly, Anusha, Sukumali and Mathali compensated by sleeping longer in standing position. Standing sleep allows the calf to suckle when it wants to without the mother having to stand up each time. Lying down and standing up for pregnant Mathali could have been more strenuous.

Analyses of individual sleep profiles (Fig. 4.4.3) showed that neonates and infants adjusted their individual sleep profiles only partially to those of their mothers and allomothers. Often the offsprings lay down earlier for a short while. The 4 mothers, who had either very young calves or were not near to parturition, normally started recumbent sleep earlier than the 3 mothers under extreme physiological stress. The adult bull Jandura was fettered in the same stable as the mothers and their offsprings during the night. But his sleep profile had two peaks, like the profiles of other bulls, and hardly followed the profiles of mothers and offsprings.

## Sleep in socially non-integrated orphans

Individual sleep profiles of most orphans reflected profiles with two peaks. It became obvious that animals standing close to each other had sleep profiles more similar to each other than animals fettered widely apart from each other. In wild elephants group members synchronized their activity patterns to a high degree (Kurt, 1992). This did not only apply to sleep but also to drinking, bathing, skin care and feeding. In the Ruhuna National Park one often had the impression that group members standing close to each other were scalping and cleaning short grasses in the same cycles. And when mating took place, other younger bulls seemed to imitate the mating bull and mounted smaller elephants. The high degree of synchronization of periods of recumbent sleep was very pronounced in circus and zoo elephants (Kurt, 1960; Tobler, 1969). The ability to synchronize activity patterns with the majority of the group seemed only to be present in socially integrated animals.

In Pinnawela it was checked, whether orphaned infants and juveniles lay down to sleep within more or less the same time (maximum: ± 5 min)

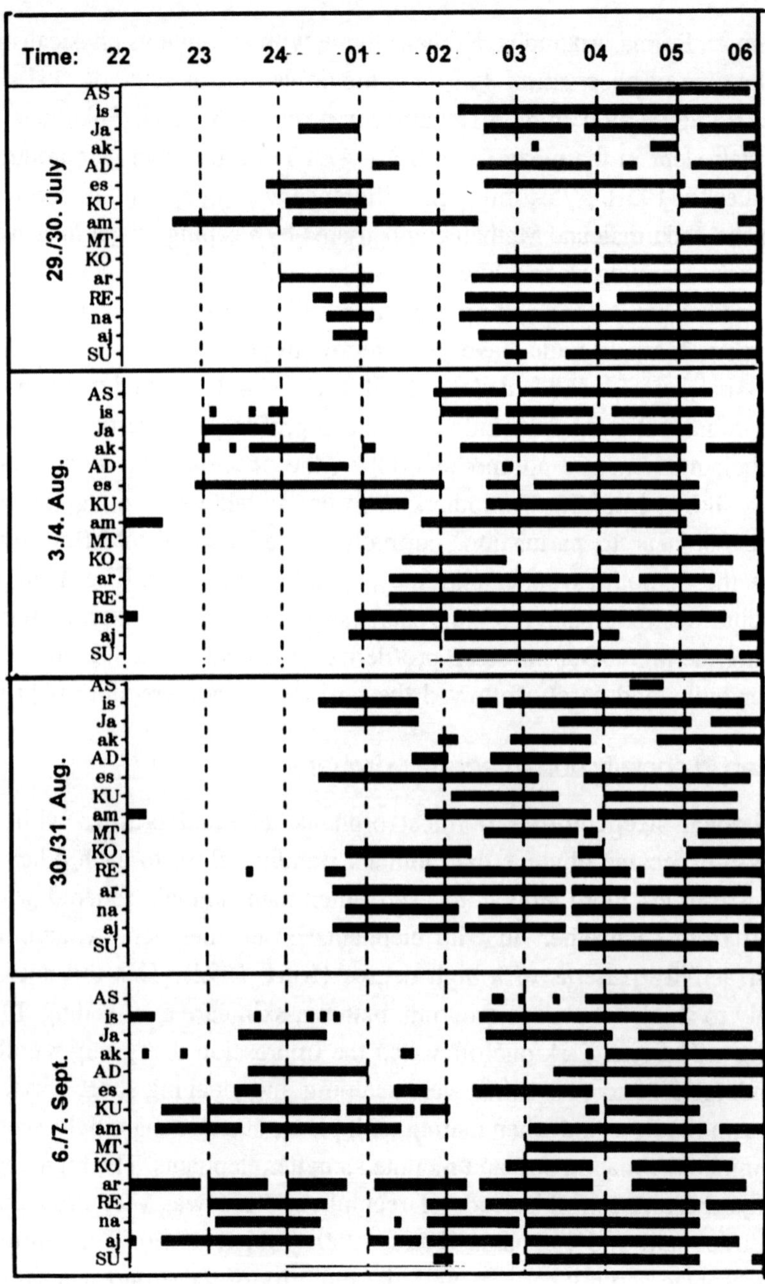



Fig. 4.4.3 Individual sleep profiles during four nights in the barn with mothers and offsprings at Pinnawela. Periods of recumbent sleep are marked with black lines. Abbreviations of names according to Table 4.1.1. & 4.1.2

as their neighbours. As a measurement for social integration the percentage of time a particular animal stood close to others or not, was considered while the herd was in the Maha Oya (see Section 5.3). A positive correlation was found between the degree of social integration and the degree of synchronized lying down for recumbent sleep. Obviously orphans which were not socially integrated not only had different sleep profiles to socially integrated animals, but also a different duration of sleep periods as shown in the following paragraph.

Juvenile males with normal body growth slept significantly less in standing position than orphaned juvenile males with retarded body growth. Normally growing juvenile females had significantly shorter periods of recumbent sleep than orphaned juvenile females with retarded body growth and a shorter mean duration of total sleep than orphaned juvenile females with retarded growth. Sub-adult females with normal growth had a shorter duration of total sleep and recumbent sleep than sub-adult females with retarded growth. Generalising these data it can be stated that orphans with retarded body growth, i.e. those, which were socially less integrated or not at all integrated slept more than captive born juveniles or socially integrated ones (Fig. 5.4.5). Therefore, prolonged periods of rest and sleep seems to be one of the behavioural strategies to cope with unnatural and stressful situations. Furthermore, it can be assumed that weaker animals need more rest.

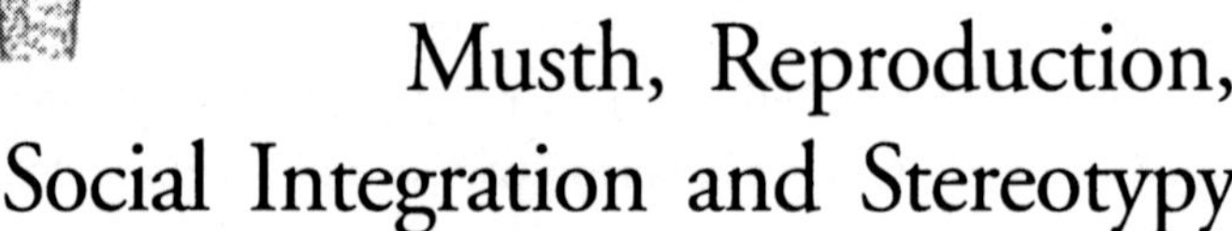

# 5

# Musth, Reproduction, Social Integration and Stereotypy

## 5.1 Musth Bulls

Both the Asian and African elephants possess a skin gland in their temples, just above the cheekbone between the eye and external ear opening. This temporal gland is periodically active in males and females. In female Asian elephants temporal gland is active for a very short period of a few hours during mating and during parturition (Kurt, 1992). In adult Asian elephant bulls, musth is considered as a period during which testes and temporal gland are extremely enlarged (Chandrasekharan *et al.*, 1992), testosterone level highly raised (e.g. Jainudeen *et al.*, 1972; Cooper *et al.*, 1990; Lincoln and Ratnasooriya, 1996), and sexual behaviour including regular masturbations, olfactory marking and aggression towards other bulls very pronounced (e.g. Kurt, 1974, 1992, 1995; Rasmussen and Schulte, 1998). The physiological changes during musth include blood acidosis, increased lipid catalysis and fluctuating, often greatly elevated serum, androgen and ketone concentrations. These alterations are characterised in the changed chemical compositions of breath, urine and secretions from the temporal gland (Rasmussen & Krishnamurthy, 2002). The chemical composition continually changes throughout the musth period. Musth is progressively assessed by increased amount of foul smelling, 2 nonanone. In young bulls the first musth periods are characterized by sporadic episodes of aggressive behaviour and fluctuating levels of serum testosterone and dihydrotestosterone, and the secretion of their temporal gland is composed of sweet smelling alcohols, ketones and esters

(Rasmussen & Riddle, 2002). The resemblance of odour profiles of secretions of the temporal gland and those of several well-known insects and their by-products provide an outstanding example of convergent evolution occurring between the chemical signals in Asian elephants and invertebrates (Rasmussen & Riddle, 2002).

However, the physiological control of musth also involves other factors than testosterone, e.g. other hormones, hormone metabolites or changes in target tissue receptors, and definitely the nutritional status of the male (e.g. Cooper *et al.*, 1990). Cortisol is markedly elevated during the musth period. This could mean that the adrenal gland actively produces steroids during musth. The hormone from the pituitary that stimulates adrenal cortisol (ACTH) production also stimulates adrenal androgen production during musth (overview in Wingate & Lasley, 2002).

The following pages summarise some ethological findings on wild and captive Sri Lankan elephant bulls, bulls from South Indian and Myanma elephant establishments as well as from European and North American zoos and circuses with special emphasis on duration and individual periods of musth, social behaviour and socio-ecological organisation of adult bulls (for detailed reports see Kurt, 1991; Kurt & Touma, 2002). Furthermore, the hypothesis that musth can be considered as synonymous with rut is discussed.

## A look at wild bulls

The previous study on wild Asian elephants in Block I of the Ruhuna National Park (103 km²) in 1967–68 revealed that sub-adult and young adult bulls (10 to 20 years) had short musth periods of a few days and musth occurred in individually known animals 2 to 3 times per year. Adult bulls (> 15 years) came into musth only once per year and their musth periods lasted about 3 to 4 weeks. Usually only one adult bull was in musth at a given time, and at least one musth bull was present in the study area more or less throughout the year (Kurt, 1974, 1992). The same was applied to bulls in the Mudumalai Sanctuary in South India. According to the recent investigations within an area of about 45 km² in the Sri Lankan Uda Walawe National Park, adult bulls in musth were found throughout the year, and most of the time only one out of 42 to 61 adult bulls was in musth at a given time.

Between 1993 to 1999 individually known bulls came into musth at the same time of the year. In wild bulls musth is obviously an asynchronous, cyclical, circannual pattern of specific physiological and behavioural characteristics, with the cyclical pattern developing progressively with increasing age.

The genesis of individual musth periods for adult bulls within a given male population is demonstrated best by the socio-ecological organisation of juvenile, sub-adult and adult bulls as it was found during the recent studies in the Uda Walawe National Park (Kurt, 2001): the distances between the respective musth bull and other bulls standing in the vicinity of 2 km correlated positively with the age of the non-musth bulls. Juvenile bulls (6 – 10 years) showed mean distances to the musth bull of 827 + 509 m (n = 18), older adults (> 20 years) distances of 1139 + 525 m (n = 33). When a musth bull was in the vicinity (< 2 km), other bulls formed significantly larger groups (2.4 + 1.6 members, n = 28), than when non-musth bull was close by (1.5 + 0.6 members, n = 33). Formations of large groups at a certain distance most probably guarantee better safety against a potentially dangerous musth bull. In case, two bulls reach musth at the same time serious fighting takes place and the loser immediately falls out of musth or has to leave the temporary home range of the superior musth bull (Kurt, 1974, 1992). In wild bull populations the asynchronous, cyclical, circannual patterns of individual musth periods seem to be reached by competition amongst adult bulls; i.e. adult bulls with established individual musth periods suppress musth in younger bulls, until the newcomers find a temporal niche for a musth period of their own. As already mentioned, the secretions of the temporal gland of sub-adult bulls as well as the secretions of adult bulls in different stages of musth can clearly be distinguished by smell.

It can be expected that bulls do not select their group partners randomly. Furthermore, different age classes are not equally distributed in the bull population. Young bulls possess higher attraction for other males than do older bulls. Older juvenile males seek group contact with males of similar age more often than with older or younger ones. Within female-offspring-groups, sub-adult males are more often seen together with males of the same class or older than with younger ones, whereas bulls older than 20 are seen together with younger adults or sub-adults more often than with other adults. Adult bulls also seem to avoid contact with other same sized bulls outside of mother-offspring groups. Group contact between juvenile and sub-adult males of the same class size is common and allows the establishment of dominance hierarchies through play-fighting as well as individual bonds between certain males. Both are important mechanisms to avoid serious fighting later, when bulls reach larger size classes. Old adults and younger bulls are found together relatively frequently in bull associations outside female-offspring-groups, as fighting between such size differences is unnecessary and therefore unlikely, and they most probably contain males with close social relationships.

In Uda Walawe, certain adult bulls formed special relationships with juvenile bulls. They provided them with sweet food by pushing down large *Tectona grandis* trees. The noise of the falling trees immediately attracted numerous juvenile males, which would have never been able to breakdown fully grown teak trees. The 'tree-breakers' always shared the sweet bast of the trees with the younger males. Tree-breaking was carried out by only a few bulls. In male groups of upto 10 members, 'tree-breakers' were found. (mean: 5.1 + 3.1; n = 15) but non-tree breaking adult males were found in significantly smaller male groups with a mean size of 2.2 + 1.6 members. In all cases observed 'tree breakers' were maknas, i.e. bulls without tusks. Furthermore, it was found that certain maknas had easier access to family units in company of numerous juvenile males than more or less solitary 'non-tree-breakers'. But a close relationship between a certain adult and certain younger bulls may also render social advantages to the younger bulls, as shown with the following example:

An adult female in oestrus rejected very aggressively the mounting attempts of a young sub-adult male until an adult bull attacked her seriously. In due course, the female allowed the sub-adult male to mount, which was immediately followed by mating with the adult bull. This as well as the observations on tree-breaking adult bulls in Uda Walawe indicate that social organisation in wild living males of the Asian elephant is more complex than as believed so far. This also concerns their organisation in space and time.

Five daily route samples of juvenile males measured between 1.5 to 6.5 km (mean: 3.5 + 2.1 km). 11 daily movements of young adults between 16 to 20 years measured between 1 to 8.5 km with a mean of 4.3 + 1.9 km, i.e. significantly longer than 14 daily movements of old adults (older than 20 years) with daily routes of 1 to 5.3 km (mean 2.4 + 1.3 km). 5 daily routes of old adult bulls in musth covered distances between 2.8 to 15 km with a mean of 8.5 + 4.5 km and were significantly longer than those of old adults not in musth. The range size in Uda Walawe occupied by 2 juvenile males during the two month observation period in 1999 had a minimum of 9.5 km². For two sub-adult bulls range sizes of 9.5 km² and 10.5 km², respectively were estimated, and for two adult tuskless males the minimum values read 9 km² and 10 km² respectively. This concludes, that these values are smaller than for temporary ranges of female-offspring-clans and they hardly say anything about the ranges used during the year (Kurt, 2001). In the Ruhuna National Park yearly ranges of adult bulls measured between 64 km² to 75 km² and in Gal Oya between 39 km² to 54 km² (De Silva, 1998, for overview).

Wild bulls in musth, scent mark their temporary positions and ranges by bringing the tip of their trunks into their mouths to collect fluid. They spray this on their temporal gland, smear the mixture of fluid and secretion over large parts of their heads and front parts of their bodies and rub their heads against trees, which are later checked by olfaction by other elephants. Furthermore, musth bulls communicate acoustically. Last but not the least wild musth bulls during the Smithsonian studies in Ruhuna National Park (Kurt, 1974, 1992) were seen more often within or in the immediate vicinity of mother-offspring-groups than males not in musth. During 1998–1999 in Uda Walawe, mating was observed 10 times. In all the cases complete copulation was carried out only by bulls in or shortly before their respective musth periods, after they had driven away most other bulls. The same was the case in Ruhuna. Before discussing the question whether musth is an entirely reproductive phenomenon, a number of observations on captive musth bulls are summarised.

## Musth in captive bulls

In captive adult bulls of Sri Lanka musth was also not a seasonal phenomenon (Fig. 5.1.1 and Table 5.1.1). Each of the bulls studied had

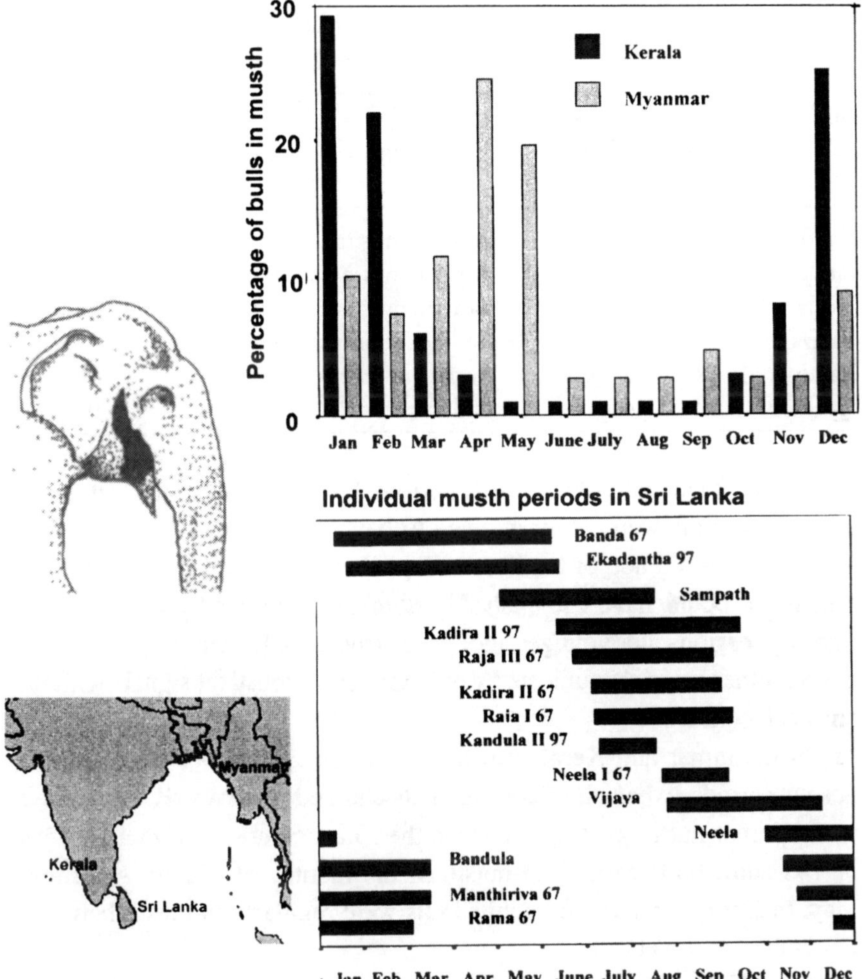


**Fig. 5.1.1** Musth periods of captive bull populations in Kerala and Myanmar and duration of individual musth periods of captive bulls in Sri Lanka (Sources: Toke Gale 1974, Chandrasekharan *et al.* 1992, Lincoln & Ratnasooriya 1996).

an individual musth period. Two of these bulls (Kadira II, Banda) had hardly changed their musth periods since 1967–68, when observed for the first time by Fred Kurt. The adult members of two bull groups in the Pinnawela Elephant Orphanage and the Dehiwala Zoo, respectively, monitored by Lincoln and Ratnasooriya (1996) and partially by our team,

**Table 5.1.1** Data on 6 bulls studied. *according to mahouts **in 1999 Jandura was in musth for the first time

| Bulls | Age in years | Age at first musth* | Situation during observation | Place of observation | Time of observation (min) |
|---|---|---|---|---|---|
| Kandula I | 17 | not yet | not in musth | Pinnawela | 525 |
| Jandura | 25 | not yet** | not in musth | Pinnawela | 465 |
| Neela | 27 | 22 years | not in musth | Pinnawela | 450 |
| Vijaya | 31 | 18 years | beginning of musth | Pinnawela | 570 |
| Kandula II | 20 | 17 years | middle musth | Kandy | 1035 |
| Kadira II | 55 | 14 years | end of musth | Kandy | 915 |

maintained their individual musth periods for at least 7 years. Although these adult bulls were never allowed to have direct social contacts, their musth periods did not, or hardly, overlap. However, at least the bulls from Pinnawela could have established a dominance hierarchy during play-fighting sessions at a younger age. Furthermore, bulls kept in the vicinity of each other can communicate by olfactory and acoustical signals without physical contact.

In Myanmar and Kerala musth periods of captive bulls concentrate during periods, when they are either discharged from work, and/or fed with optimal nutrition (Fig. 5.1.1). In the forest camps of Myanmar 55% of 148 adult bulls came into musth in the months of March, April and May. In Kerala 85% of 160 adults bulls were regularly in musth between December and February.

In southern Asia captive bulls are normally not made to work during musth, but are fettered at specific secure places with heavy chains. In most zoos bulls are kept in a smaller enclosure during musth than when not in musth, in modern zoos they are kept with the herd. In many cases the musth periods of captive bulls last longer than those of wild bulls. In Myanmar the duration of musth was 3 to 80 days with a mean of 33 days (n = 210; Toke Gale, 1974). In temple elephants of Kerala musth periods lasted for 30 to 150 days with a mean of 68 days (n = 127; Chandrasekharan *et al.*, 1992). The State Elephant of the last Maharaja of Mysore (Karnataka) was in musth during 9 months every year (Stracey, 1963). The White Elephants in the Chitralada – Palace of Bangkok are in musth nearly permanently. Asian elephant bulls in western establishments can also have extremely long musth periods of 1 to 8 months (e.g. Rübel & Tanner, 1993). Between 1984 to 1988 one bull kept in the zoo in Columbus (Ohio, USA) had musth periods of 124 to 187 days (Cooper *et al.*, 1990).

According to Buddhist and Hindu belief musth of temple and state elephants mean fertility and wealth for the country and the ruling families (Kurt, 1993). Accordingly, they are well-fed during their musth periods despite their extreme danger to humans. The same applies to fighting and war elephants. They were almost permanently in musth and led close to their opponents blindfolded (e.g. Meyer, 1929). Musth also means danger, mainly for the mahout of the bull concerned. He is often selectively attacked and hence, often absents himself during the musth period of his bull (Jainudeen *et. al.*, 1972a; Kurt, 1993; Dickerman *et al.*, 1997). Asian elephant lore predicts a number of treatments for renitent musth bulls and shortening of their musth periods, such as low diet, drugs, hard work (before the expected onset of musth) or more or less permanent painful flogging with especially sharp *ankuses* or spears wounding sensitive parts of the body like ears, eye-edges, temples, nail beds, penis and perineum (Jainudeen *et al.*, 1972b.).

For captive bulls, several models were developed which divide the musth period into two, three or four parts (Fig. 5.1.2): only two are used in the present context. In the first part of the musth period, the bulls seem to be aggressive and strongly sexually aroused. In the later parts of the musth period, they remain aggressive but become sexually inactive and greatly increase the frequency of olfactory marking behaviour. Between the 20th of July to the 10th of August 1997, Chadi Touma (2001) and his team studied the behaviour of 6 captive bulls aged between 7 to 55 years, kept singly on chains during 66 hours by focal animal sampling method and continuous recording (Table. 5.1.1). Three of these bulls were in musth, one at the beginning, one in the middle and one at the end of the respective individual musth periods. The 3 bulls in musth were continuously kept in chains and received several portions of food and water during the day. The 3 bulls not in musth were brought to the water and fed only in the evening with a large portion of food.

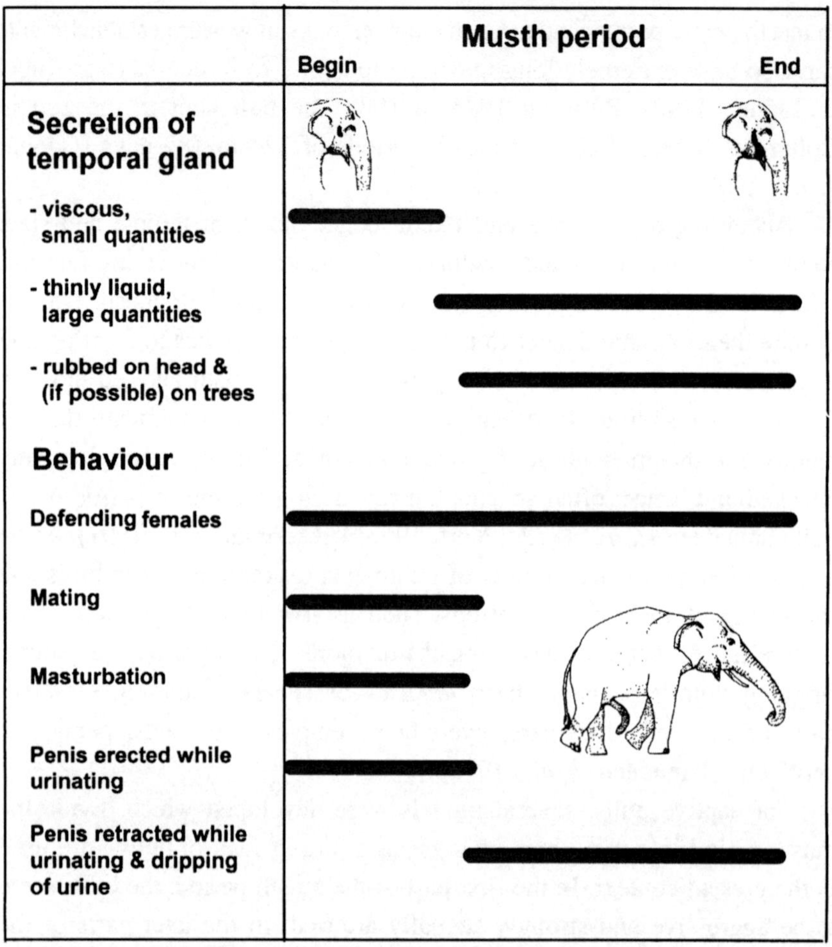

| | Musth period | |
|---|---|---|
| | **Begin** | **End** |
| **Secretion of temporal gland** | | |
| - viscous, small quantities | | |
| - thinly liquid, large quantities | | |
| - rubbed on head & (if possible) on trees | | |
| **Behaviour** | | |
| Defending females | | |
| Mating | | |
| Masturbation | | |
| Penis erected while urinating | | |
| Penis retracted while urinating & dripping of urine | | |

Fig. 5.1.2   Some characteristics of the musth period

The data revealed that musth bulls spent less time in feeding than bulls not in musth, however, they spent significantly more time in stereotypic behaviour, so-called weaving (Fig. 5.1.3). The hormonal changes during musth most probably induce a reduction of food intake. The increase of weaving could be due to the fact that musth bulls were permanently chained and hardly able to make one or two steps to either side. Furthermore, it could be postulated that weaving was triggered by lack of social partners (see Section 5.4). In wild bulls social behaviour increases enormously during the period of musth (Kurt, 2001), however in captive musth bulls social contacts are reduced to zero due to chaining.

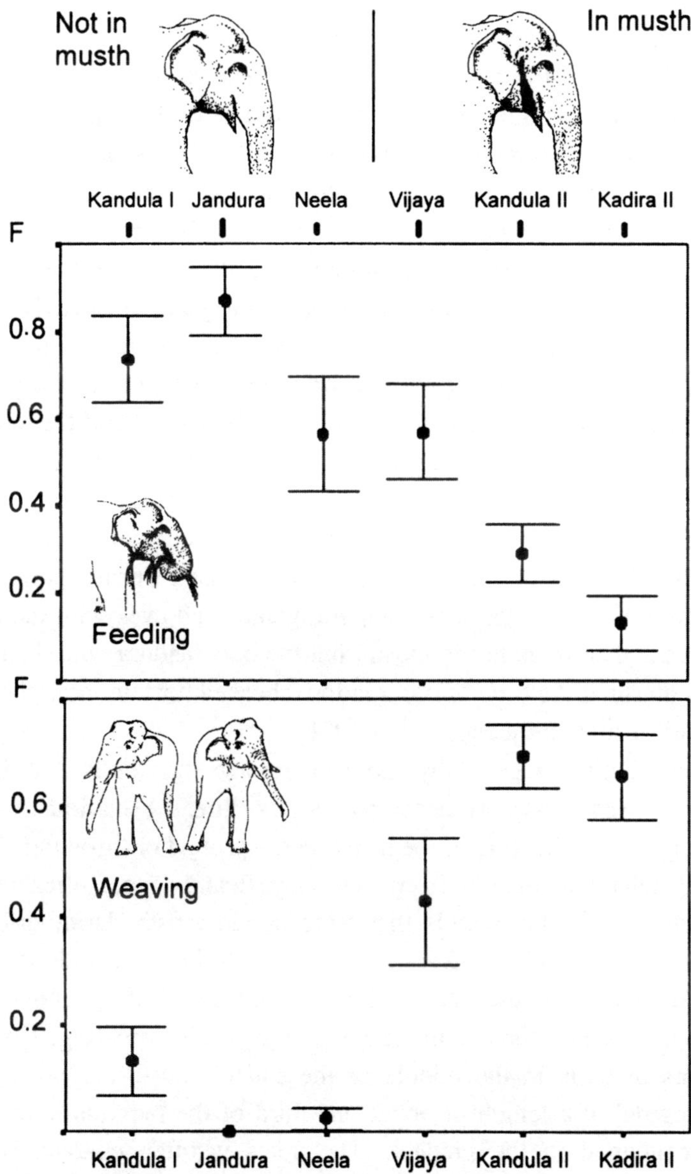


**Fig. 5.1.3**   Mean frequencies (acts/min; according to the method of focal animal sampling and continuous recording) and confidence intervals of 95% of two behavioural elements in six captive bulls observed during 66 hours. Top: Feeding, bottom: Stereotypies. Significant differences between musth bulls and non-musth bulls were found for the element stereotypy (Source: Kurt & Touma, 2002).

Musth is, amongst other, characterised by swollen temporal gland which produces a strong smelling secretion. The secretion is often mixed with fluid and then smeared over the head and front parts of the body. This behaviour was regularly observed in the 2 bulls that were in the middle or at the end respectively of their individual musth periods (Fig. 5.1.4). The temporal glands of the other 4 bulls were not or hardly swollen and did not produce secretion. The fluid to mix the secretion stems from the *Bursa nasopharyngialis*, a muscular pocket, lying just behind the root of the tongue and formed by muscles of the larynx (Tennent, 1867; Shoshani, 1998). To bring the fluid from this organ into the mouth, the head is raised, the mouth slightly opened and the muscular system of the cheeks pressed inwards. Then the fluid is taken out of the mouth with the tip of the trunk. These behaviour patterns termed 'Raise head and trunk in mouth' were observed significantly more often in the 3 musth bulls than in the three bulls not in musth. Furthermore, musth bulls more often touched their temporal glands and eyes with the tips of their trunks than bulls not in musth, but the bull Jandura, who had never been in musth at the time of observation, checked his temporal gland and eyes with a high frequency (Fig. 5.1.4).

Musth bulls are generally more aggressive than bulls that are not in musth. Aggression is accompanied by threat behaviour such as spreading of ears, beating of the trunk on the ground or throwing objects. These 3 behaviour patterns were seen significantly more often in bulls in musth than in the 3 bulls that were not in musth. Urine-dribbling, another typical musth behaviour, was seen only in the 2 bulls that were in the middle or at the end of their individual musth periods. The 4 other bulls erected their penis more or less while urinating. However, the penis of bulls in the middle or the end of musth are not erected, but 'hanging' at a length of about one third of the maximum and seem to be contained in the prepuce. The penis is often characterised by greenish rings, which could either stem from vascular congestion or, more probably, from algae. During this period urination seems to be almost blocked and only possible by more or less continuous urine dribbling. Since captive musth bulls are permanently chained and not brought to a daily bath, prepuce and hind legs are permanently soaked with urine, which leads to painful inflammation, wounds and, in due

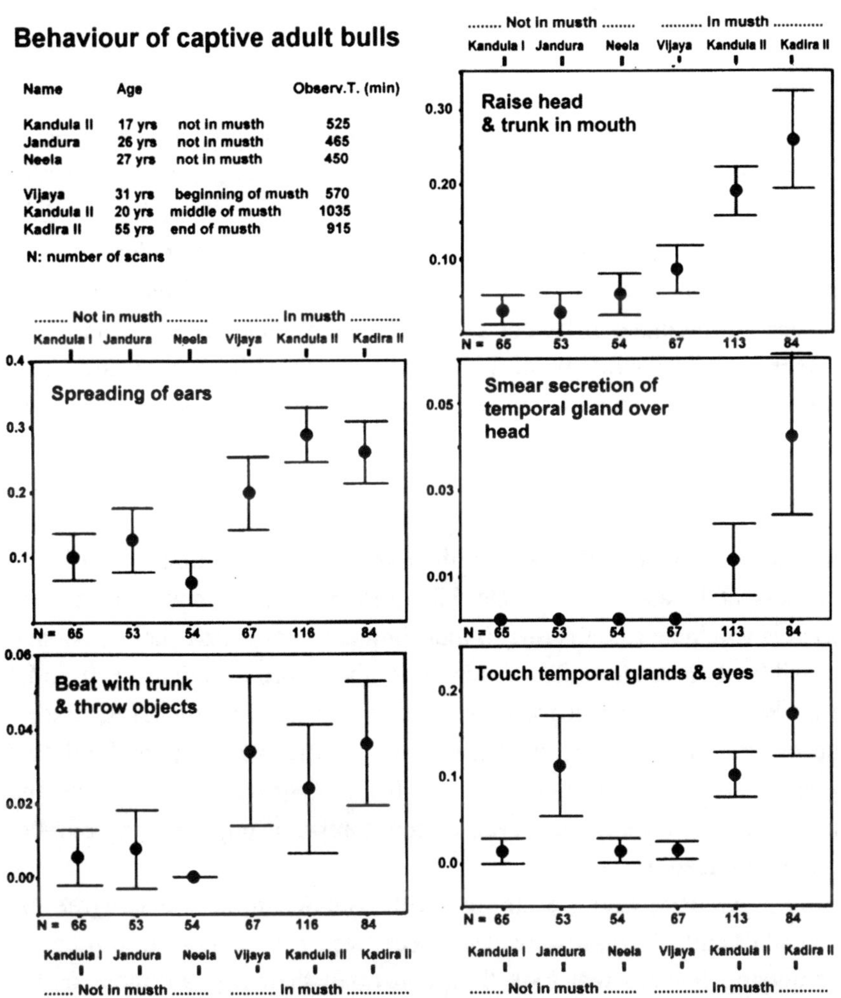


**Fig. 5.1.4** Mean frequencies (acts/min; according to the method of focal animal sampling and continuous recording) and confidence intervals of 95% of five behavioural elements in 6 captive bulls observed during 66 hours. Significant differences between musth bulls and non-musth bulls were found for the elements 'Raise head and trunk in mouth', 'Spread ears', 'Beat trunk on ground and/or through with objects' (Source: Kurt & Touma, 2002)

course, most probably to a further acceleration of aggressive behaviour (Jainudeen *et al.,* 1972 a,b.).

## Musth and rut

In wild bulls individual musth periods are most probably reached by competition amongst adult bulls, and the physiological fitness of musth is established by optimal body condition due to high food intake. After gradually separating from their mother-offspring-units, before puberty bulls group in socially open aggregations and use different ecological niches following a different feeding strategy than females and their offsprings. The strategy of bulls reads 'high risk-high reward' (e.g. Sukumar, 1992). Due to their body size, strength and independence from vulnerable neonates and infants, they can find their daily rations in areas inaccessible or too dangerous for mother-offspring-units such as steep slopes, swamps, or often well protected paddy fields or sugar cane plantations, where bulls may run the risk of getting wounded or killed by farmers and poachers (e.g. De Silva, 1998). Furthermore, much more food is available to bulls due to their immense body size, their tusks and, possibly, their special relationship with other more experienced 'friends', from whom they can learn how trees are broken and debarked; and, if necessary, bulls can invest more time to find and prepare food than females, due to their independence from offsprings. That means that bulls can use, if necessary, even climax forests or high growing bamboo groves as feeding grounds.

Researchers, who are familiar with wild Asian elephants, agree that sub-adult and adult bulls always appear 'well-fed' compared to pregnant or nursing females that look thin, and often their ribs are more or less visible (e.g. Kurt, 2001). However, bulls at the end of their musth periods may look similar. This is due to some marked metabolic changes during musth. Musth bulls extremely decrease their food intake due to both the impact of testosterone as well as due to totally changed daily activities, which are dominated by testing and defending females, bathing and scent marking and extremely increased roaming, leading to a drop in body condition and even a considerable drop of muscle mass. However, the same phenomenon occurs also in captive bulls that are kept under restricted conditions without increased social contacts or free movement. Their increased triglyceride levels, elevated lipase activity and increased ketones (Rasmussen & Perrin, 1999) are highly suggestive of fat breakdown as it happens during starvation. The most extreme possible interpretation of this evidence is that adrenal activity actually triggers musth (Wingate and Lasley, 2002). But there is more evidence that adrenal activity actually causes musth.

According to our hypothesis, musth periods are established by competition amongst bulls, and a once established musth period is maintained as long as possible during the life span of an adult bull. This fact can only be the product of a highly structured social organisation amongst bulls, in which everyone knows 'who is who', 'who is with whom' and 'who is where', and which is guaranteed by chemical and acoustical signals transmitted over large distances (e.g. Hart *et al.*, 2001), as well as outstanding cognitive abilities (e.g. Hart *et al.*, 2001). Therefore, it is obvious to assume that a certain bull knows when his established musth period is imminent and then he "has an internal drive to increase his body condition as a store against the upcoming losses which will occur during musth" (Wingate & Lasley, 2002, p.154). He will enter a short phase of obesity. But over-nutrition also occurs in certain captive bulls (e.g. state and temple elephants) and stimulates adrenal activity, since cortisol is a glucose counter-regulatory hormone and would be stimulated under conditions of elevated blood glucose. This is because the ACTH that would stimulate cortisol production also stimulates adrenal androgen production (Wingate & Lasley, 2001). So far it is not clear whether musth is controlled only by adrenal gland or to a certain degree directly by testicular activities. Most probably by both, as is the case in rut of certain deer. In *Cervus* and some *Odocoileus* species antler cycles (shed antlers mark the ability for rutting), are controlled by adrenal as well as directly by testicular hormones (Bubenik, 1966).

Increased frequencies of sexual, antagonistic and marking behaviour advocate the opinion that musth in adult Asian elephant bulls is synonymous to rut as known from other gregarious ungulates with complex social systems, especially camels, sheep and deer (Eisenberg *et al.*, 1971; Poole, 1987; Kurt, 1992; Kurt and Touma, 2002). Musth bulls as well as rutting camel stallions, rams and stags have increased testosterone levels, and are more aggressive towards other males. All stay in mother-offspring-units more often than males that are not in musth or rut and they frequently test females. Their conspicuous vocalisations is lacking in non-rutting males. Their body postures with raised heads and stretched forelegs conspicuously signals their presence, they frequently rub body secretions, either directly or indirectly by wallowing, on their body and they rub these secretions on trees and rocks. They frequently urinate which stains the inside of their hind legs or their chests and flanks and leaves scent trails in their environment.

In deer for example, antlers of stags are shed during rut and certain skin glands are active, male deer are in physiological, ethological and social respects fit for reproduction. In several South Asian deer species such as *Axis axis, A. porcinus* or *Cervus unicolor* reproduction takes place throughout the year, although the reproductive capability of individual males last only for periods much shorter than 12 months (e.g. Kurt, 1978). The same applies to elephant bulls. However, deer stag do not mate outside their rutting season. For deer living in non-tropical ranges, rut is triggered photoperiodically and their testes are inactive and not producing sperms (e.g. Kurt, 2002). Whether the same is the case in tropical deer species has been discussed for a long time (e.g. Bubenik, 1966). In tropical deer female choice favours the larger stags with largest antlers and in wild African and Asian elephants females prefer bulls in or shortly before musth as mating partners. Established musth bulls have proven longevity and the ability to find enough food to afford the luxury of momentary obesity and in due course musth. Both are important selective qualities for a long lived species living in an environment with unpredictable food resources. However, one could argue that musth and rut are not that closely related, since captive elephant bulls are known to reproduce successfully when not in musth. The following pages will first bring some examples and later shortly discuss whether bulls have certain rut-like periods during which their reproductive capacities are more pronounced.

In Pinnawela, first studies on musth bulls by Lincoln and Ratnasooriya (1994) revealed that mating was only successful by the two breeding bulls, Neela and Vijaya, shortly before the onset of their individual musth periods. The present study again compared, observed or estimated mating times with the musth periods of Neela and Vijaya, which in the meantime had been established. The available data from Pinnawela refer to gestation periods of 625 to 650 days, i.e. 21 to nearly 22 months. Accordingly, the time of successful matings of the mothers of the 12 Pinnawela-born elephants that are being referred to in the study can be estimated quite accurately. Four matings took place in October, two in November, one in December and one in January, i.e. eight successful matings happened in the periods, which were (later) found to be the periods when Neela and/ or Vijaya were in musth. However, 4 matings must have taken place, when neither Neela nor Vijaya were in or shortly before musth. The Pinnawela data do not fully confirm the hypothesis mentioned above. But Pinnawela elephants were far from being kept under natural conditions. Therefore, the question whether musth is necessary for successful reproduction to extensively kept timber elephants, that live at least when not working in a more natural environment, where they meet with captive as well as wild conspecifics was addressed.

In Asian timber camps females are mated to a high degree by wild bulls and it is said that these bulls are in or shortly before musth. When wild musth bulls are close to an elephant camp, captive bulls are kept within the camp and, if necessary, protected by fires from aggressive visits by the wild counterparts. But females are brought to the forest after work as usual. In Myanma timber camps it is an old tradition that captive adult bulls are allowed to roam freely at night shortly before their annual musth periods begin and mate with the captive females. About 50% of the captive born elephants have been fathered by such captive bulls (Kurt & Mar, 2003). Although in Myanmar musth and reproduction take place throughout the year, neither of them are equally distributed throughout the year but show prominent peaks. The highest percentage of musth bulls is found in April, i.e. shortly after a period of rest of 3 to 4 months at the beginning of the rains and the working period. The highest percentage of parturitions takes place in January, i.e. in the middle of the dry season when work ceases temporarily. The time span between the two peaks is little less than 9 months or, if one measures over two consecutive years, about 21 months, i.e. the average length of the estimated gestation period for timber elephants. It can be assumed that during the period of rest, reproducing male and female timber elephants are capable of building up enough resources to reach oestrus or musth status respectively. However, these findings are no strict proofs that only males in musth reproduce successfully.

## Is musth 'a must' for successful reproduction?

Addressing the question whether musth is a necessary precondition, 'a must' for successful reproduction to the population of bulls kept in zoos, it was found that this is not the case. For example Maxi, the bull of the zoo in Zurich mated successfully during musth and when not in musth (Kurt, 1992). This bull is doubtlessly an exception, since most captive bulls in musth are kept under extreme restrictions and not allowed to meet with other elephants. Furthermore, many breeding bulls in western zoos are extremely young and have not even reached the age when first signs of musth become obvious (Haufellner *et al.*, 1993, 1996, 1999, 2000). In 144 successful matings in European zoos, 24% of the bulls were between 9 to 15 years, i.e. before first signs of musth are visible, 26% were between 16 to 20 years and 50% older than 20 years, i.e. at an age when most wild bulls have established their individual musth periods. These data from zoos make it clear that musth is definitely not a necessary precondition for mating success. However, it must be stressed that in all of the 144 recorded successful matings in European zoos none of the bulls had a smaller body size than his mate (Fig. 5.1.5). As long as no older male was present, a 9 or 10 year old bull could reproduce with a female of 8 or 10 years but not with an adult one of 20 or 30 years. In contrast, a 45-year-old bull like the famous Siam in the zoo of Paris mated successfully also with females that were hardly 5 or 6- year-old. All data from reproducing zoo elephants strongly advocate the opinion that female choice is always in favour of the strongest and most dominant available mating partner.

In European and North American zoos and circuses parturitions take place throughout the year. A high percentage of births occur in the months of April and May with 34 parturitions (11.1%) each and relatively few offsprings are born in January and September with 18 parturitions (5.9%) each, but no pronounced breeding season can be found. The same applies to wild Asian elephants in South India and Sri Lanka (Kurt, 1991). Nevertheless, the dates of birth of the offsprings of the 12 most successfully breeding bulls in European and North American establishments

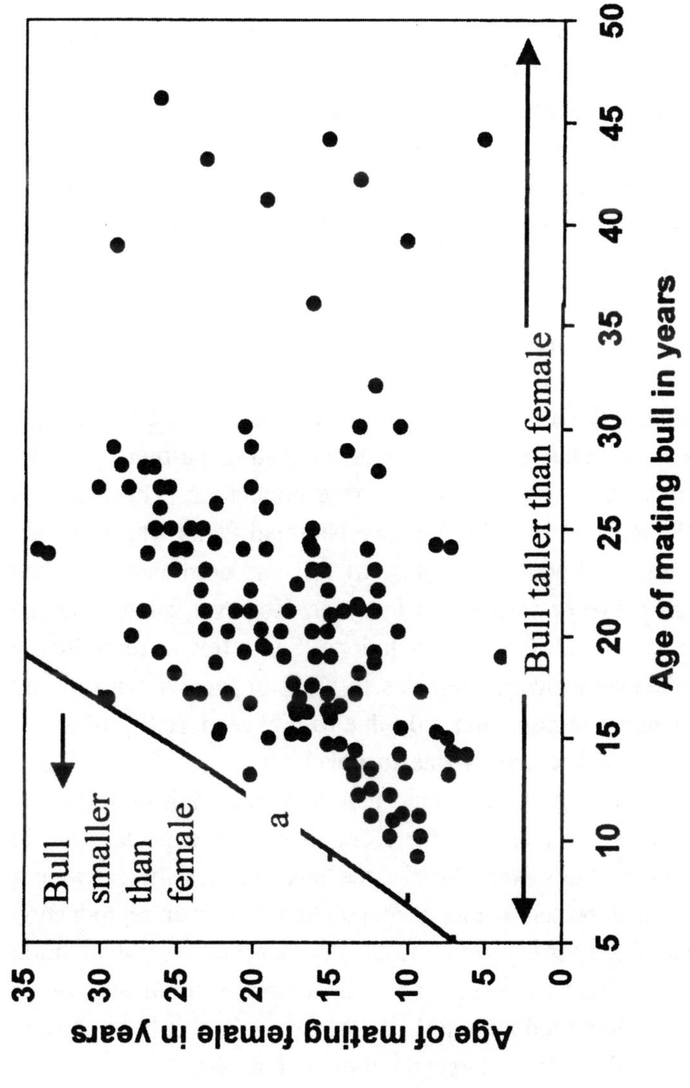

**Fig. 5.1.5** Comparison of mating ages in zoo kept males and females. Data from European Elephant Group (1995–2003)

were investigated (Fig. 5.1.6). Each of these males had fathered seven to 20 offspring (mean:10) with 2 to 8 females (mean:5) and the time span of male reproduction lasted (so far) 7 to 35 years (mean:15 years). Dates of birth of the offsprings of each of these bulls were not equally distributed over the year, as expected, but seemed to be more or less clustered. The individual reproduction period for each of the 12 bulls was defined as the shortest possible period during which 90% of their offsprings were born. These individual reproduction periods lasted between 111 to 245 days with a mean of 183 days (6 months) and the individual reproduction periods of these bulls correlated neither with the number of offsprings, nor with the number of females or the length of the time span of successful breeding (Fig. 5.1.6). According to these findings one can postulate the existence of individual breeding periods during which a bull reproduces more successfully than during other periods of the year. Unfortunately, data on musth periods are not available.

The wild Asian female elephants that are being referred to in this report preferred old adult bulls in musth as mating partners, i.e. the strongest and most dominant members of the population. These studies took place in the Ruhuna and Uda Walawe National Parks where the sex ratios of sub-adults and adults was close to 1:1, and dominant bulls had a relatively high presence throughout the year. However, such balanced sex ratios are exceptional in wild populations. Sex ratios in reproducing age classes are skewed towards females in most of the Sri Lankan and South Indian elephant regions, not only due to higher mortality of males in middle age classes but mainly due to poaching. In the Periyar Tiger Reserve of Kerala for e.g. it was found that only one adult bull per 100 reproducing females survived and therefore, it can be expected that in regions where many bulls were lost in the last decades there was not always a bull in musth present within the population. However, reproduction can still continue if one assumes that musth is not a necessary precondition for successful reproduction. Even under the extreme circumstances of one adult bull per 100 reproducing females the few bulls would be capable of mating successfully with a large number of females.

If a parturition interval of 5 years is assumed, a bull can mate with 20 females per year or one or two per month under the assumption that there is no pronounced breeding season. To answer this question, data

# Birth of offsprings from 12 European & North American bulls

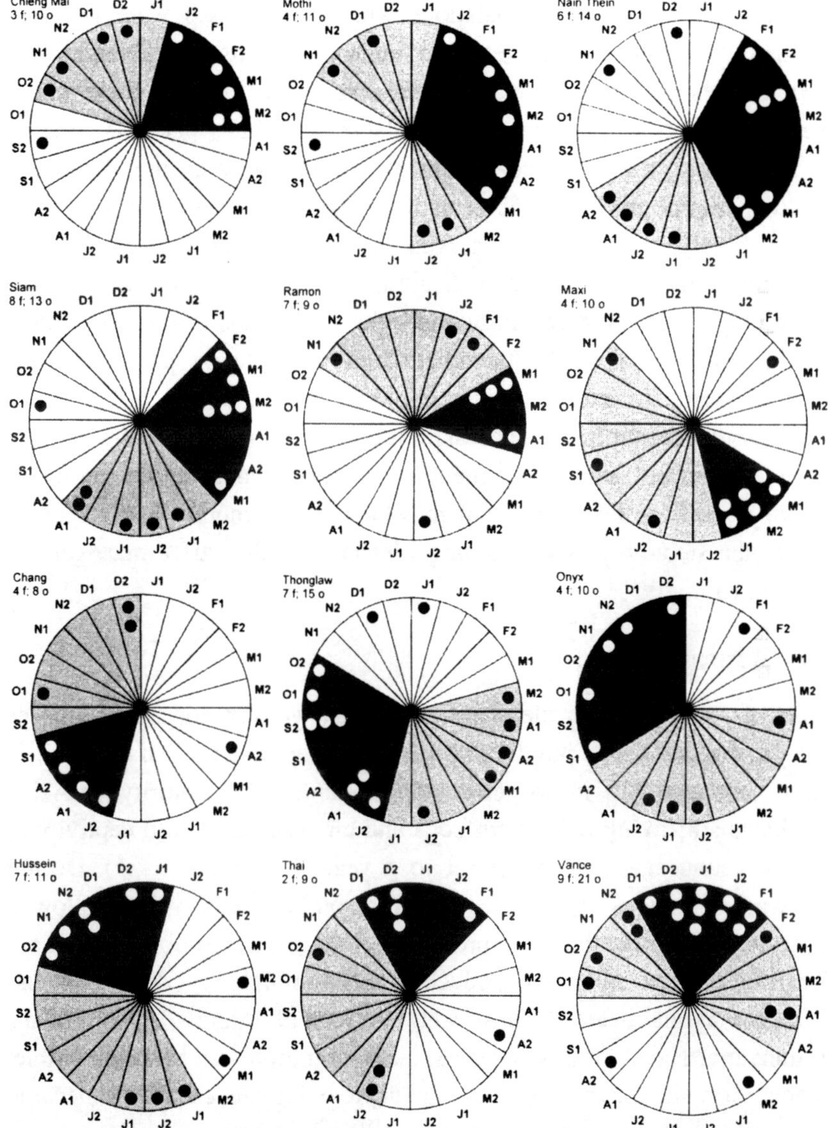

**Fig: 5.1.6** Parturition of offsprings from the 12 most successful zoo bulls in North America and Europe. The shortest period of 50% of all parturitions are marked black, the shortest of 85–90% of all parturitions are marked grey.

from population census in several Sri Lankan and South Indian populations were compared and the number of neonates per 100 females in reproductive age, or, in populations where age classes were not well defined, the number of 'calves' (i.e. neonates, infants and juveniles) per 100 females in reproductive age was calculated. Finally, these values were compared to the number of available adult bulls per 100 females. In both cases, a negative correlation was found between the abundance of adult males and the number of offsprings. The fecundity of females drops sharply, when the sex ratio in reproductive age classes becomes less than 1: 6. These findings indicate that successful reproduction takes place only when there are sufficient older adult bulls in the population.

In wild Asian elephants musth bulls seem to be paramount for reproduction. In captive Asian elephants musth seems not to be a necessary precondition for successful reproduction. But from the point of view of welfare and management of captive elephants the period of musth is still an unsolved problem which shows more clearly than all others the dubiousness of keeping elephants under strict management rules in captivity.

## 5.2 Reproduction

Most of the few Asian elephants that reach western zoos and circuses today, originate from jungle camps, where reproduction, if desired, is more common than in western establishments. The majority of Asian elephants still living in western zoos and circuses came into captivity at a very young age, i.e. between 1 to 5 years (see Section 7.1). During the kraal operations upto the early 70s in southern Asia, which followed large-scale deforestation and produced elephants for timber extraction, the babies were cheap by-products, readily bought by American and European animal dealers. Those who survived often grew up as orphans in unnatural social groups, where social integration was hardly possible. When growing older, many of the captive female elephants became obese. Reproduction rate was and is still low in zoos and circuses due to the rarity of bulls, the relatively short reproduction period of female zoo elephants between 7 to 32 years, and the high neonate mortality due to stillbirth and infanticide.

In a few Asian establishments captive elephants show higher reproduction rates. At the Pinnawela Elephant Orphanage 12 females gave birth to 22 offsprings between 1983 to 2003 (Fig. 5.2.1). 12 neonates survived, i.e. the mortality rate at a young age was extremely low. In the following pages some details concerning age and body measurements of reproducing and non-reproducing sub-adult and adult females are given. Furthermore, pregnancy periods and duration of parturition are discussed and, finally, a description of the birth of a calf in August 1998 is given.

## Relative body weight of mothers, pregnancy period and birth weight

Shoulder height and thorax circumference of most elephants in Pinnawela were known, hence, body weight could be estimated. A number of orphaned females showed retarded body growth. The same animals were socially more or less isolated. Body size does not only have an impact on social status but also on the reproductive status of a female.

In female elephants the ability to reproduce is obtained when an animal reaches a certain size and thus has enough body resources of its own for pregnancy and later for lactation (c.f. Kurt & Mar, 1996). This means that for puberty and for the first pregnancy, it is not the actual age alone of the female elephant that is decisive, but also her physical condition which can be expressed roughly in relative body weight. Relative body weight is the quotient of the effective body weight and the shoulder height (W/S). However, it must be stressed that seen from a strict statistical point of view, relative shoulder height is not an unequivocal value. Hence, the following data should be considered as a valuable 'rule of thumb' as shown in the following samples from Pinnawela.

Anusha, an orphaned female with retarded body growth, was born from wild parents in 1949 and until January 1986 lived in the Dehiwala Zoo, where she never reproduced. Then she came to Pinnawela, where she became mother for the first time at the age of roughly 40. In 1967, i.e. at the age of 18, her body measurements were taken for the first time. Her shoulder height measured 219 cm and her thorax circumference 298 cm, her estimated body weight was 1910 kg and her relative weight 8.7 kg/cm (Kurt & Nettasinghe, 1968). In 1997–98 the respective values read 207 cm for shoulder height, 321 cm for thorax circumference, 2090

## Development of families in Pinnawela (1983–2003)

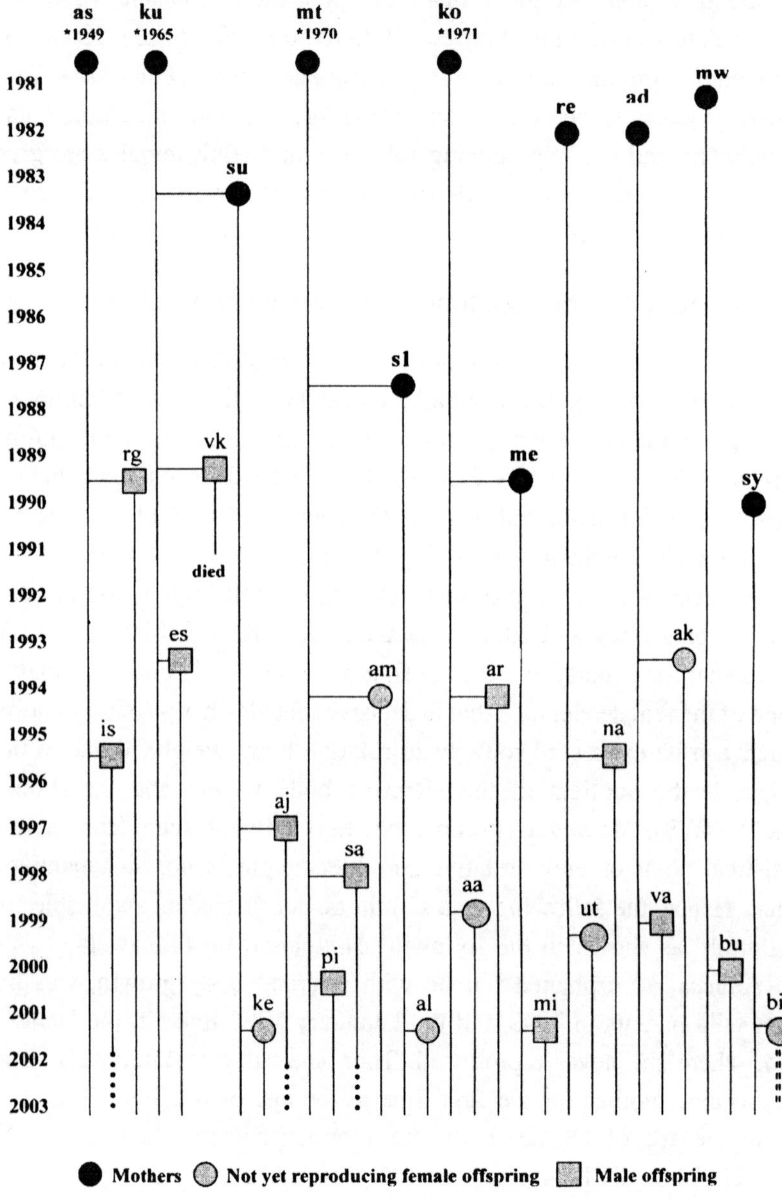



**Fig. 5.2.1**   Development of mother-offspring groups at Pinnawela between 1983 to 2003

kg for body weight and 10.1 kg/cm for relative body weight, i.e. she took part in reproduction only after she had reached a relative body weight of 9 or more kg/cm.

The same applied to the orphaned Kumari, born from wild parents in 1965. In 1997–98 her relative body weight was 9.3 kg/cm. She became mother at the relatively late age of 18 years. The females Anuradha (born 1982), Rejina (born 1982), Komali (born 1971) and Mathali (born 1970) had normal body growth and in 1997–98 their relative weight measured 10.2 kg/cm, 9.7 kg/cm, 10.9 kg/cm and 11.6 kg/cm respectively. They gave birth to their first calves between 11 to 18 years. The captive born Sukumali (born 1983) parturiated for the first time at the age of 13 years. In 1997–98 her relative weight was 9.4 kg/cm. The 4 females Ninja (1985), Mauri (1983), Ranmali (1982) and Mahaweli (1981) showed retarded body growth. In 1997–98 their relative weights were 6.2 kg/cm, 6.9 kg/cm, 7.9 kg/cm and 8.4 kg/cm respectively. Although these females were of 12 to 18 years, they had never reproduced so far. Concluding from these data, it can be said that in Pinnawela females with normal body growth parturiated for the first time at an age between 11 to 18 years, and females with retarded body growth between 18 to 40 years, if at all. Reproductive status was reached only by females with a relative body weight of at least 9 kg/cm. In some females kept in western zoos relative weight was found to be more than 12 kg/cm. This value seems to be the upper threshold value. In cases where the relative weight of a female was more than 12.0 kg/cm the animal never reproduced (Kurt & Mar, 1996).

The comparatively low relative body weights of females in Pinnawela caused relatively short durations of pregnancy of 19.5 to 20.6 months, low birth weights of 54 to 70 kg, and relatively small shoulder heights of the neonates. Whereas in zoos, duration of pregnancy often lasts longer, sometimes upto 24 months. Accordingly, birth weights are heavier and shoulder heights of neonates taller, as duration of pregnancy correlates positively with the size of the neonate and also with the probability of stillbirth. Stillbirths happened after pregnancies of 619 to 718 days (mean 666.6 + 37.2 days, n = 5), the birth weights measured between 93 to 165 kg (mean 127.2 + 20.7 kg, n = 17) and the shoulder height between 84 to 165 cm (mean 92.1 + 7.4 cm, n = 8). Surviving newborns were born after pregnancies of 510 to 718 days (mean 615.5 + 47.9 days, n = 32), had weights of 49 to 159 kg (mean 90.5 + 27.2 kg, n = 33) and shoulder heights between 76 to 106 cm (mean 84.4 + 8.1 cm, n = 19). The differences between surviving newborns and stillbirths are significant.

Stillbirths also occur after extremely long parturition of 7 to 40 h. In Asian jungle camps as well as in Pinnawela parturition is fast, i.e. within one or one and a half hours. As with wild elephants, they take place at night (Toke Gale, 1974; Krishnamurthy, 1992; Kurt, 1992). In western zoos and circuses, babies are born during day time also, but the parturition normally starts between 21:00 h to 07:00 h. Apart from one case, in Pinnawela births took place either between 04:00 h to 08:00 h or between 20:00 h to 23:00 h. Parturition lasted about 90 min.

## Record of a birth and social integration of the neonate

The team had the chance of witnessing a birth during the studies in Pinnawela. This birth took place on August 7th, 1998. This event is described in the following lines (c.f. Dastig, 2002; Dastig *et al.*, 1999):

The 28-year-old female Komali stopped eating at 7 a.m. and became increasingly restless, shifting her weight from one side to the other. Komali pressed her head against the outer wall of the stall, simultaneously lifting one of her hind legs. She also took turns, swatting at her genitals with the tuft of her tail and kicking at them with one of her forefeet. After a while, a thigh-sized protrusion in the perineum became visible, and slowly sank lower. In addition, sticky white, mucoid discharge was expelled from the vulva. Komali began voiding small amounts of faeces and urine. During contractions, the abdominal wall was tense and the back arched upwards, while at the same time, Komali raised the left and right hind leg alternately. During that time, she kicked her abdomen with one foot and tried repeatedly to remove the iron bars in front of her and the chains on her feet with her trunk. She then pressed one of her front legs against the wall and began, during the continually accelerating contractions, to sink onto her carpal joints, pressing one hind leg against the concrete floor, shifting her weight forward and, standing and stretching one hind leg backward. In this phase, the reddened and edematous mucous membranes of the vulva became visible.

After the keeper had cleaned her stall at 8 a.m., the contractions came at the following intervals: 8:06 a.m. with a length of 75 sec, 8:10 a.m. (20 sec), 8:15 a.m. (45 sec), 8:21 (20 sec), 8:24 (44 sec) and at 8:25 a.m. the calf was born. In the final moments of the birth, the thigh-sized protrusion moved very quickly in the direction of the genital opening. At 8.23 a.m. its water burst and blood mixed with yellowish mucoid amnion fluid was expelled in phases. Finally, the milky white foetal membranes began to protrude from the vulva under which about 15 cm of the front legs became visible. Thus, the calf was born in front end position. Two seconds later, the following contractions completed the expulsion phase. During the birth, after the head had come out, the calf turned and landed on the floor on its side. As the neonate hit the floor, the amnion sac broke. The calf then immediately tried to get onto its belly into sternal position, and Komali turned to her calf and began to feel it with her trunk. The veterinarian carefully removed the remaining fetal membranes from the neonate's head. After a few minutes, the calf tried to stand up. The afterbirth was expelled at 10:25 h. The parturition took 1 h and 25 min in total. The neonate, named Ashokamala, had a shoulder height of 95 cm, a thorax circumference of 105 cm and an estimated weight of about 75 kg. This was the third time Komali gave birth at the Pinnawela Elephant Orphanage.

Already during the parturition the other members of the mother-calf-group had tried to reach over towards Komali, but were hindered by chains and keepers. After parturition all members of the mother-calf-group inspected the newborn. Especially the 8-month-old Salia was highly interested in the small Ashokamala. Later, when Ashokamala started to explore her environment, she was never attacked by any other elephant. Socially she was integrated into the group from the very beginning of her life. The same must have applied for the other captive born neonates, such as Salia (son of Mathali), Arjuna (son of Sukumali) and Isuru (son of Anusha).

In summer 1998 Salia, Arjuna and Isuru were regularly observed during the morning and the evening for a total of 68 h. They often played with each other. Salia drank not only from his mother, but also from his unrelated allomother Rejina. Food was taken mainly close to these 2 females or close to his 4-year-old sister Amali. Arjuna was not only suckled by his mother Sukumali, but also by his grandmother Kumari. Isuru was suckled only by his mother Anusha. He fed not only next to her, but 50 or more metres away, close to his 9-year-old brother Rangiri.

## 5.3  Social Integration and Socialisation

Both the Asian and the African elephant are highly social animals that display a great amount of maternal care (Lee & Moss, 1986; Moss, 1988) and also an impressive amount of allomaternal care (Asian elephant: Gadgil & Nair 1984; Rappaport & Haight, 1987; African elephant: Douglas-Hamilton, 1975; Wilson, 1971; Lee, 1987; Garaï, 1997). Information on the social behaviour of the Asian elephant is sparse (wild: Kurt, 1974; captive: Eisenberg, McKay & Jainudeen, 1971; McKay, 1973; Gadgil & Nair, 1984; Rappaport & Haight, 1987; Garaï, 1992) but there is some indication that there are similarities in the social organisation to that of the African elephant, although there may well be some differences.

Mothers and their offsprings form the core of the elephant society, with usually several related females and offsprings forming a larger family unit, the female members of which form strong lifelong bonds with each other (Douglas-Hamilton, 1975; Moss, 1981, 1988; Moss & Poole, 1983). Both male and female calves grow up in the security of this family unit, protected and cared for by their female relatives. Before reaching puberty  young males leave gradually the maternal group to form lose associations with other males of differing age (Kurt, 2001a).

In view of the relative long period of physical maturation in elephants, one can assume that a growing elephant has much to learn and has to acquire social skills for its future role and position in elephant society. For females, this could be mothering, leadership, allomothering, integration into and being a useful member of the existing society; for males this includes social integration into the dominance hierarchy, behaviour during and out of musth towards other bulls, acquisition of successful matings, as well as habitat use for both sexes. Each social society is made up of an intricate network of relationships and a process of interactions with other members, the mechanisms of which must be learned. Therefore, the loss of its family must be a traumatic experience for a very young elephant and particularly for a young female elephant, which hardly ever leaves its maternal group. It has been shown that young African elephants suffer a fair amount of stress upon losing their families during culling operations (Garaï, 1997) and there is anecdotal evidence that some very young orphaned individuals that are deprived of the opportunity of bonding with older females or other juveniles do not recover from the trauma for years.

A very good measure for social bonding is spatial proximity (Kummer & Kurt, 1963; Lamprecht, 1973; Hinde, 1977; Garaï, 1992). In a normal undisturbed group of elephants neonates will be near their mothers and allomothers. Infants and juveniles upto the age of about 8 years will tend to be in the vicinity of their mothers, even though these are not always their nearest neighbours (Asian: Gadgil & Nair, 1984; African: Lee, 1987). In the African elephant females tend to stay near their mothers and younger siblings, whereas young males increasingly spend more time with same age peers (Lee, 1987). Neonates and infants that have to be taken care of are at the centre of elephant society (Kurt, 1974; Gadgil & Nair, 1984; Garaï, 1992) African orphaned juvenile elephants will attempt to form a social organization similar to a normal one, however, this is possible only where the basic mother-offspring core is present (Garaï, 1997).

The stress and distress that the orphans must have experienced prior to be taken care of at the orphanage can only be assumed. The central question of the social study was to assess whether the orphans are capable of forming a social unit in which they can fulfill their social and emotional requirements and how the stressful experience has affected them physiologically and psychologically. A restriction in behavioural complexity and deprivation of social and emotional requisites can be experienced as stressful and frustrating, resulting in atypical behaviour patterns such as stereotypy (Sachser & Lick, 1991; Marriner & Drickamer, 1994; Mason, 1991; Kurt & Hartl, 1996, see Section 5.4), aggression or even apathy (Garaï, 1997).

## Preferred nearest neighbour

Data sampling took place during July-August 1997 with some additional focal sampling during August 1998. In 1997 there were 56 elephants (25 males and 31 females) at the orphanage (see Section 3.1). In August, 1998 there were a few changes in the group, the new neonate male Salia had been born to Mathali, the orphan Anura had died and Thilaka had been moved to the Dehiwala zoo. Sumana had been removed from the herd and was being trained. The adult males did not form part of the study as these were held separately. A total of 48 elephants, therefore formed part of the social study, and were individually identified by their size, shape of head, body and ears, pigmentation pattern, tail tip features and other characteristics. Individual recognition when the elephants mingled in the river was a prerequisite in order to assess 'nearest-neighbour-pairs' and socialisation partners as well as focal animal behaviour patterns.

Nearest-neighbour-pairs were sampled at 5 min intervals during the time they were in the Maha Oya. The nearest neighbour had to be within 3 m radius to the focal animal. If the neighbour was further than 3 m away it was considered to be standing 'alone'. This rather small distance arose from the fact that the elephants were herded together by the mahouts to stand in a cluster for the tourists. Within this restriction however, there was a pattern as to which elephants stood next to each other. With the help of cluster analysis (Morgan *et al.*, 1976) the similarity value (*S*) was calculated, whereby $S_{ij} = 1000 * [ N_{ij}/ (N_i + N_j)]$, with $N_{ij}$ being the total times individuals i and j were nearest neighbours of each other, $N_i$ the amount of samples taken with individual i; $N_j$ the amount of samples taken with individual j. The multiplication factor 1000 was chosen for practical reasons.

Based on these data a maximum spanning was drawn (Fig. 5.3.1) to illustrate the associations between the individual elephants. Each individual has been given at least one connecting line to another individual. The strongest relationships were indicated by the shorter lines and the higher similarity values, the weaker associations were indicated by longer lines and lower values. The diagram clearly showed that the strongest associations were between mother-infant pairs: Kumari and Esala (ku-es), mother and son; Rejina and Nandimithra (re-na), mother and son; Komali and Menika (ko-me), mother and daughter; Komali and Aruna (ko-ar), mother and son; Anuradha and Anuradhika (an-ak), mother and daughter; Anusha and Isuru (as-is), mother and son; Sukumali and Arjuna (su-aj), mother and son. Certain other pairs showed rather high similarity values (above 110): Esala and Singharaja (es-sg), a Pinnawela-born juvenile and a male orphan; Soma and Anusha (so-as), an adolescent orphan female and the oldest female; Sandalee and Arjuna (sd-aj), a young

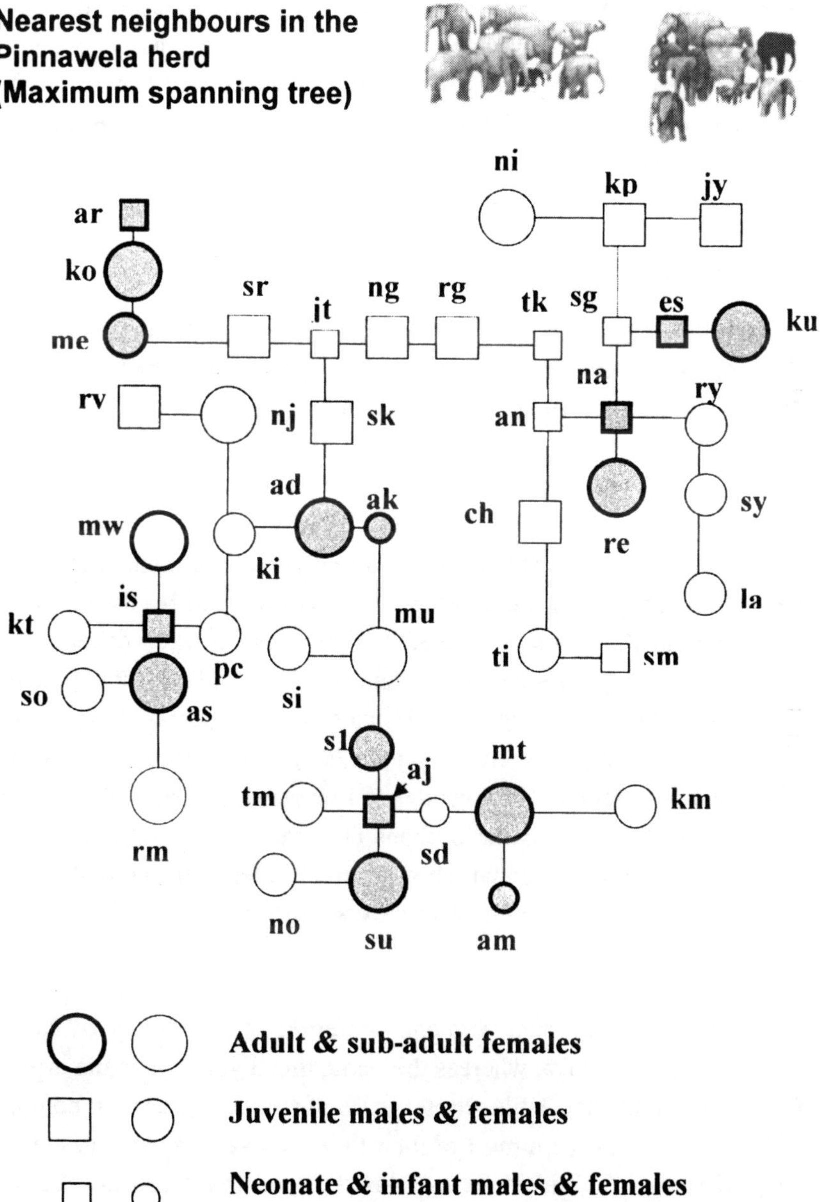


**Fig. 5.3.1** Maximum Spanning Tree of the Pinnawela herd (without adult and sub-adult bulls). The closer two individuals are, the stronger the similarity. Gray: Mothers and captive born offspring. (Names given as abbreviations according to Table 4.1.1 & 4.1.2)

juvenile female orphan and a male Pinnawela-born calf. Other individuals had very low values compared to any other elephant and this showed that they were socially and spatially at the periphery of the group. This holds particularly for the females Kamani (km), Noni (no), Kanthi (kt), Lasanda (la) and Nikini (nk) and the males Jayathu (ju) and Kapila (kp). These elephants are discussed further on.

## Preferred social sub-group

In the following analyses the data were classified into categories, which differ slightly from the ones used elsewhere in this book, but which allowed comparison with other similar studies on African elephants (Garaï, 1997); calves (0 to 1.9 years), young juveniles (2 to 4 years), old juveniles (5 to 9 years), adolescents (10 to 15 years) and adults (15< years). The time spent alone by either orphans or Pinnawela-born elephants was measured and analysed statistically. For individual preferences for certain age or sex sub-groups, the relevant total was divided by the number of individuals in that class (Fig. 5.3.2). There were marked differences in the choice of sub-groups between Pinnawela-born and orphaned juveniles. The Pinnawela-born young juvenile males (2 to 4 years) chose an adult female and the orphans chose same age peers as preferred partners. Pinnawela-born female old juveniles (5 to 9 years) preferred the company of adult females whereas the orphans preferred same aged females. The orphaned female calf, Sandali showed much interest in the male calves, whereas the Pinnawela-born male calves, Isuru and Arjuna preferred an adult female.

There were some individual differences in the choice of partners. The young male orphan, Sumana preferred the company of the old juvenile female Thilaka, whereas the remaining 3 young juvenile orphans, Singharaja, Anura and Jatila, chose males of similar age as their company. Old juvenile males spent most of their time with same aged males, except Rangiri, son of the oldest female Anusha, and the 3 orphans Sanka I, Kapila and Ravana spent more time near females than near any other elephant. One female Ramnya, chose to be near the males and, judging by her coquettish behaviour interest in the bulls, she was possibly in oestrous during both observation periods.

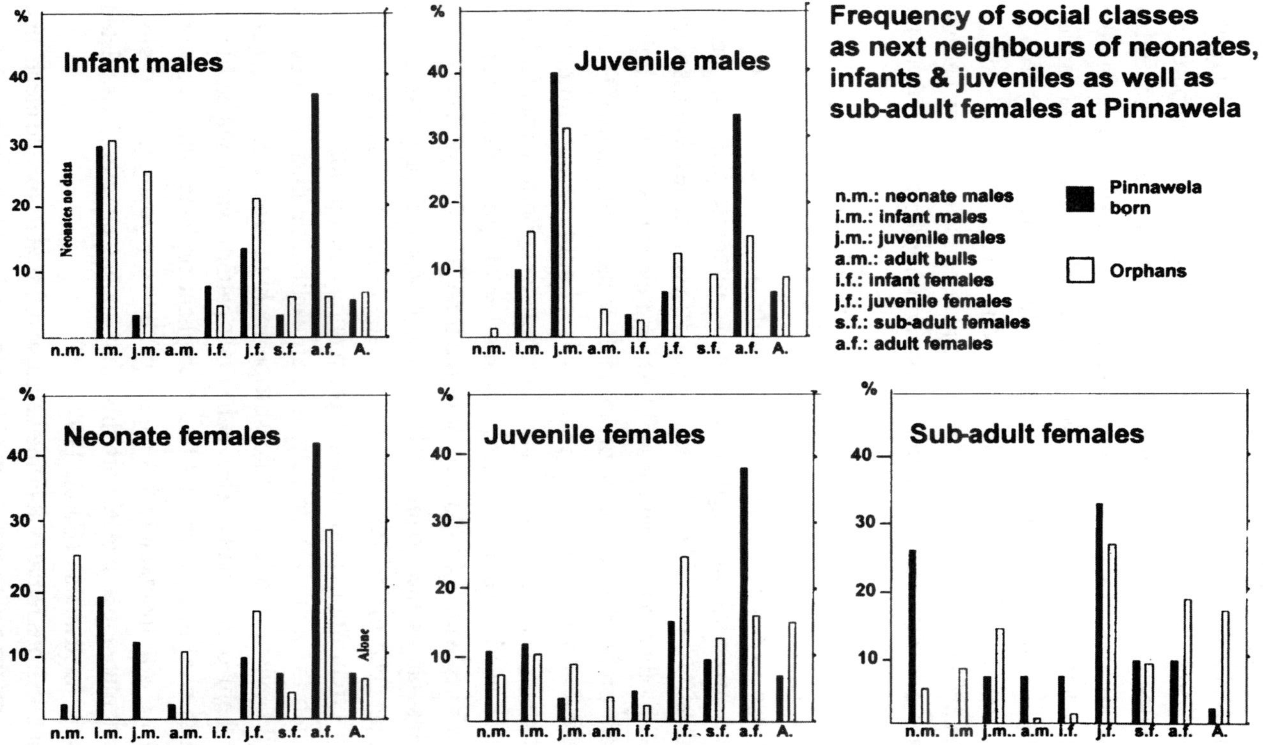

**Fig. 5.3.2** Choice of partners for infant and juvenile bulls, neonate, juvenile and sub-adult females according to their social status (Pinnawela-born or orphans). The columns show their respective choices, in percentage of total for each possible choice.

All orphans spent more time just standing alone than did the Pinnawela-born juveniles. Specifically the females Noni, Thilaka, Punchi, Kirimenika, Nikini, Amali and Ninja, and the males Sanka I and Jayathu spent more time alone than with a preferred partner. If one compares the physical condition of these orphans, it is apparent that they were all in inferior condition and showed retarded growth. Particularly Noni, Thilaka, Kirimenika and the male Sanka I were too small, thin and had far less body weight compared to similar aged elephants. (Table 5.3.1). Nikini and Ninja were visibly in a sickly condition. Amali, who was born in Pinnawela to Mathali, was visibly dejected and depressed, possibly because her mother was chained the whole day, even during the time spent in the river.

Studies on wild African elephants showed that males upto the age of about 4 years will tend to prefer the company of an adult female, then from the age of about 5 years they increasingly spend more time with same aged juveniles, although their second preference will still be an adult female (Garaï, 1997). After the age of 10 years males will associate mainly with other males. Female wild African elephants upto the age of 9–10 years prefer the company of an adult female, and thereafter start to seek the company of younger individuals whom they can allomother, but still remain near an adult female. The Pinnawela-born Asian elephant juveniles showed a similar pattern; all mother-offspring pairs were in close proximity with each other most of the time. The Pinnawela orphans differed insofar as the young males below 4 years already associated with similar aged males. Possibly these young orphaned males had been forced by circumstances to become prematurely independent and were slipping into the normal male pattern at an earlier age than is necessary in the wild, where mothers would still be looking after them and even suckling them occasionally. This behaviour pattern was comparable to that seen in wild in a group of African elephant orphans which had one adult female with them, and where some of the very young males (below 4 years) attempted to associate with this adult female, but others associated with similar aged males (Garaï, 1997).

The nulliparous females at Pinnawela also differed from wild elephants, as many spent time with similar aged females, and some either spent much more time comparatively or else never associated with younger individuals.

Table 5.3.1  Comparison of inferior physical conditions and socially deviating behaviour for 21 orphans. X: characteristics present. For full forms of abbreviations of elephants name see Table 4.1.2. & 4.1.2

| Name (see. Section 3.1.) | Too small or too thin | Sick, bad condition | More time alone than with preferred partner | Equal time alone as with preferred partner | Socially peripheral, no ties to adult female | Socially different to others | Extreme stereotypy |
|---|---|---|---|---|---|---|---|
| no | X | – | X | – | X | X | X |
| so | X | – | – | – | – | X | X |
| nJ | X | X | X | – | X | – | – |
| nk | X | X | X | – | X | – | – |
| ti | X | – | X | – | X | – | – |
| sh | X | – | X | – | X | – | – |
| ki | X | X | X | – | – | – | – |
| mw | X | – | – | X | – | – | – |
| mu | X | – | – | X | – | – | – |
| rm | X | – | – | X | – | – | – |
| sk | X | X | X | – | – | X | – |
| kp | X | – | – | X | X | X | – |
| ju | X | – | X | – | X | – | – |
| rv | X | X | – | X | – | X | – |
| sm | X | – | – | – | – | – | X |
| kt | – | – | – | – | X | X | – |
| la | – | – | – | X | X | – | – |
| pc | – | – | X | – | X | – | – |
| sy | – | – | X | – | X | – | – |

The 4-year-old orphaned female Sandali, spent most of her time with one of the calves, although she attempted to stay near an adult female. She was displaying premature allomothering tendencies. Interestingly, those older juvenile females that showed allomothering tendencies (Thammena, Punchi, Kamani, Soma, Ninja, Nikini and Ranmali) towards younger individuals, only did so towards the Pinnawela-born calves, and not to one of the orphans. Again, this behaviour was comparable to that seen in some of the wild African orphan groups (Garaï, 1997), where certain individuals tried to look after the youngest individual in the group, even if an adult female was already looking after it, and not after other similar aged, solitary individuals. In Pinnawela the calves Isuru, Arjuna and in the second year Salia, had at least 3 allomothers each. Even Sama I (daughter of Mathali) allomothered Isuru (son of Anusha) and, surprisingly, not her own 4-year-old sister Amali, who was always alone.

It appears that adaptive behaviour in elephant society is to protect the youngest individual, irrelevant of whether there are other females taking care of it, and not to try and protect as many young as possible. Although this behaviour seems contra-productive to the whole group, from an evolutionary view it makes sense. In a normal family the youngest individual will be the most vulnerable, and a calf's survival chance increases with the number of allomothers (Lee, 1987), and as the allomothers are generally related to the calf, in terms of propagating their genes it makes sense to help that calf. In the Asian elephant there may be an additional cause for helping a mother raise its offspring: it has been suggested that Asian elephants might help look after calves of older females for reasons of reciprocity (Trivers, 1971; Gadgil & Nair, 1984; Garaï, 1992). At Pinnawela the mothers were the most experienced. They invested in their calves which might pay off in the future. The evolutionary approaches would be genetically fixed, and may explain why females in a group look after the youngest individual and not display altruistic behaviour towards solitary juveniles.

## General activity patterns

A comparison between the social and physical conditions clearly showed that the relatively thinner, smaller and sickly elephants were socially isolated from the rest of the group. Most of these elephants slept longer

(Section 4.4) and also displayed more stereotypic behaviour (Section 5.4). Deprivation of socialisation factors could also lead to a restricted and impoverished activity pattern. In order to answer this question, in 1998 the activity patterns of orphans and Pinnawela-born same aged individuals were compared. Focal sampling was done when the elephants were in the Maha Oya twice a day. 5 males and 12 females were observed continuously for an hour at a time. Each individual was observed equally during the morning and the afternoon over several days. Total observation time was 105 h. Following activities were recorded at 1 min intervals (Table 5.3.2): 'standing (doing nothing) alone'; 'standing near partner' (within 3 m); 'standing total'; 'exploring (actively searching with the trunk), standing or walking'; 'manipulating and object', for example a stone or piece of wood, which was held in the trunk longer than three seconds and not eaten. 'Feeding' of edible plants found in the water. 'Sand-or mudbathing' was recorded when the elephants threw sand on themselves or crossed the river to wallow in the sand or mud along the edge of the water; 'bathing' was recorded when the elephants lay down in the river; 'social' included any type of deliberate contact (with the trunk or other body part, or visual contact) between two or more elephants and 'social play' was recorded when two individuals sparred (pushing with the trunk base, climbing up on each other with the front legs, friendly chasing) or tumbled over each other in the water. Any type of rhythmical head and body swinging or nodding behaviour was recorded as 'weaving' (see Section 5.4) The values were calculated as percentage of the total observation time.

The very thin orphaned male Ravana spent significantly more time in standing and doing nothing, than the same aged healthy looking Nilagiri, who spent 43.4% time standing alone. Despite this he was a more social elephant than Nilagiri, comparable to the 5-year-old Singharaja and Esala. However, atypically Ravana directed 78.8% of his social interactions and play towards calves, whereas Nilagiri mostly sparred with his favourite partner Rangiri, as would be expected from a young male of that age. Although the orphan Singharaja showed a similar activity distribution to the same aged Pinnawela-born Esala, he displayed stereotypic behaviour in the river. This behaviour had not been noted during the previous year.

Table 5.3.2    Actogram during the stay in the Maha Oya river expressed in percentage of total for each individual. For abbreviations of elephants name see Table 4.1.1 and.1.2 Sta: standing alone; Stp: standing with partner; Stto: standing total; Exst: exploring standing; Exw: exploring walking; Exto: exploring total; Man: manipulating object; Wa: walking; Fe: feeding; Smb: sand or mud-bath; Ba: bathing; Sopl: social and play; We: weaving; nd: no data.

| Na | Sta | Stp | Stto | Exst | Exw | Exto | Man | Wa | Fe | Smb | Ba | Sopl | We |
|---|---|---|---|---|---|---|---|---|---|---|---|---|---|
| es | 2.6 | 18.4 | 21.0 | nd | nd | 3.9 | 27.9 | 15.4 | 6.7 | 0.0 | 0.0 | 25.1 | 0.0 |
| sg | 2.2 | 19.3 | 21.5 | nd | nd | 3.4 | 16.7 | 16.4 | 4.9 | 0.9 | 0.1 | 20.4 | 15.7 |
| ng | 0.0 | 34.7 | 34.7 | nd | nd | 19.0 | 14.1 | 25.4 | 0.7 | 0.7 | 0.4 | 14.9 | 0.1 |
| sk | 0.0 | 34.9 | 34.9 | nd | nd | 20.9 | 7.7 | 23.4 | 0.7 | 0.9 | 0.9 | 10.6 | 0.0 |
| rv | 22.7 | 29.6 | 52.3 | nd | nd | 17.4 | 0.6 | 6.2 | 0.5 | 0.9 | 1.4 | 28.9 | 0.0 |
| am | 3.7 | 16.3 | 20.0 | 55.9 | 4.1 | 60.0 | 2.0 | 7.5 | 2.8 | 2.7 | 0.0 | 5.0 | 0.0 |
| s2 | 3.3 | 55.4 | 58.7 | 14.2 | 0.6 | 14.8 | 0.4 | 2.0 | 4.2 | 0.5 | [2]19.4 | [2]21.5 | 0.0 |
| ki | 16.0 | 54.7 | 70.7 | 9.9 | 0.1 | 10.0 | 0.5 | 13.1 | 2.1 | 0.7 | 0.1 | 2.8 | 0.0 |
| me | nd | nd | 19.7 | 50.9 | 2.0 | 52.9 | 3.5 | 9.7 | 0.0 | 0.0 | 0.0 | 14.0 | 0.2 |
| ry | 0.0 | 3.8 | 3.8 | 22.3 | 0.0 | 22.3 | 0.2 | 14.9 | 2.3 | 0.0 | 0.0 | 56.5 | 0.0 |
| no | nd | nd | 14.6 | 22.8 | 7.2 | 30.0 | 0.7 | 6.2 | 0.0 | 0.0 | 0.0 | 1.1 | 47.4 |
| pc | 7.7 | 30.7 | 38.4 | 19.7 | 2.7 | 22.4 | 13.2 | 2.6 | 2.5 | 0.8 | 0.0 | 25.9 | 0.0 |
| sl | 3.8 | 7.7 | 11.5 | 13.9 | 0.0 | 13.9 | [2]54.5 | 3.4 | 3.4 | 0.6 | 0.0 | [2]24.2 | 0.0 |
| nJ | 1.0 | 7.6 | 8.6 | nd | nd | 55.9 | 6.0 | 4.6 | 0.0 | 0.8 | 9.4 | 14.7 | 0.0 |
| m | 0.8 | 3.4 | 4.2 | nd | nd | 53.8 | 6.9 | 8.3 | 0.8 | 3.1 | 11.2 | 11.7 | 0.0 |
| mu | 11.8 | 35.3 | 47.1 | 20.1 | 6.3 | 26.4 | 7.9 | 5.9 | 3.3 | 0.9 | 0.7 | 7.8 | 0.0 |
| mw | nd | nd | 35.2 | 48.3 | 3.0 | 51.3 | 4.0 | 2.5 | 1.3 | 0.0 | 0.0 | 5.7 | 0.0 |

[2]Manipulation and social behaviour occur simultaneously.

The females Nikini, Ninja and Majuri were similar aged but there were slight differences in their activity patterns. Majuri stood significantly more than Ninja, and showed much less exploring than the other two. However, during this second year Majuri seemed to have strengthened her bond to Shanti, and although she showed relatively less social behaviour than either Nikini or Ninja she persistently stood next to Shanti, indicating her interest or maybe concern in this extremely small elephant. The exploration done by both the sickly Ninja and Nikini was nearly as stereotyped as picking up of small stones and dropping them again. Both showed hardly any interest in anything happening around them. Mahaweli was another elephant to show very little interest in anything going on around her. She never even followed the group when they crossed the river for mud-wallow. Although she was already 17-year-old, she did not have a calf and displayed no allomothering tendencies, and during 1997 very little and in 1998 no interest in any other elephant. Her dejectedness seemed even greater in the second year. Her body weight was far below that of the same aged mothers Anuradha and Rejina.

In contrast, the 8-year-old Ramnya displayed an extreme interest during both years in older bulls, who repeatedly tried to mount her and chased her along the river bank, showing that she was very likely in oestrus both times. The orphan Punchi showed a great amount of social behaviour compared to the Pinnawela born Menika. During 1998 Punchi was still interested in the calf Isuru, but was social towards many other elephants and spent far less time alone than during 1997. Sama I was often occupied with manipulation and social behaviour. Together with a number of other elephants she spent most of her time in the river by a large log, picking off pieces of bark and wood, which led to much social activity between these elephants. Sama I showed no interest in her younger sister Amali. Due to her severed forefoot Sama II could not walk around easily and stood most of the time at the same spot. She hardly ever moved from Sukumali's side and when the latter crossed the river for a mud-wallow, Sama II bellowed. Sama II was particularly interested in Sukumali's calf Arjuna and played and bathed with him in the water. (lying down in the water was possibly a relief for her good front leg). She spent significantly more time in social behaviour than Menika, Kirimenika or Noni, indicating her integration into the group despite her disability. Amali and Anuradhika were both Pinnawela-born and showed similar activity patterns, however, Amali gave a dejected impression, even when 'exploring', similar to Ninja and Nikini, she just picked up and dropped stones showing no interest in anything. She always stood at the same spot and very seldom went to her mother who was chained to a rock, or her elder sister Sama I. Amali only socialised for brief moments when another elephants happened to pass her. Noni weaved during nearly half of the observation time and significantly more than Menika, who only weaved for 30 seconds. In addition Noni spent significantly less time in social behaviour than the same aged Pinnawela-born Menika. The relatively high percentage of 'exploring' was due to the fact that she was repeatedly led to the shore by the mahouts to be photographed with tourists and this activity was sampled as 'exploring'. Otherwise, when on her own she showed hardly any interest in anything around her. Noni's back legs were slightly deformed, possibly a result of her stereotypic behaviour. Similarly, the underweight and peculiarly shaped Kirimenika showed very little interest in anything, spending significantly more time standing than either Sama I or Menika and that of 22.6% standing alone. Kirimenika had problems with swallowing food and her body was too broad and her back flat, possibly due to constant bloating. The activity patterns of the above few selected individuals illustrate that the physically inferior animals indeed show differences to other elephants, with one exception, Sama II who appeared socially integrated into the group.

## Social Integration

The elephants in physical inferior conditions were socially less or not at all integrated into the group (Table 5.3.1). Specifically Noni and Kirimenika were totally asocial, i.e. did not show any social behaviour at all towards any other elephant. An exception was Sama II who, in spite of her impediment, displayed a large amount of social behaviour and even allomothered one of the calves, mainly Arjuna, Sukumali's calf. Even when she was chained for the night in the stable, she displayed much social bonding towards her neighbour, the male Suranimala. During 1997, before she was allowed to join the other elephants in the river, she was very attached to this male. During the initial period when she was introduced to join the group in the river, the mahouts used Suranimala to help Sama II. In 1998, she had obviously formed a bond with Sukumali and her calf Arjuna. There were differences in the social integration patterns between the females that were in physically inferior conditions. Soma, Ranmali, Ninja and Nikini started showing allomothering tendencies towards the calves only during 1998. Soma in fact, never left Isuru's side, whereas she had been always alone or near an adult female during the first year. One may assume that these elephants were slipping into the roles of allomothers and thus formed part of the social unit.

Mahaweli, Thilaka, Noni and Kirimenika were not integrated into any social unit. The same was true for Majuri and the relatively very small Shanti during 1997, but luckily these two females managed to form a bond in 1998. Kamani and Punchi, who had been socially isolated in 1997 (although of normal physical appearance) showed allomothering tendencies in 1998, showing their ability of social integration. Although they were physically in good condition but there were differences in the social integration ability of the similar aged orphan Sandali and the Pinnwela-born Amali. Sandali showed a great interest in the calves, whereas Amali was very alone and appeared dejected, probably because her mother was chained all the time and her sister Sama I showed no interest in her. This could have been a result of management, as Sama I was not housed with her family.

The physically weaker males also showed differences in social integration ability. Kapila and Jayatu kept much to themselves, whereas Ravana spent much time with calves in 1998, maybe he had found a social niche, as due to his physical weakness he was probably low down in the dominance hierarchy. Sanka I had no social partner but rather kept near the females. In contrast, the physically normal orphan males Chula, Tharaka, Nilagiri, Jatila played much with other same age males.

The obvious question is: are the elephants in inferior physical conditions because of social stress (Sachser & Lick, 1991) resulting due to their low hierarchal positions, or are they socially not integrated because of their weak conditions? In general, elephants are very considerate towards sick family members. There are indications from South African translocated juveniles that originated from culling operations, that they grow less quickly than their wild counterparts, a hypothesis that needs to be tested though.

This could imply that the stress they experienced during the cull, capture and subsequent translocation affected them physically. Many of the Pinnawela orphans showed retarded growth and weight values, which could be the result of trauma due to loss of their social environment, capture and transport, experienced at an early age (Sapolsky, 1990). A sick animal is probably not in the position to fight for social integration and a position in the group. Orphaned African elephants will form a dominance hierarchy (Garaï, 1997), but acquiring a higher position needs physical strength. The physically inferior animals will therefore be at the bottom of the hierarchy, which immediately eliminates many social partners, unless such an individual can attach itself to an adult female, but again these chances will be limited and the position will have to be acquired by strength and stamina.

One way for a young female to attach herself to a more dominant female would be to allomother its calf (Dublin, 1983). A young male however, will have to find his place in the male dominance hierarchy and fight for it, or else be at the bottom of the social periphery. In conclusion, one may predict that physically inferior orphans will be at the lower end of the dominance hierarchy, a situation which in return may well have adverse influence on their physical condition due to social stress, as long as there is no older female to protect them. But as has been stated above it is not adaptable behaviour for elephants to protect a non-kin or older juvenile. For this reason, the orphans are left to handle the situation alone, but have possibly not yet acquired the necessary skill to cope with social stress induced by dominant animals in addition to their situation of being an orphan. The psychological effect must not be underestimated and depression and dejection can lead to apathy. It needs strength to overcome this apathy and/or outside influence in the form of another individual, therefore, for a depressed animal it would be unlikely to initiate a social interaction and possible relationship. On the other hand, a healthy animal will not necessarily initiate social contact with a weak animal which does not respond.

The question posed at the outset of the study, whether the elephants at Pinnawela are capable of forming a social unit in which each individual may fulfil its role, can be answered as follows: Assuming that the mothers and offsprings form the centre pillar of the social structure, it can be expected that some of the younger females will attempt to fulfil their roles as allomothers, which will be of advantage to them when they have their own offsprings. These females were socially integrated, and this might have induced a feeling of security. Whereas, other females had less opportunity or, for reasons stated above, were not capable of social interactions. As these elephants could not fulfil their roles, this might have left them with additional frustration and insecurity to cope with. What the long-term psychological effects are cannot be ascertained at present. Some of the young males formed play groups and were thus able to practice acquiring and maintaining a higher position, a vital social skill when they grow up (Poole, 1987). Those elephants that were socially not integrated in either female or male society and were in physically inferior conditions might well be deprived of certain psychological requisites and learning opportunities necessary to ensue a normal social structure and individual development.

One may further assume that these elephants have more difficulty in coping with stress and the negative experience of their capture and loss of family. Surely, they would feel less secure than the socially integrated elephants. Whether an individual is capable of achieving integration will depend on its health, stamina, opportunity, but also largely on the individual character. The Pinnawela-born elephants are really in a better position, except that housing and management practices can also have a negative influence and impede a normal development. This appeared to be the case with Amali, whose mother was constantly chained and therefore, Amali was deprived of this important maternal learning partner. The same applies to Menika who was chained separately from her family at night and developed stereotypic weaving after being separated from her mother and before being allowed to join her again in the morning.

Stereotypic movements are ritualised behaviour patterns derived from searching behaviour and develop through lack of social contact. If this hypothesis deduced in Section 5.4 is true, then the elephants which stand in the Maha Oya should show a negative correlation between social and exploration behaviour. Indeed, this was found to be confirmed by a comparative analysis. In this comparison the values of 3 elephants were excluded: Singharaja and Noni, both of whom also stereotyped in the river, as well as the values of the obviously very sick Kirimenika. Amali,

Ninja and Nikini had the lowest values for social contact, but the highest values for exploratory behaviour. For example Ravana, Punchi and Esala had high social values, but low exploratory behaviour values (see Fig. 5.4.5b). If one follows this thought further and pools the values of 'exploring' and 'walking' together as 'exploratory behaviour', and adds the values of 'standing with partner' to 'social behaviour', the negative correlation between 'exploring' and 'social' is significant.

## 5.4  Stereotypies

Stereotypies are repetitive invariant behavioural patterns that have no obvious goal or function (c.f. Mason, 1991). Stereotypies occur regularly in intensive farm animal husbandry systems characterised by extreme spatial confinement and monotonous environment (c.f. Wechsler, 1991). They are frequently observed in primates reared in isolation lacking appropriate social partners, and they have been described in a wide variety of zoo animals, such as bears, cats or ungulates including elephants (Hediger, 1950; Morris, 1964; Meyer-Holzapfel, 1968; Boorer, 1972; Carlstead *et al.*, 1991). In captive Asian and African elephants stereotypies have been described as intentions to stepping  forward and backward and accompanied by rythmical body and trunk swaying, either in a forward-backward movement or sideways while the animal is standing on the spot, accompanied by head-nodding and often by trunk-swinging. These stereotypies are known as 'weaving'. Next to weaving, other forms of stereotype behaviour are also known among elephants. Certain zoo elephants walked stereotype laps, and elephants kept in narrow pens developed complex stereotype series of fore-steps , side-steps and back-steps. However, data given in this chapter concentrate on the most prominent form of stereotypies in elephants i.e. weaving.

Circus people sometimes state that weaving is essential for blood circulation – a rather extraordinary conjecture, since in none of the numerous studies on wild elephants weaving is mentioned, although zoo speakers occasionally tell that weaving occurs also under natural conditions. Weaving is always associated with intensive keeping systems, i.e. more or less permanent chaining in small stalls, where living conditions deviate fundamentally from the natural environment in which the animal's behavioural organisation has been shaped by evolution. Accordingly, stereotypies, also those in elephants, are generally considered as indicators of poor welfare (e.g. Kiley-Worthington, 1990; Gruber *et al.*, 2000). Although weaving in captive elephants is a well-known phenomenon, neither the genesis of weaving nor the external stimuli leading to the stereotypy have been described so far.

This chapter does not only deal with the elephants of Pinnawela. Data from Asian elephants living in temples (Kandy), jungle camps (Tamil Nadu) and in the Swiss National Circus have been included. Most of the 43 animals monitored (Table 5.4.1) came into captivity at a very young age. All of them were fettered to chains or ropes for at least a few hours everyday and the space, which they could reach physically, measured between 8 to 16 m². In this monotonous environment, important parts of the specific natural environment of the species were missing, such as bath, wallow, scratching trees and often adequate social partners were also missing. In the following pages, firstly stereotypies are described with special reference to movements of legs, head, trunk and ears. The number of stereotype movements per minute was considered as frequency of a stereotypy. Then the influence of keeping conditions on stereotypies, possible environmental and social factors that could trigger this behaviour are looked at. Lastly, the curability of stereotype behaviour and the consequences of social isolation are discussed.

## How does stereotype behaviour evolve?

In young elephants, which were fettered daily for 14 to 20 h, weaving evolved from complete forward and backward steps and exploratory trunk movements. With age and time spent in captivity the movements were increasingly reduced to weaving and swaying forwards and backwards on the spot and/or rhythmical nodding of the head and swinging of the trunk (Fig. 5.4.1). 7 of 10 young elephants (not older than 5 years) were weaving either with complete forward steps or by raising at least one forefoot, but 9 of 10 elephants over 20 years and having been in captivity for a considerable time, kept their feet on the ground and only swayed their bodies. During the forward steps, head and trunk were lifted and stretched forward, during the backward steps they were kept low. At a later stage, head and trunk were swinging diagonally to the longitudinal axes of the body or were nodding up and down, as found in

**Table 5.4.1** Form and frequency of stereotypies in 44 elephants according to age (A, years), sex (S, m = males; f = females) and place (C; Pi = Pinnawela, C = Circus Knie; K = Kandy). Movements of head and trunk (HT): 1 =l linear, parallel to body axis; 2 = oval, ± parallel to body axis; 3 = lateral, vertical to body axis; 4 = nodding. Movements of forelegs (FL) and hind legs (HL): 1 = both legs make a complete fore step; 2 = one leg makes a complete fore step, the other one is only lifted up; 3 = one leg is lifted; 4 = weight is reduced from 1 leg; legs are hardly moved. Frequency = number of acts per min.

| Name | A | S | C | HT | FL | HL | Frequency (n) |
|---|---|---|---|---|---|---|---|
| *Neonates* | | | | | | | |
| Kumari II | 1.5 | f | Pi | 1 | 1 | 1 | 17.0 ± 0.6 (36) |
| Sapumali | 1.5 | f | Pi | 3 | 3 | 4 | 25.9 ± 0.5 (26) |
| Surangani | 1.5 | f | Pi | 3 | 4 | 4 | 26.8 ± 0.3 (25) |
| *Infants* | | | | | | | |
| Anuradhika | 3 | f | Pi | 1 | 2 | 3 | no data |
| Nandimithra | 3 | m | Pi | 1 | 3 | 3 | no data |
| Aruna | 3 | m | Pi | 1 | 2 | 3 | no data |
| Ma-Palai | 3 | f | C | 1 | 1 | 1 | 23.7 ± 1.6 (78) |
| Ma-Palai | 3.5 | f | C | 3 | 3 | 4 | 28.8 ± 3.7 (148) |
| Sigharaja | 4 | m | Pi | 3 | 4 | 5 | no data |
| Sumana | 4 | m | Pi | 1 | 4 | 4 | 18.7 ± 1.0 (45) |
| *Juveniles* | | | | | | | |
| Kanchana | 6 | f | K | 3 | 4 | 5 | 26.9 ± 0.8 (23) |
| Minjak | 7 | f | C | 2 | 4 | 5 | 20.4 ± 1.5 ( 49) |
| Kirimenika | 8 | f | Pi | 1 | 3 | 4 | no data |
| Menika | 8 | f | Pi | 1 | 3 | 5 | 19.7 ± 1.4 (65) |
| Seela | 8 | m | K | 4 | 3 | 5 | 22.2 ± 1.2 (40) |
| Ramnya | 8 | f | Pi | 3 | 1 | 4 | 22.3 ± 1.1 (38) |
| Punchi | 9 | f | Pi | 1 | 3 | 4 | 17.4 ± 2.0 (30) |
| Noni | 9 | f | Pi | 3 | 4 | 4 | 24.0 ± 1.8 (64) |
| Sarna 1 | 9 | f | Pi | 3 | 3 | 5 | no data |
| Ravana | 9 | m | Pi | 4 | 5 | 5 | no data |
| Karnauti | 9 | f | C | 3 | 4 | 5 | 23.3 ± 2.2 (52) |
| Mangala | 9 | f | K | 2 | 3 | 5 | 14.7 ± 1.1(23) |
| Soma | 10 | f | Pi | 2 | 2 | 4 | no data |

[Table 5.4.1 Contd.

Contd. Table 5.4.1]

*Sub-adults*

| | | | | | | | |
|---|---|---|---|---|---|---|---|
| Singharaja II | 12 | m | K | 4 | 5 | 5 | 21.3 ± 1.0 ( 48) |
| Ninja | 12 | f | Pi | 2 | 3 | 5 | no data |
| Seetha | 13 | f | K | 2 | 4 | 5 | 11.7 ± 1.1 (51) |
| Sukumali | 14 | f | Pi | 1 | 3 | 4 | 19.9 ± 2.1 (57) |
| Siam | 14 | m | C | 2 | 4 | 4 | 13.4 ± 0.4 ( 34) |
| Rejina | 15 | f | Pi | 1 | 4 | 5 | no data |
| Nardi | 15 | m | M | 3 | 5 | 5 | 24.5 ± 0.9 ( 36) |

*Adults*

| | | | | | | | |
|---|---|---|---|---|---|---|---|
| Ranmali | 16 | f | Pi | 3 | 3 | 3 | 25.3 ± 1.4 (41) |
| Mahaweli | 17 | f | Pi | 3 | 3 | 3 | no data |
| Ekadanta | 20 | m | K | 3 | 4 | 5 | 19.3 ± 0.8 ( 48) |
| Kandula II | 20 | m | K | 3 | 3 | 5 | no data |
| Seetha II | 25 | f | K | 3 | 4 | 5 | 20.7 ± 1.5 (29) |
| Komali | 26 | f | Pi | 2 | 4 | 5 | no data |
| Kadira | 27 | m | Pi | 3 | 2 | 3 | no data |
| Raja II | 27 | m | K | 1 | 4 | 5 | 10.6 ± 0.8 (44) |
| Mathali | 27 | f | Pi | 3 | 5 | 5 | 20.0 ± 1.9 (66) |
| Vijaya | 31 | m | Pi | 4 | 5 | 5 | no data |
| Kumari | 32 | f | Pi | 1 | 5 | 5 | 16.9 ± 1.6 (42) |
| Raja III | 45 | m | K | 3 | 5 | 5 | 11.1 ± 0.9 (48) |
| Rosa | 49 | f | C | 3 | 5 | 5 | 17.2 ± 0.6 (45) |
| Raia | 50 | m | K | 2 | 4 | 4 | 11.8 ± 0.4 (49) |

6 of 10 elephants older than 20 years. In the early periods of an evolving stereotypy the longitudinal body axis, pointed to a goal, unreachable for the fettered elephant (e.g. a potential social partner or food). Later on, this characteristic would disappear.

In young elephants the frequency of weaving was about 30 movements / min, and in old ones about 10 movements / min. The frequency of steps in moving wild elephants followed the same pattern. The frequency of forward weaving was found to be between a slow walk and fast walk of wild elephants, but the frequency of nodding and sideways weaving was higher than the frequency of steps in fast walking. The data presented so far indicate that weaving in captive elephants could be considered as ritualised appetitive behaviour as proposed earlier for stereotypies in several domesticated and zoo animals (e.g. Morris, 1964; Cronin *et al.*, 1984).

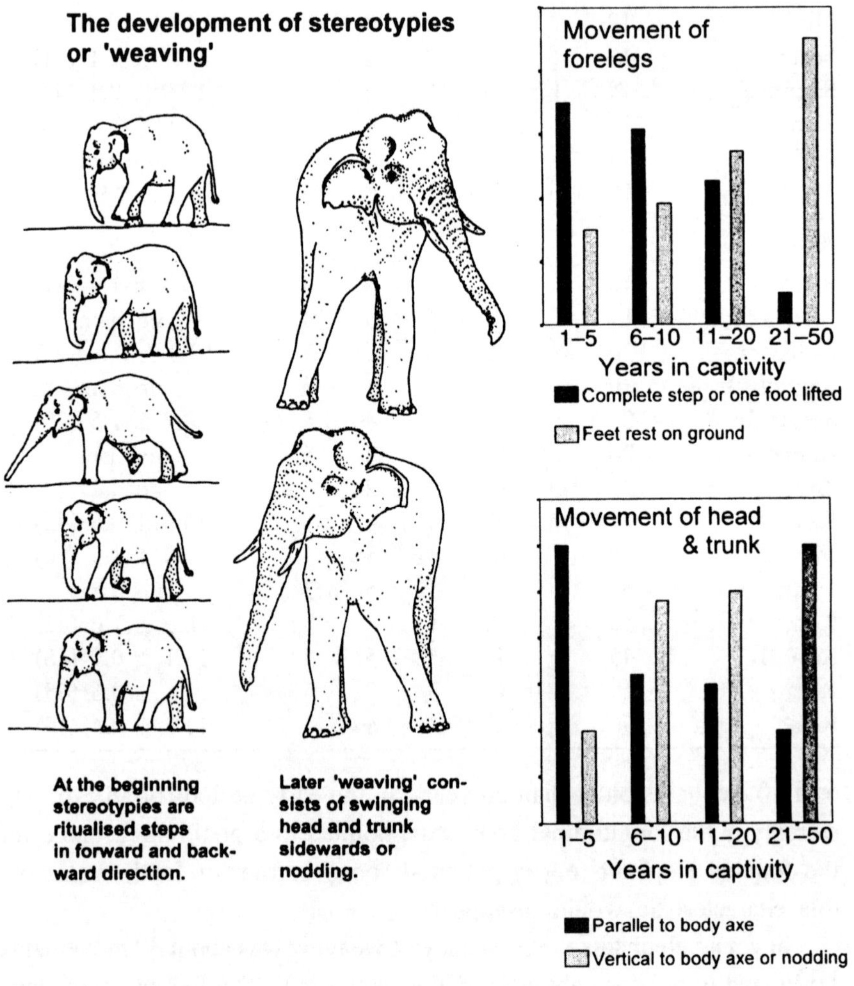



**Fig. 5.4.1** Development of stereotypy in 44 Asian elephants. With increasing age and time in captivity forward steps are reduced and head and trunk are increasingly moved vertically to the longitudinal body axis or 'nodding' appears.

Appetitive behaviour can be observed when animals are motivated to perform species-specific patterns of behaviour, but cannot perceive the adequate releasing stimuli (Craig, 1918; Lorenz, 1937; Tinbergen & Tinbergen, 1972). Accordingly, stereotypies are specific for certain situations and predictable in time and space.

## How do activity patterns change under restrictive keeping conditions?

Elephants captured in 1968 in the *khedda* of Kakanakote were brought singly to the Pilkana ('elephant stable') and fettered between large trees. First they tried to escape, attacked humans and dogs or explored their stands as far as possible. Next to such phases of aggression and exploration, there were periods of resignation, i.e. the newly captured animals ate less than already tamed conspecifics, they rested often or tried to sleep in the recumbent position (Kurt, 1992). During the process of breaking in and taming, the behaviour of flight and aggression disappeared, and gradually weaving replaced pulling of the ropes and exploration.

The daily activities (6. a.m. to 6. p.m.) of wild elephants in the Ruhuna National Park (1968–69) and newly captured ones after the *Khedda* operations in Kakanakota (1968) were compared (Fig. 5.4.2). Activity patterns of adult captive bulls stem from Sri Lanka and Karnataka. All captive animals studied were fettered and had to replace locomotion with weaving. Wild infants took bath and rested more than fettered ones and social behaviour was much more pronounced. However, captive neonantes and infants spent much more time in weaving than wild ones with locomotion. Chained sub-adult and adult females spent less time in feeding and social behaviour but wove more than wild ones walked. In chained sub-adults and adult bulls, feeding was the most prominent activity, and resting occured more often than in wild ones. Skin care and social behaviour were less pronounced than in wild bulls and stereotypies took less time than locomotion in wild ones. These patterns change fundamentally when bulls are in musth.

In fettered musth bulls weaving was the main activity and appeared for a longer period than extreme locomotion in wild musth bulls. Captive musth bulls spent more time in feeding, skin care or social behaviour than

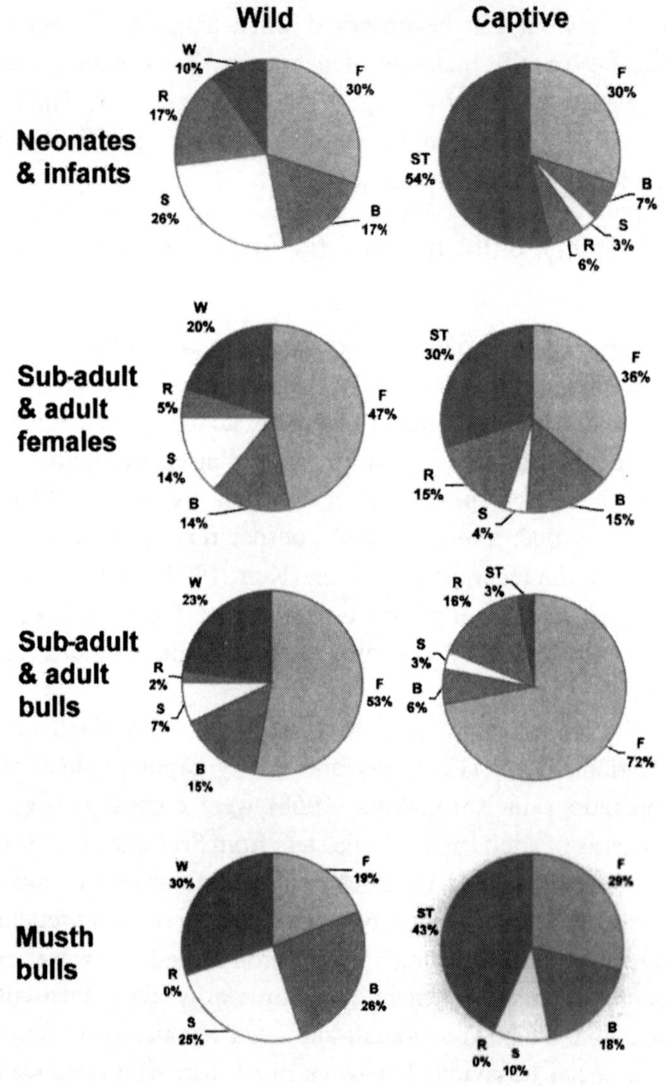



F:  Feeding % food preparation
B:  Bathing & Skin care
S:  Social behaviour including acoustical & chemical patterns

R:  Rest & sleep
W:  Locomotion
ST: Stereotypies or weaving

**Fig. 5.4.2**  Comparison of daily activities of wild (Ruhuna N.P.) and newly captured elephants (Kakanakota-*Khedda*; musth bulls in South India and Sri Lanka). Values are given as percentage of the time between 6 a.m. to 6 p.m.

those not in musth. But they did not rest. Also, wild musth bulls did not rest during the day, but they spent, as could be expected, much more time in social behaviour than captive ones.

In 1998, activities of 8 neonates (3 neonates growing up with their mothers and 5 orphaned neonates) at the age of 1 to 2 years were monitored in Pinnawela. The captive born neonates were never chained, and the orphans were kept for about 22 hours daily with one foreleg and one hind leg fettered. They were allowed to roam free in a small paddock for 2 hours. In the paddock the orphans spent almost twice as much time exploring the area than captive born neonates, but they showed reduced manipulation (of food plants), social behaviour and playing activities (Fig 5.4.3). Orphans tied with ropes in their first months in captivity often pulled on their fetters and tried to free themselves. Occasionally, they rested while lying down and exploratory behaviour appeared quite often, but social behaviour was reduced. Orphaned neonates, having lived in captivity for at least 3 months and mainly tied with ropes, had already established their stereotypies and spent more than 50% of the day time in weaving. Manipulation of objects, pulling at chains, resting in the recumbent position and social behaviour were reduced. Obviously, stereotypies had changed the pattern of daily activities and displaced exploratory and appetitive behaviour to a certain degree.

## What are the environmental factors which trigger stereotypies?

In many privately-owned Sri Lankan elephants as well as in circus elephants, weaving appeared under different conditions and not necessarily when the animals were fettered. Circus elephants wove whenever they had to wait for something or were expecting something. They increased the frequencies of  weaving before they were fed, or when they were expecting water or being cleaned, or at night, when the neighbours on either side were lying in such a way that the animal concerned found no place to lie down for its own recumbent sleep. Circus elephants wove when waiting for rehearsals and show time. Elephants observed during the Kandy *Perahera* wove before they were fed or when other elephants were led away. About 40% of them wove even while marching slowly in the Buddhist processions. Naive observers used to say that during the *Perahera*, elephants would dance on music.

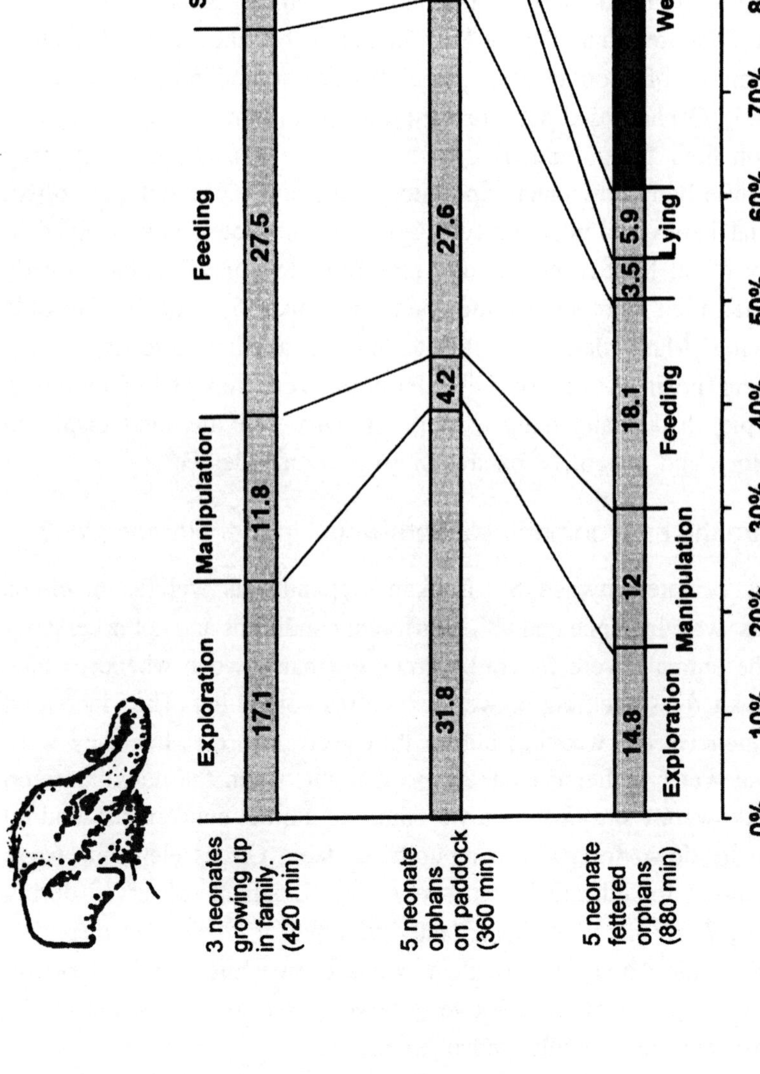


Fig. 5.4.3   Comparison of daily activities of three Pinnawela-born and 5 orphaned neonates. Captive born neonates were never fettered, orphans were either fettered or kept on a paddock.

Such data made it clear that elephants regularly kept in chains immediately started to weave when they had to wait, could not reach a certain goal, or when they did not know how to cope with a situation. In such elephants stereotypies are 'frozen' (Hediger, 1950). As manifold as the stimuli triggering weaving might be, they should not be seen as reason for the origin of weaving. The array of stimuli in frozen stereotypies in older animals go back to the fact that, once stereotypies are established, they are easily transferred from one situation of conflict to another (Kiley-Worthington, 1977). The weaving which is typical for elephants evolved in a period of at least 2 to 3 months, during which no owner could keep his elephants without food or water, but he could keep them without adequate social partners and movement. Hence, in a species with highly developed social behaviour, it can be hypothesised that the process forming the stereotypy is first of all triggered by permanent lack of social partners and restricted movements and not by occasional, relatively short periods of waiting for food and water or acoustic and other disturbances which elephants quickly become accustomed to. Restricted movements seem to be responsible for forming the stereotype weaving. Captive elephants with more space still develop stereotypies but they finally end up with more complicated sequences of forward, backward or sideward steps.

## Are stereotypies formed by the lack of adequate social partners?

The hypothesis that stereotypies are (1) ritualised forms of appetitive behaviour and are formed (2) by lack of adequate social partners could be shown in the Pinnawela herd. If the first hypothesis is correct, elephants frequently displaying exploratory behaviour should have had little social contacts. When the Pinnawela elephants were allowed to move freely in the Maha Oya, a significant negative correlation was found between the frequency of exploratory behaviour and the amount of social contacts. Furthermore, it was highly significant that, when fettered, socially integrated elephants stereotyped less than socially isolated orphans. These observations advocated for the second hypothesis, stating that stereotypies are first of all triggered by lack of adequate social partners. Generalizing the data, one found that elephants born in Pinnawela and grown up in more or less close contact to their relatives stereotyped less than orphans, and socially integrated orphans less than socially isolated orphans.

In socially integrated elephants stereotypies always appeared in social contexts. Already 90 to 60 min before release in the morning, mothers increased the amount of time spent on weaving, and 5 to 20 min before the actual release they wove almost permanently (Fig. 5.4.4). Similar to these mothers, the captive born juvenile female Menika also increased the amount of time spent on weaving before release and reached 100% 45 minutes before she was allowed to join her family. Her night quarter was close to those of other juveniles but at a distance of some 50 meters from the stable of her mother, her younger brother and the other elephants with whom she spent the day. She was shackled about half an hour before her family was brought to the stable in the evening. Until her relatives had passed her to be chained in their stables, Menika wove almost permanently (Fig.5.4.4).

Sub-adult and adult elephant bulls led a more or less solitary life but they were highly social, shortly before and during the period of musth. Bulls kept were intensive systems, where they were kept in chains when not working, were found to stereotype very rarely. During the period of musth weaving became a prominent pattern of activity.

## Can stereotypies be cured?

In Pinnawela, socially integrated elephants stereotyped only when separated by chaining from their social partners, and they stopped weaving when the preferred partners were within the reach of their trunk or when they were released and could go close to the preferred partners. However, in certain elephants, release from chains was hardly a therapy to abolish their stereotypies. They weaved even when they were not in chains and some of them weaved more or less permanently, when not sleeping, drinking or walking. These particular elephants had retarded body growth and were socially isolated. Social isolation leads to reduced food intake and retarded body growths, and retarded body growth hinders socialisation. In these elephants the search for adequate social partners by continuous appetitive behaviour was finally replaced by stereotyped behaviour as the goal of their appetitive behaviour, i.e. adequate social partners could not be found at all (Fig. 5.4.5a).

Seen in such a way, weaving must be considered as one of the several symptoms of a complex, slowly developing serious pathological

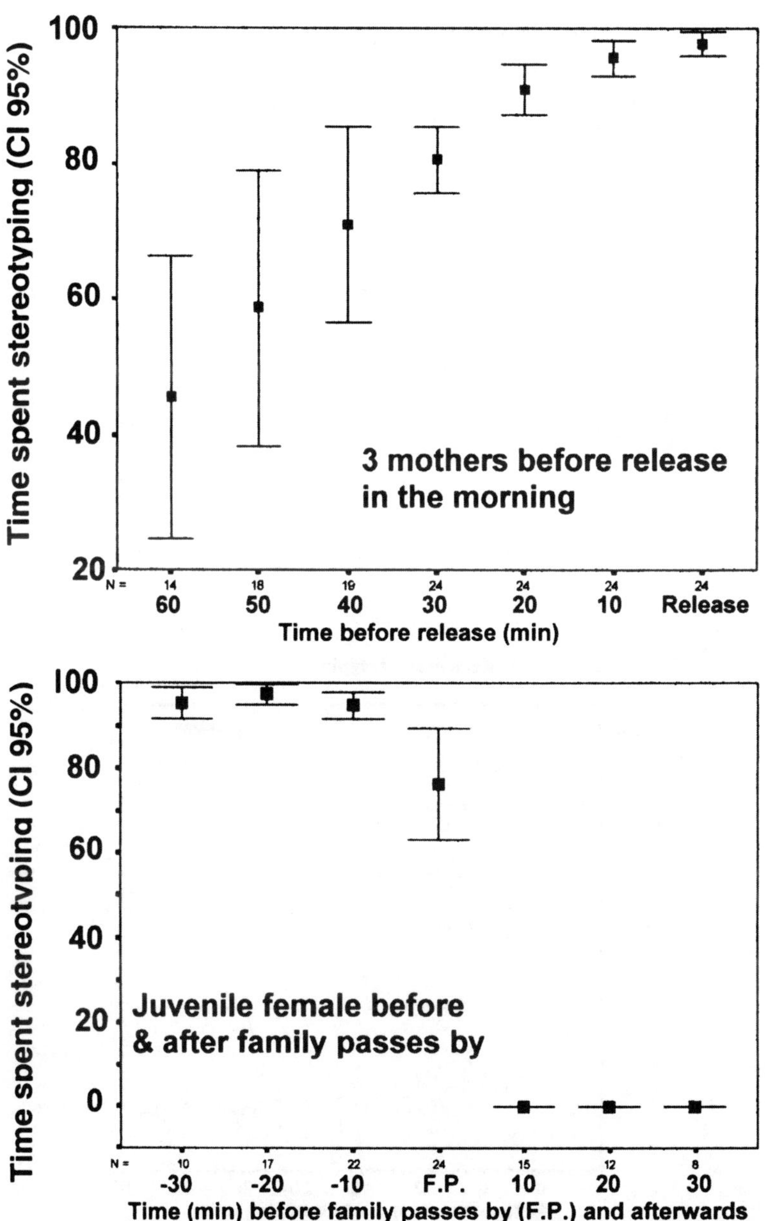



Fig. 5.4.4 Time spent stereotyping (%) of 3 mothers (Kumari, Mathali, Sukumali) before release in the morning (top) and time spent stereotyping (%) of Menika in the late afternoon before and after her family passed and went into the night quarter. Interval of confidence: 95%.

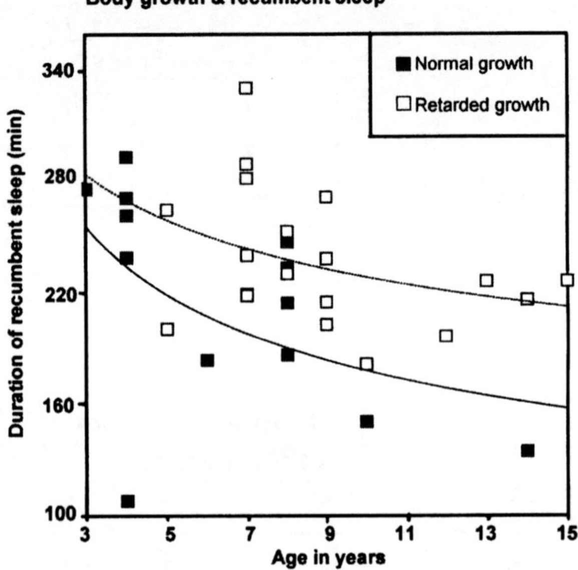


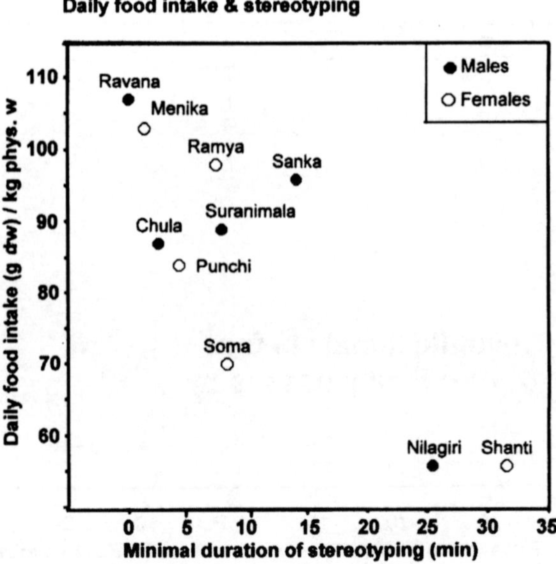


**Fig. 5.4.5(a)** Duration of recumbent sleep compared to age and body growth (top). Minimum duration of stereotypies at night of young elephants compared to the daily food intake in dry weight per physiological body weight (bottom) in gram.

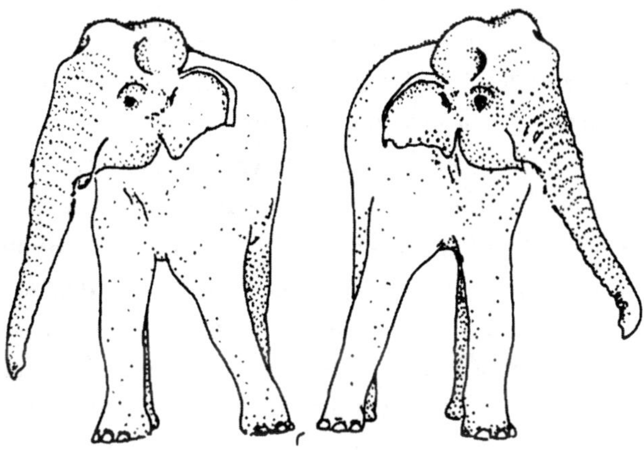

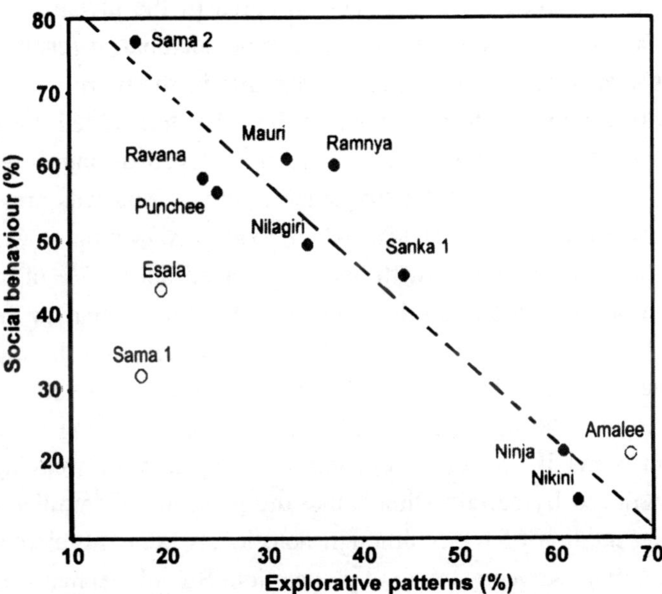


**Fig. 5.4.5(b)** Comparison of all social behaviour patterns and all explorative behaviour patterns in percentage of all activities shown during the stay in the Maha Oya

process, which starts with social isolation and delayed body growth at a young age and is later followed by delayed puberty. According to the endorphine-hypothesis by Cronin *et al.*, 1985) stereotyping animals (e.g. domesticated sows) produced endorphines which suppress aversive or missing stimuli, which again lead to more intensive stereotyped behaviour. But after these few initial experimental results, later experiments could not confirm this theory (Staufacher, 1998). According to another theory, the coping-hypothesis of Wiepkema (1987), stereotypies are supposed to suppress stress and, in due course, harmful accidents. In this theory stereotyped behaviour is considered as a strategy to reach an equilibrium of motivations, when other coping-strategies such as aggression, flight, helplessness or exploration fail. If the coping-hypothesis was correct, then the weaving of elephants would always be curable and would only reappear in the moment of new coping situations. But this is not the case.

As already mentioned, socially integrated elephants weaved only when separated from social partners. In Sama II, the juvenile female in Pinnawela had lost a forefoot in a land mine accident, it resulted in the lack of social partners and triggered weaving. In the morning, when her neighbours were released from their chains, she stereotyped during 29% of the time. After her own release, she stereotyped during 30%. Later, weaving occurred in different frequencies: When food was present, the time spent on weaving was 9%, when food was lacking it was 28%; when familiar keepers were with her, she weaved during 7% of the time, and when her kin passed on the way from the grassy area  to the water close to her, the value increased to 21%. When they were brought back in the evening, Sama II immediately stopped weaving, and after she was allowed to join the herd permanently, she stopped weaving completely.

As in Sama II, in other elephants the frequency of weaving could also be reduced by certain stimuli like the presence of familiar keepers or the presentation of food, although notoriously weaving elephants are able to combine weaving and food preparation. But all elephants observed interrupted weaving when they slept, drank, defecated and urinated. In zoo and circus elephants, which had been kept more or less permanently in chains for years or even decades, the frequencies of stereotypical behaviour were reduced when they were kept in paddocks without chains. Under such improved living conditions, more stimuli were found for exploration, comfort or social behaviour. But in most cases, these zoo and circus elephants never completely gave up their stereotypies, but increased the frequencies again after they had sufficiently explored the new environment (Kiley-Worthington, 1990; Schmid, 1995). The reason for this could be the fact that adequate social partners or other species-specific parts of the environment were absent, or that the stereotypies were already frozen.

The last example concerns a tusker of 15 years, who had spent about 10 years in a south-Indian temple, before he was moved to a jungle camp in Nardi (Mudumalai). He never showed social behaviour towards the other elephants present, but immediately started weaving when he was standing alone or waiting for his food rations. This elephant was restricted in his movements all his life and had therefore lost out on socialisation. The other timber elephants also had developed stereotypies when they were broken in and trained. But they never wove in the jungle, where they had to work during the day, and where they found most of their fodder during the night. Here, they also met tame and wild conspecifics. Obviously, the stereotypy of the former temple tusker had already been frozen.

## What are the ultimate consequences of social isolation and stereotypies?

The persistence of stereotypies, even in new environments with adequate social partners, is best explained by the psychopathological model of Dantzer (1986). According to this hypothesis, stereotypies are formed by frustration of certain highly-motivated behaviours which cannot be performed under restrictive keeping conditions. Accordingly, the animals react with more appetitive behaviours. Again, success fails to appear. There is no negative feedback from the central nervous system, and accordingly, appetitive behaviours are not blocked. And what is more: the lack of a consumatory act leads to a positive feedback. The animals once more intensify appetitive behaviours, which finally leads to a severe disturbance of the neuronal centre which steers appetitive behaviours. The pattern is finally 'burnt' with permanent 'scars' in the central nervous system. According to Dantzer's hypothesis, stereotypies must be considered as a consequence of a pathological process in the central nervous system.

According to our opinion, stereotypies in Asian elephants are a symptom rather than a cause of a serious pathological process which starts with social isolation at a young age and with retarded body growth, and is then followed by delayed puberty. In the end, the pathological process can lead to the killing of social partners and offsprings due to lack of socialization (Kurt and Mar, 1996). Although stereotypies have no social function, they can cause further pathological processes, because elephants with persistent stereotypies weave almost permanently, even without chains, and they do not change their positions. That means they permanently force their extremities in a specific configuration, which may lead, in due course, to deformations of hooves and soles, and later to deformations of the skeleton of the extremities.

Abnormally growing hooves and soles as well as pathological deformations of the skeleton of the extremities belong to the most common injuries of Asian elephants kept in zoos and circuses (e.g. Ruthe, 1961; Salzert, 1972) and have been attributed to different causes, but so far never to persistent stereotypies, although the relation is obvious in many cases. In such cases, these elephants should be given opportunities for daily movements. However, the best therapy against social isolation, stereotypies and other pathological processes would be to keep Asian elephants in permanent family groups in a species-specific environment.

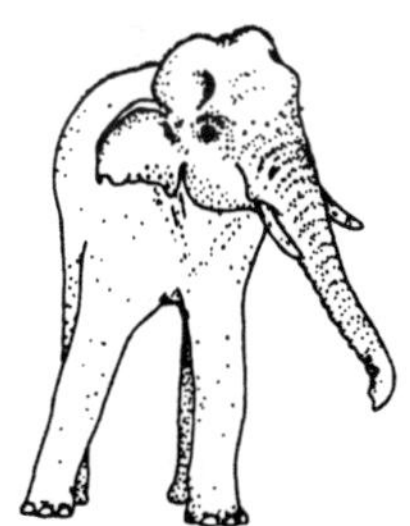

# 6

# Captive Elephants and Conservation

During the meeting of the International Union for Conservation of Nature and Natural Resources (IUCN) Asian Elephant Specialist Group in Phnom Penh in May 2002, the AsESG formed a 'Captive Elephant Taskforce'. Some AsESG members met again at the Workshop on Captive Elephant Management in Trichur (Oct., 2002) and again the Chairman of AsESG, Professor Dr. Raman Sukumar, asked for ideas concerning the aims of such a taskforce. Together with Khyne U Mar, the former Head of Elephant Research at the Myanma Timber Enterprise, we drafted a few ideas for *Gajah* the paper of AsESG and for the Smithsonian Institution of Washington ('Ethics and Elephants'). They included the following thoughts:

The AsESG as well as other agencies should stop to speak of 'domesticated' elephants. Misuse of the term 'domesticated' leads to the fact that many captive elephants were, and still are, treated as cattle and managed accordingly.

One should ask what are the aims of IUCN and the IUCN Specialist Groups as compared to the aims of animal welfare organisations and zoos which are concerned with management problems of captive elephants, as discussed in the following pages. Then some very brief overviews are given on existing and forthcoming guidelines on the management as well as on the estimated sizes of captive populations. Different aspects of the significance of captive populations are broadly discussed according to the various keeping systems defined in Section 3.2. Finally, some suggestions are made concerning further activities of  the AsESG and other conservation agencies in the management of captive elephants.

## What are the aims of IUCN, modern zoos and animal welfare organisations?

The aims of IUCN are conservation of biodiversity with emphasis on the maintenance of threatened species and their genetic diversity and the maintenance of ecological processes. The Asian elephant is considered as a threatened species. Furthermore, *Elephas maximus* must be considered as a keystone species playing an important role in ecological processes (e.g. dispersion of seeds, acceleration of soil ecology in cooperation with insects and other invertebrates) in their habitat. The Asian elephant can also be considered as a flagship species promoting conservation of numerous less popular plant and animal species surviving in its habitat (Leader-Williams & Dublin, 2000).

In 1993 modern zoos formulated the 'World Zoo Conservation Strategy' and started to take responsibility for ex-situ conservation of certain animal species including the Asian elephant. Accordingly, zoos want to keep elephants in as natural ecological and social conditions as possible. This means that the environment must be provided to ensure species-specific behaviour patterns, e.g. food preparation, or social possibilities for forming a dominance hierarchy, family units and successfully rearing offsprings. The World Zoo Conservation Strategy aims to maintain a natural socio-ecological set-up, refers to results of field research on wild populations and aims through environmental enrichment to reduce the frequencies of abnormal behaviour as well as the increase of behavioural diversity, the range or number of behaviour patterns, the positive utilisation of the environment and the ability to cope with challenges in a more normal way. Elephants kept in near-natural conditions convey more understanding of the species and its requirements to the visitor and insight into the necessity to protect the species. Furthermore, near-natural conditions promote captive breeding and, hence, add to the conservation of genetic and behavioural diversities.

During the last years, animal welfare organisations have helped to raise elephant welfare standards. A considerable number of new elephant welfare organisations have been formed. They campaign, train and provide resources to improve living conditions of captive elephants in western as well as several range countries including India, Sri Lanka, or Thailand. Animal welfare activities refer mainly to the fitness (e.g. health, body condition) of the individual, but hardly touch socio-ecological considerations to maintain captive populations living in near-natural conditions.

In traditional elephant keeping establishments, the captive animals are not able to enjoy what Webster (1984) terms 'the five freedoms' (freedom from malnutrition, from thermal and physical discomfort, from injury and disease, from fear and stress, as well as freedom to express most genetically adapted normal patterns of behaviour). Nevertheless, assessment of the fundamentals of animal welfare organisations shows that their objective is tantamount to the fundamental basis of the maintenance of the Asian elephant in captivity. Taking the above mentioned definitions into consideration, Conservationists have to assist all serious animal welfare activities, but, according to the aims of IUCN, conservation agencies have furthermore to promote the aims formulated by the World Zoo Conservation Strategy. The goals of the strategy should accordingly be valid for any conservation minded facility keeping captive Asian elephants.

## Existing guidelines for the management of captive Asian elephants

The earliest works concerning care of captive elephants have already been mentioned in Section 3.2. In the nineteenth and early twentieth century British civil servants recorded many local practices, improved them, if necessary, and formulated a number of guidelines to regulate capture, care, daily food rations, working times and working loads of timber and army elephants in India and Myanmar (e.g. Sanderson, 1907; Evans, 1910; Milroy, 1922; or Ferrier 1947, Krishnamurthy & Wemmer, 1995a, 1995 b). State-run elephant establishments had guidelines on elephant husbandry which were later prescribed as Department Standing Orders and are still implemented. A system of record keeping right from the early days of elephant management under colonial rule provides valuable data on the performance of individual animals. Several forms of reportage are routinely practised. These records also provide guidelines, particularly to the new entrants to the administrative set-up, for the proper usage of working elephants under their custody.

In India the Central Zoo Authority has prescribed standards and norms for all zoos in the Recognition of Zoo Rules (1992). But no such standards and norms exist for the care of hundreds of privately owned elephants. In the last 7 years 'Standards and Norms for Elephant Owners' have been drafted and, for Kerala, the Captive Elephant Management Act by Jacob. V. Cheeran and his colleagues was enforced in 2003 (Ghosh, 2005). These new guidelines are paramount to improve living conditions of captive elephants. However, the fact that only few captive elephants are breeding and that few live in a social and ecological environment, and/or in physical and psychological condition conducive to breeding has hardly been recognised.

In North America and Europe a considerable number of management guidelines have been formulated by governmental and non-governmental organisations. Many of them are rightly criticised by elephant experts or welfare organisations as being not strict enough in guaranteeing the minimum species-specific requirements. However, most of these guidelines consider the Asian elephant as a species with highly evolved social behaviour and therefore, prescribe to keep the species in groups. In Europe for example, a radical change came with the 'EEP Management for Elephants in EAZA Institutions' (2000). Amongst many other points, it was approved, "that matriarchal family units are developed and kept together, i.e. the intention to keep female offspring within their family group during their life" (e.g. Dorrestyn, 2001).

In 2002 a 'Review of the Welfare of Zoo Elephants in Europe' commissioned by RSPCA and carried out by Ros Clubb and Georgia Mason of the Oxford University summarised results from a few rather traditional zoos and found poor welfare of most zoo elephants and demanded improvement of the elephants' social and physical environments. They concluded: "The factors responsible for the poor welfare of zoo elephants should be empirically investigated as a matter of urgency. The zoo elephant population should be frozen, i.e. breeding and importation should be ceased until these factors have been identified. Then only zoos that solve these problems should be allowed to keep elephants in the future".

The Oxford study has been fundamentally refuted by the European Elephant Group (EEG), an association monitoring captive elephants in practically all European establishments since 30 years, based on the fact that it lacked information from numerous large and modern zoos where elephants are living under almost optimal conditions. EEP published a detailed report on 'Elephants in European Zoos and Safari Parks' in 2002.

In *Management Guidelines for the Welfare of Elephants* Miranda F. Stevenson (2002) compiles all the requirements for a near-natural keeping of elephants in modern zoos. The Elephant Management Policy Statement of Federation of Zoos in Great Britain and Ireland defines: "Elephants must only be kept in zoos as part of an overriding conservation mission so that they are in actively managed breeding programmes ... . Their presence must enable progressive educational activities and demonstrate links with field conservation projects and benign scientific research, leading to continuous improvements in breeding and welfare standards. Zoos must exercise a duty of care so that standards of husbandry practices, housing, health and welfare management are humane and appropriate to the intelligence, social behaviour, longevity and size of elephants... ".

## How many Asian elephants live in captivity?

Accurate numbers of captive Asian elephants in the range countries (Bangladesh, Cambodia, India, Indonesia, Lao PDR, Malaysia, Myanmar, Nepal, Sri Lanka, Thailand, Vietnam) are hardly known. According to the most recent estimates, (Table 6.1.1) their total population (including zoo and circus elephants in the range countries) is about 14,000 animals (Baker & Kashio, 2002). The largest population is found in Myanmar (40% of the total range countries population), followed by India (26%) and Thailand (19%). Outside the range countries at least 870 Asian elephants are kept in zoos and circuses (European Elephant Group, 2003). Therefore, the total captive population includes about 15,000 elephants or 25% to 33% of the total present population of the species (Baker & Kashio, 2002; Sukumar, 2003). The overall captive population, which still drains from wild ones, dwindles more or less rapidly due to several reasons such as low reproduction and high (juvenile) mortality, and mainly the restriction of capturing operations and the lack of employment opportunities, as logging was banned in many south Asian countries.

In Vietnam, for instance, the number of captive elephants dropped from 600 in 1980 to 165 in 2000 and the present population consists mainly of old animals (Cuong *et al.*, 2002). In Cambodia the number of captive elephants dropped from an estimated number of 300 to 600 some 20 years ago to 162 in 2000 (Lair, 1997; Dany, 2002). In Sri Lanka privately owned and temple elephants numbered 532 in 1970, 344 in 1982 and 186 in 2002 (Jayewardene, 2002). However, in Sri Lanka's Pinnawela Orphanage captive propagation started very promisingly in 1982 and until now 22 offsprings have been born and only one has died. In Nepal, the captive population dwindled continuously from 328 in 1903 to 47 in 1973 and later increased to 171 in 2002. In Indonesia a captive population of about 400 Sumatra elephants is considered as stable due to regular capturing operations in the last years (Hutadjulu & Janis, 2002). In Indonesia as well as in Malaysia keeping elephants in captivity was abandoned about 100 years ago and later again activated with the assistance of mahouts from Thailand. The present small population of Malaysia consists of 36 elephants and is said to increase. India's captive elephant population is considered as stable (Bist *et al.*, 2002), and Myanmar's population seems to be decreasing slowly (Aung & Nyunt, 2002). In Europe the captive population of nearly 500 dropped to 450 in 2000 and within the next 20 years it is expected to breakdown to 120, although captive breeding has been extremely successful in the last years (Kurt, 2001). In North America the total captive population of about 300 at present will drop to approximately 10 elephants in 50 years and be demographically extinct without continued importation or a drastic increase in birth rate (Wiese, 2000).

In Europe it can be expected that the present population of about 240 elephants in 88 circuses (90 African, 150 Asian) will disappear until 2020 (EEG, 2001). The population of 300 Asian elephants (2005) in 83 European zoos and safari parks will drop until 2015 to about 180 in 25 to 30 establishments. Then a stabilisation or even a slight growth of population can be expected (EEG Database).

In all range countries, considerable shifts of captive elephants from forested to urban areas are reported. In India for example, restriction of logging operations in the north-east and the Andaman Isles imposed by the Supreme Court in 1996 resulted in a great exodus of captive elephants in these regions, while their number has increased in Kerala from about 250 in 1983 to more than 700 at present with a male-biased population structure, and the city of Jaipur (Rajasthan) with about 100 captive elephants with a female-biased population structure has presently become a major elephant centre (Bist *et al.*, 2002). Most captive elephants all over Thailand suddenly became unemployed after the government banned logging in 1989. Unemployment and, in due course, starvation are at the root of many problems with the captive population of Thailand. Many elephants are brought to the cities for begging, tours or are engaged in illegal logging (Lohanan, 2002). They are given amphetamines in order to speed up work. The unworkable elephants are sent to slaughter houses and their meat is sold (Mahasavangkul, 2002).

The multiple and diverse deployment of captive Asian elephants requires various keeping systems (Kurt, 1995) as already defined in Section 3.2. The intensive keeping system has been taken over by western circuses and in former times also by zoological gardens. The keeping system in traditional zoos as well as in the Pinnawela Elephant Orphanage in Sri Lanka can be considered as intermediate between extensive and intensive, for example, during the day the elephants are kept free in their paddocks but during the night they are shackled or otherwise kept solitary in small cages. In very modern zoos alternative keeping systems are now evolving very fast according to the guidelines of the World Zoo Conservation Strategy. In these systems elephants live in appropriate family groups which are paramount for the processes of learning and socialisation (e.g. Garaï, 2002 a; Kurt, 2001, 2002; Stevenson, 2002) and only sub-adult and adult bulls are kept singly, if necessary. A further alternative keeping system is found in Elephant Transit Homes (e.g. Uda Walawe National Park in Sri Lanka). In the Transit Home of Sri Lanka, young elephants are kept in a so-called semi-natural environment with minimal contacts with humans (e.g. bottle-feeding). This is done to avoid imprinting processes, which could later complicate the release of the captive animals into the wild.

## Significance of captive Asian elephants for man and their individual fitness and population dynamics

Captive elephants have an economical significance for owners and keepers/mahouts. They may be important as a driving power for forestry applications. Furthermore, they may have a religious significance (mainly in Buddhist and Hindu countries) or a social significance for non-religious ceremonies. Captive elephants can be an attraction in urban tourism, they can be paramount for education (including eco-tourism) concerning life sciences and conservation. Last but not the least, captive Asian elephants can still have an ecological significance in selective timber harvesting and as keystone species in the maintenance of biodiversity and ecological processes.

For each of these above categories the degree of significance has crudely been estimated and given in 4 figures (0: insignificant; 1: low, 2: medium and 3: high significance). Relative low total significance (total less than 10 points) is found in elephants kept in transit homes, urban tourist centres, circuses or traditional zoos (Table 6.1.2). Relative high significance (total $\geq$ 10 points) is found in elephants kept in temples, in the Pinnawela Elephant Orphanage, modern zoos, timber and tourist camps (Table 6.1.2). It must be stressed that relative high significance has different reasons. Those living in jungle camps reach high values mainly due to their value for local people, education and ecology, while those living in temples reach a relative high significance due to their economical, religious and tourist value.

Living conditions can be defined by the individual fitness, the food offered, the daily activity as well as aspects of population dynamics such as social behaviour, social structure of the population and reproductive performance (Table 6.1.2.). Individual fitness depends on the degree of care by veterinarians and keepers/mahouts and becomes visible in body condition. Both aspects have been characterised by numbers: 1: low, 2: medium, 3: high. The food offered to elephants can be monotonous (1), divers (3) or intermediate (2). Monotonous food consisting of the leaves, branches and stems of 1 to 3 plant species is common for elephants kept in temples and urban tourist centres. But also in most of the traditional zoos and circuses, food can be considered as monotonous. The same applies to the baby elephants kept at the Transit Home in Uda Walawe. Beside the daily portion of milk, they live on only a few wild growing grass species. Elephants living in jungle or forest-based extensive systems find the most diverse diet. In modern zoos as well as in the Pinnawela Elephant Orphanage the diversity of food can be considered as intermediate.

**Table 6.1.1** Estimated population sizes of wild and captive Asian elephants in the range countries according to 1:Jayawardene (2002) and own data; 2: Bist *et al.* 2002); 3:Kharel (2002;4: Islam (2002); 5: Aung & Nyut 2002); 6: Daim (2002); 7: Lohanan (2002); 8: Norachack (2002); 9: Dany *et al.* (2002); 10: Cuong *et al.* 2002); 11: Suprayogi *et al.* (2002). Ownership can be by government (G) or private (P). The keeping system (Ks) is either extensive (E) or intensive (I). Captive elephants are used for forest and wildlife work (fw), eco-tourism (et), ceremonies and urban tourism (cut), begging (beg), display (dis) and illegal logging (ilog). The captive population is recruited from last legal captive operations (L.leg.capt), from neonates and infants found as orphans (orp), from captive breeding (br), illegal capture (ill cap) or by purchase or donation (PD). +: regular, (+) rare, - absent. Population structure and dynamics depend on sex ratio (=: equal, i.e. at least 1 male: 4 females, # unequal, i.e. one sex is extremely predominant), age structure (Y: mainly young animals, O: mainly old animals, =:± normal) and captive breeding +: regularly; (+) rare; (-) absent. The present population trend can be increasing (inc), decreasing (dec) or stable (stab). nd: no data.

| Country | Wild pop. (min–max) | Capt. pop. | Owned by (n) | Ks | Use for fw | et | cut | sh | beg | dis | ilog | Population recruit from: L.leg capt | orp | br | ill cap | PD | sex rat. | age str | cap br | T |
|---|---|---|---|---|---|---|---|---|---|---|---|---|---|---|---|---|---|---|---|---|
| 1. Sri Lanka | 3500 | 286 | G (±100) | I | – | – | – | + | – | + | – | 1995 | + | + | – | – | = | y | + | me |
|  |  |  | P (186) | I | – | (+) | + | – | (+) | – | – |  | – | – | (+) | + | = | O | – | dee |
| 2. India | 28100- ± 29200 | 3500 | G (562) | E | + | + | – | – | – | (+) | – | 2000 | (+) | + | – | – | = | = | + | dee |
|  |  |  | P(±2938) | E,I | (+) | (+) | + | – | + | – | – |  | – | (+) | (+) | + | # | # | – | stab |
| 3. Nepal | 100 | 171 | G (77) | E,I | + | + | – | – | – | – | – | 1973 | – | + | – | + | = | = | + | ine |
|  |  |  | P (94) | I | – | + | – | – | – | – | – |  | – | – | – | + | = | = | nd | mee |

[Table 6.1.1 Contd.

Contd. Table 6.1.1]

| No. | Country | | | Sample | Src | | | | | | | | Year | | | | | | | | Trend |
|---|---|---|---|---|---|---|---|---|---|---|---|---|---|---|---|---|---|---|---|---|---|
| 4. | Bangladesh | 200 | 93 | G (17) | E | + | − | − | − | − | − | − | 1974 | − | (+) | − | − | = | = | (+) | dee |
| | | | | P (76) | I | − | − | + | − | + | − | | | − | − | − | + | = | = | − | de |
| 5. | Myanmar | 4000 | ±5700 | G (±2700) | E | + | (+) | − | − | − | (+) | − | 1994 | − | + | − | − | = | = | + | dee |
| | | | | P (±3000) | E | + | − | − | − | − | − | − | | − | + | + | − | = | = | + | dee |
| 6. | Malaysia | 1200-1500 | 36 | G (8) | E | + | (+) | − | − | − | (+) | − | 2000 | (+) | − | − | + | # | = | − | ine |
| | | | | P(28) | I | (+) | (+) | − | − | − | + | − | | − | − | − | + | = | = | − | ine |
| 7. | Thailand | 2250 | 2500 | G (125) | E,I | − | + | − | + | − | + | − | 1975 | − | + | − | − | = | = | + | dee |
| | | | | P(2375) | E,I | − | + | + | + | + | + | + | | − | (+) | + | + | #.= | #= | (+) | dee |
| 8. | Laos PDR | 2100-3200 | 864 | P (864) | E,I | + | − | − | − | − | (+) | + | nd | − | − | + | + | nd | nd | − | dee |
| 9. | Cambodia | 300-600 | 162 | P (162) | E,I | + | − | + | − | (+) | (+) | (+) | 1999 | − | − | + | − | nd | O | − | dee |
| 10. | Vietnam | 2000 | 165 | P (165) | E,I | (+) | (+) | + | − | − | − | − | 1960 | − | (+) | + | − | # | O | − | dee |
| 11. | Indonesia | 2100-2700 | 418 | G (391) | I | (+) | (+) | (+) | + | − | + | − | 1999 | − | − | − | − | O | y | − | stab |
| | | | | P (27) | I | (+) | − | − | + | − | + | − | | − | − | − | + | nd | nd | − | stab |
| | Total | 45850 | ±13895 | | | | | | | | | | | | | | | | | | |
| | | 49250 | | | | | | | | | | | | | | | | | | | |

**Table 6.1.2** Evaluations of different keeping systems according to economical, non-profitable and ecological significance, as well as to fitness, population dynamics and significance of conservation of the Asian elephant.

| Significance | Keeping systems | | | | | |
| --- | --- | --- | --- | --- | --- | --- |
| | Qualifications: 0: absent;  1: low; 2: medium; 3:high | | | | | |
| | Temple | Circus | Trad. zoo | Pinna-wela | Forest camp | Mod. zoo |
| *Economical significance* | | | | | | |
| - For owner | 3 | 3 | 3 | 3 | 2 | 3 |
| - For mahout/keeper | 1 | 1 | 2 | 2 | 2 | 3 |
| *Non-profit significance* | | | | | | |
| - Religious processions | 3 | 0 | 0 | 0 | 0 | 0 |
| - Urban tourism | 3 | 1 | 1 | 3 | 0 | 2 |
| - Education | 1 | 1 | 2 | 3 | 2 | 3 |
| *Ecological significance* | | | | | | |
| - Forestry | 0 | 0 | 0 | 0 | 3 | 0 |
| - Keystone - function | 0 | 0 | 0 | 0 | 3 | 0 |
| Individ. fitness elephants | | | | | | |
| - Staff, vet. care | 2 | 1 | 2 | 2 | 3 | 3 |
| - Body condition | 2 | 1 | 2 | 2 | 3 | 3 |
| - Food | 1 | 1 | 1 | 2 | 3 | 2 |
| - Daily activity | 1 | 1 | 1 | 3 | 2 | 3 |
| *Population dynamics* | | | | | | |
| - Social behaviour | 0 | 0 | 1 | 3 | 3 | 3 |
| - Social structure | 0 | 0 | 1 | 3 | 3 | 3 |
| - Reproduction | 0 | 0 | 1 | 3 | 3 | 3 |
| *Signif. for conservation* | | | | | | |
| - Promotion conservat. | 1 | 0 | 1 | 3 | 2 | 3 |
| - Reprod. potential | 0 | 0 | 1 | 3 | 3 | 3 |
| - Chances of pop. surv. | 0 | 0 | 0 | 3 | 2 | 3 |
| - Natural biodiversity | 0 | 0 | 0 | 0 | 3 | 0 |
| *Total points* | 18 | 10 | 19 | 38 | 40 | 40 |

For elephants kept in extensive systems daily life includes, besides work, the search for food, skin care, socialising and sleeping at a species-specific place. There are no, or hardly any, stereotypies. Their daily species-specific activity is considered as medium (2). Elephants kept in extensive and intensive systems are brought twice a day to certain waterways by their mahouts, where they are scrubbed and massaged with the shells of green coconuts or bread-fruits, stones or bark. Elephants kept in intensive systems are fed once a day or repeatedly with prepared fodder. When they are not working they are often idle. Stereotypies occur regularly. The daily species-specific activity is considered as low (1). The same applies to elephants kept in traditional zoos, circuses and the Transit Home of Uda Walawe. However, a high degree (3) of diverse daily activities is found in modern zoos and the Pinnawela Elephant Orphanage.

Species-specific social behaviour is very rare or practically absent (0) in elephants kept in intensive systems. But it is relatively high (3) in the Pinnawela Elephant Orphanage, in extensively kept elephants as well as in modern zoos. Low social activities (1) are found in traditional zoos as well as in the Transit Home. Low or even absent social behaviour is a result of the respective management systems, where intraspecific interactions are either not allowed by owners and keepers/mahouts or hardly possible because of a monotonous social structure of the captive groups concerned (Fig. 6.1.1). In most intensively kept populations social structures can hardly be recognised (0) since only members of certain age and sex classes are kept (e.g. mainly sub-adult and adult bulls in temples, or older females in circuses, or juveniles in many urban tourist resorts. Groups in traditional zoos consist mainly of middle aged or old females and rarely include bulls or younger members. Therefore, their social structure can be considered as comparatively low (1). The same applies to elephants in the Transit Home of Uda Walawe, where only orphaned neonates, infants and juveniles are kept. The populations in the Pinnawela Elephant Orphanage, as well as those in jungle based camps are socially well structured (3).

Reproduction is practically absent in intensive keeping systems and of course in the Transit Home. But reproductive performances are relatively high at the Pinnawela Elephant Orphanage and in extensively

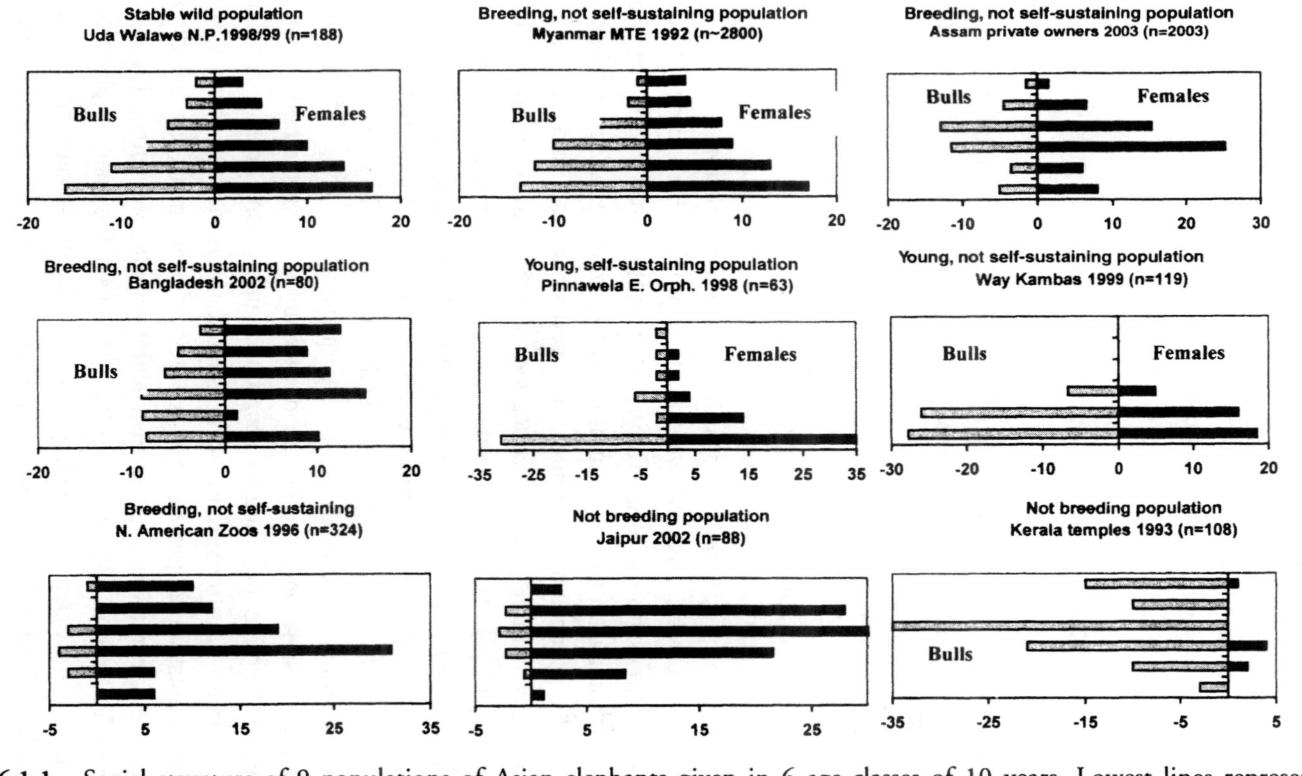



**Fig. 6.1.1** Social structure of 9 populations of Asian elephants given in 6 age classes of 10 years. Lowest lines represent animals between 0 to 10 years. Highest lines represent animals between 51 to 60 or more years. Given as per cents of the total population.

kept populations. The total of crude evaluations of different parameters of living conditions and population dynamics is low (<10 points) in intensive keeping systems as well as in traditional zoos and the Transit Home, but is significantly higher in the Pinnawela Elephant Orphanage, in modern zoos as well as extensive keeping systems (Table 6.1.2).

## The significance of captive Asian elephant populations for conservation

The significance of captive elephants for economy, religion, education as well as for the natural environment is the root cause for the ex-situ conservation of the species. The same applies to the parameters of living conditions and population dynamics. To assess crudely the significance of captive elephants for the conservation of the species, it must also be taken into consideration that captive elephants can promote conservation efforts, that some of the populations can add to the natural biodiversity and that captive propagation is possible in some populations.

There is a wide array of ways, how captive Asian elephants can promote conservation efforts. Riding elephants enable researchers to carry out ecological and behavioural studies in wild areas. Zoo elephants living under near-natural conditions are paramount for fund raising as well as conservation oriented education in urban areas. In protected areas, riding elephants are an important asset for wildlife and conservation research as well as for eco-tourism. It must be stressed, however, that more and more captive elephants are misused for shows which have little or nothing to do with the species-specific behaviour or good taste and aim to entertain the crowds in a very questionable way. Often such shows are possible only due to the most cruel training methods.

Several captive populations reproduce on a more or less regular basis. It must be stressed, however, that there is no self-sustainable captive population at present due to lack of well managed breeding programme. Nevertheless, there are smaller populations with a high (3) potential for reproduction such as the Pinnawela Elephant Orphanage, modern zoos and extensive keeping systems. In intensive keeping systems reproduction is practically absent (0), due to several reasons. The first one often mentioned, refers to the lack of economical interest of owners to breed elephants. But many intensively kept elephants are neither physiologically nor psychologically able to reproduce. They show serious retardation in body growth; they never go through a species-specific socialisation process and tend towards infanticide, or they are kept under such tremendous physical and psychological suppression by their keepers/mahouts that neither their sexual organs nor their sexual behaviour can develop normally (Hildebrandt *et al.*, 2000). Many females kept in traditional zoos become so obese that reproduction is impossible after the age of 30 years (Hildebrandt & Göritz, 1995; Kurt & Mar, 1996).

The chances of survival of captive populations are practically zero in intensively kept elephants, and very insignificant for those kept in traditional zoos, where the mortality rate is extremely high at a young age (Fig. 6.1.2). However, survival chances are relatively high in the Pinnawela Elephant Orphanage, modern zoos as well as in extensive keeping systems (i.e. Sukumar *et al.*, 1997), with low mortality rates at a young age. Finally, the significance of captive populations increases when they add to the natural biodiversity of the species' habitat. Summarising the single values of qualifications for different keeping systems (Table 6.1.2), the following conclusions can be drawn: Elephants living in intensive keeping systems, circuses or traditional zoos have low significance for conservation of the species (total < 20 points). But elephants living in the Pinnawela Elephant Orphanage, modern zoos and extensive keeping systems have a relative high significance for conservation. The same may apply to Transit Homes.

The relative high number of Asian elephants currently in captivity, estimated to enclose 25 to 33% of the total present population of *Elephas maximus,* advocates for a conservation orientated management of captive Asian elephants. However, the adequate management of captive Asian elephants can only be reached by cooperation of many different disciplines such as wildlife experts, veterinarians, elephant owners and managers, and accordingly, of several governmental agencies and NGOs. First of all, every attempt should be made to ensure that all captive Asian elephants meet the maximum welfare requirements.

It is doubtlessly more important for species conservation to maintain large wild populations in their natural habitat than to maintain considerable captive populations. However, the pressures on a number of wild populations by human-elephant conflicts, poaching of young elephants and other

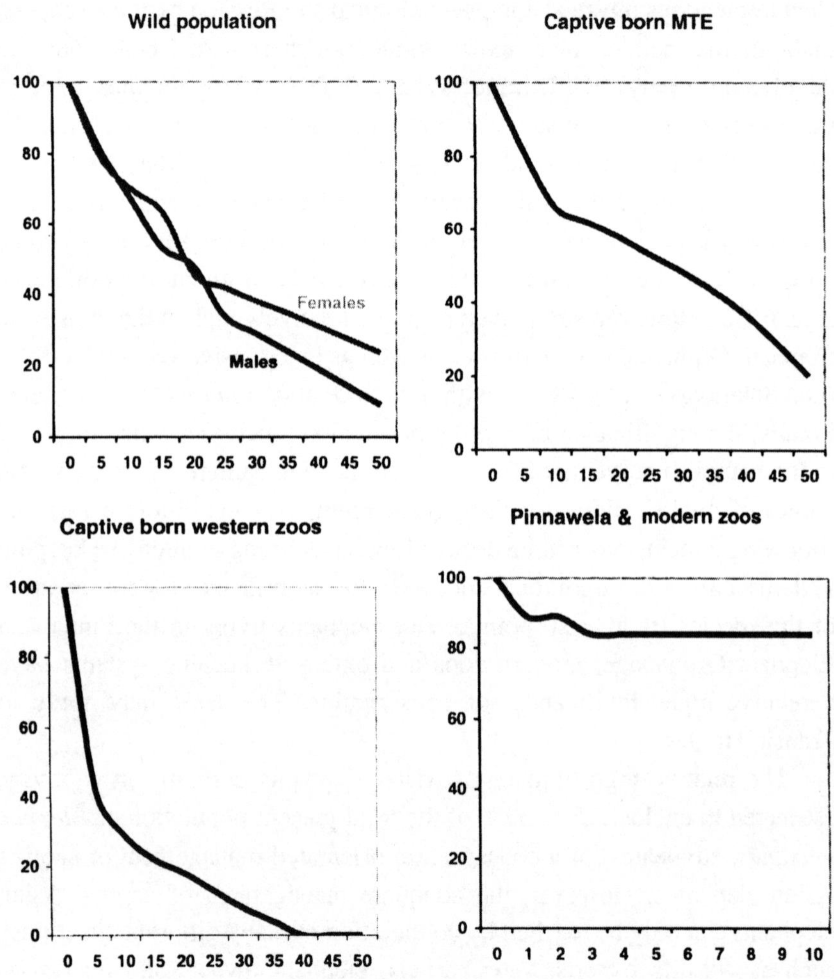


Fig. 6.1.2   Survival rate according to age of wild (adapted from Sukumar 1991 and Kurt 2001) and captive populations (Kurt 2001, Mar 2001, 2002). 'Modern zoos': Emmen (Holland) and Hagenbeck - Hamburg (Germany). Age is given in years.

disturbances by man (e.g. destructive agricultural practices, fuel wood extraction, etc.) seem to be the reasons that will cause continuous inflow of wild elephants into captive populations in the years to come, and in some populations can cause even extinction. Taking this fact into consideration governmental and non-governmental conservation agencies have to fight for appropriate elephant keeping facilities where elephants are kept according to modern welfare guidelines. They must also take conservation into consideration by promoting captive propagation of genetic and behavioural diversities and conservation minded education and promotion. This goal is best reached by maintaining jungle based establishments, improving elephant facilities in zoos, exerting control over intensive keeping systems, upgrading the knowledge on elephants in decision makers and mahouts/keepers, and developing alternative keeping systems.

## Maintenance of jungle based establishments

About 60% of the captive elephants in the range countries are kept extensively and still live in or close to jungles (for overview see Baker & Kashio, 2002). Following the logging bans in several countries, it must be expected that these populations will dwindle, while intensively kept populations in urban areas will increase accordingly. Jungle based captive elephants are important for nature conservation as well as for captive breeding, as given in a few examples: of 198 privately owned elephants registered by the Assam Forest Department in 2003 only 56 (28%) were captured from the wild (Ashraf *et al.*, 2002). In the Jaldapara Wildlife Sanctuary (115 km²) of West Bengal, 67 captive elephants of the Forest Department are breeding so successfully that a further increase in population would affect the resources, and birth control is becoming a major issue (Mathew, 2002). In the forest camps of Tamil Nadu and Karnataka captive females have an annual fecundity rate of about 0.1 per adult female during the twentieth century and 0.16 per adult female between 1969 to1989. Given the relative high survival rate at a young age, this population would grow strongly, with the higher fecundity rate observed in recent years (Sukumar, 2003). In Myanmar the annual fecundity rate of timber elephants ranks below 0.1 per adult female, and their large population is declining slowly under the prevailing natality (average 3.1% of the population) and mortality (3.3% of the population) rates (Aung & Nyut, 2002). However, this decline cannot only be blamed on relative low breeding success; some changes in the survivorship rate of younger animals, improved healthcare and additional food rations given to the elephants, mainly to pregnant and nursing females, could help to achieve a stable population size, as it may have existed already during British rule of the timber camps in former Burma at the beginning of the twentieth century, (e.g. Williams, 1950).

In the jungle camps of Karnataka and Tamil Nadu upto 90% of the extensively kept females reproduce (Sukumar, 2003). Of the 57 females in the national parks and reserves of Nepal 16 (28%) have parturiated between 1979 to 2000. This figure is still higher than in European zoos, where in 2005, 225 females were kept, but only 22% of them bred (EEG Database). However, this low fecundity rate is also due to lack of suitable males. In the city of Jaipur, where elephants traditionally live in intensive keeping systems, only two parturitions have been recorded during the last 100 years (March, 2003), and in the privately owned as well as in the temple elephants of Kerala and Sri Lanka captive propagation is practically absent. But it is not only the relative high reproduction rate that makes jungle camps potentially important for conservation of *Elephas maximus* in future. Many babies that are being born in jungle camps have been fathered by wild bulls.

Seen from the point of view of conservation of genetic diversity, many jungle camps function as gene traps, i.e. rather small captive populations harbour a relative diverse gene pool. In European zoos the situation is vice versa: for a total female population of about 225 there are only 16 actively breeding bulls and in several cases have already reproduced with their daughters.

But captive breeding is not the only reason to maintain jungle based elephant camps. Timber elephants play divers ecological roles as already mentioned above. And they can continue to have a high significance for conservation, also in regions where selective logging has been banned. With their extreme cross-country mobility they are unique mounts for wildlife officers, researchers and eco-tourists. They are used mainly in India and Nepal as irreplaceable partners in the management and studies of the Bengal Tiger, the Great Indian Rhinoceros or wild elephants. Since antiquity, specially trained hunting elephants, so called *koonkies*, carry elephant catchers into the herds of wild elephants for *mela-shikar*, i.e. the noosing of selected animals. *Koonkies* still play an important role in taming, training and translocation of wild elephants in India, Malaysia, Myanmar or Thailand. The tradition of using *koonkies* for taming and training of newly captured elephants has been lost in Sri Lanka.

In Sri Lanka a rich fauna and flora survives in a total protected area of 8,220 km² or 12.5% of the country's surface and a relative large wild elephant population of about 3,500 wild elephants, which create considerable human-elephant conflicts (De Silva, 1998). Captive elephants could be used, as in other South Asian countries, for patrolling in protected areas and for protecting men and crops from elephant raids. Today eco-tourism on elephant back, is evolving in Sri Lanka. In Thailand it is already established. But not all former working elephants are suitable for tourism. They can be used for diverse conservation work. Thailand harbours 100 national parks and other nature reserves. If only 4 elephants are assigned to patrol in each reserve, 400 elephants could survive in a near-natural habitat (Salwala, 2002).

Conservation of captive Asian elephants in jungle camps also means employment of local, often tribal people who would be taking care of the conservation of elephants. The importance of their traditional knowledge of elephants and biodiversity of the natural elephant habitat can be paramount for conservation, applied research and local, conservation minded politics.

## Improvement of facilities in zoos

Keeping of elephants in western zoos and circuses has been copied from the intensive system in South Asia. In the traditional circuses, elephants are chained when not at rehearsal, show or on parades. There is direct contact between elephants and the men looking after them. Next to the whip, the *ankus* is the tool used to guide the animals, and even some Asian commands are still used. In Europe, circus elephants are old, and it can be expected that the present circus population will disappear soon, since it is unlikely that circuses will obtain permits for new and younger animals, although many circuses now keep their elephants in electro-fenced paddocks during the day. In India, with more than 100 captive elephants, the circuses are one of the biggest buyers of elephants and they are continuously looking for replacements for their old animals (Bist *et al.*, 2002). Worldwide, circus companies face tremendous criticism from animal welfare activists for frequently subjecting their elephants to pain and cruelty.

In traditional zoos, elephants are chained during the night in small stables and kept during the day in often very small enclosures. In Germany for example, the directive given by the Forest Ministry in 1996 permitted 4 elephants to be kept in an area of 650 m², the size of the penalty area of a soccer field. In 2000, the rules were changed to 3,000 m², i.e. the size of 12 tennis courts. Out of 138 European zoos and safari parks with elephants (both species), 118 enclosure sizes are known. In 30 (25.4%) zoos the size is 1,000 m² or less, 63 (53.4%) are between 1,000 to 5,000 m² (the size of a hockey field), 10 (8.5%) are between 5,000 to 10,000 m² and are about the size of a soccer or rugby field, 8.5% are between 10,000 to 50,000 m² and 4.2% are larger than 50,000 m². The present trend is towards large enclosures which include bath, mud-wallows and scratching trees. Bulls older than 10 years are no longer chained. Today, only 25% of European zoos with elephants, chain females and young animals during night time. In the modern concept of zoos chains are replaced by boxes, and direct contact between keeper and elephant is replaced by the system of 'protected contact' or 'no contact' (European Elephant Group, 2003). Increasingly, elephants are being considered to be captive wild animals like rhinos, hippos or tigers. However, this does not apply to most zoos and zoo-like establishments in the range countries, where elephants live under more or less antiquated conditions.

Even in world famous zoos like Dehiwala (Sri Lanka) or Yangon (Myanmar), elephants are still kept in the same manner as at the beginning of the twentieth century when these zoos were built by Europeans, i.e. the animals are permanently chained when not brought to their daily bath and used for shows or riding. In the Pinnawela Elephant Orphanage elephants are chained between 6 p.m. to 8 a.m., i.e. for 14 hours per day (Chapter 3). The grassy area where 65 to 70 elephants are kept during the day without chains measures about 93,000 m² (Tilakaratne & Santipillai, 2002). But since this area lacks an elephant proof fence, the animals are permanently herded together in dense clusters by a handful of mahouts and hardly able to completely use this area with the size of about 12 rugby fields.

Keeping elephants permanently or for long periods in chains is cheap, but always and everywhere accompanied by serious accidents (e.g. Haufellner *et al.*, 2002, 2003). In elephants, chaining causes foot problems such as uneven wear of toenails and soles, and bed - or pressure sores. More or less permanent chaining leads to notorious 'weaving'. This stereotypy is caused through lack of space and inadequate social partners and can be considered as a symptom of social isolation (Section 5.4) which finally leads to abnormal social behaviour including infanticide. Social isolation in more or less permanently chained elephants is furthermore enforced, by the fact that they are lined up according to sex and size – one of the many attributes leading to the belief that captive elephants are domesticated animals. Under such circumstances, any particular elephant is, if at all, allowed to have direct social contact with only two conspecifics, i.e. the neighbours to the right and left. In most cases they are the wrong partners. The members of a group of wild elephants organise themselves in a typical spatial pattern in which neonates (0 – 2 years) and infants (3 – 5 years) stay close to their mothers and/ or allomothers, and juvenile females close to neonates and infants (Section 5.3). Most members of the Pinnawela herd tried to follow this pattern when not chained. In wild herds juveniles were encouraged by sub-adult (10 – 15 years) and adult (> 15 years) females to care for neonates and infants and accumulate the necessary experiences to care for their own offsprings later in their lives.

A survey done at the Pinnawela Elephant Orphanage revealed that 74% of foreign visitors do not approve of the chaining of elephants (Kurt, 2001). Visitors to western zoos rarely spend much time at elephant facilities as long as the animals are chained and weaving or 'misbehaving' in other ways. In that case, visitors prefer, for e.g. animals in aquariums, which they had not planned to see before their visit. Elephants only trigger attention when they take bath, play, care for their offsprings or mate as shown in a survey done by Stolba and Müllers (1990). Zoo and circus elephants kept in miserable conditions in many cases, certainly do not attract initiatives for the preservation of the species. The Zoo Conservation Strategy is as follows: in future, zoo elephants and elephant enclosures must promote "the public awareness of the necessity of nature conservation, careful use of natural resources and the development of a new balance between man and nature" (IUDZG , 1993). Improvement in the keeping conditions is urgently needed in South Asian zoos. In cities like Kuala Lumpur, Bangkok, Yangon or Colombo, decisions are made for the future policies of the countries. Here, a new generation of decision-makers is emerging who will soon decide the fate of the last Asian elephants. Modern zoo biology along with the Asian traditional knowledge can make the changes necessary in order to ensure the survival of elephants in zoos.

## Control over intensively kept populations

In many densely populated countries of South Asia, relatively large numbers of wild elephants have survived most probably due to the fact that the elephant is venerated as a symbol of Buddha and also as a Hindu God. "Thus, elephants kept in temples or participating in cultural or religious festivities can reinforce this sentiment of sacredness among the people" (Sukumar, 2003). With the exception of one kept in Punjab, all the other 192 official temple elephants of India are in the southern states, mainly in Kerala (Bist *et al.*, 2002). However, this small fraction of only 5% of the country's captive population is greatly outdone by privately owned elephants hired out to temples for religious ceremonies. Impressive elephant processions today attract large numbers of Indian and foreign tourists. Accordingly, the number of colourful festivities have increased in South India as well as in Sri Lanka.

Captive elephants have become a highlight for urban tourism, also as riding animals for tourist, for visiting cultural centres such as the Amber Ford in Rajasthan or Angkor Wat in Cambodia. In several places captive elephants are forced to play football or polo. In the cities of Thailand elephant shows mimic attacks of war elephants and the noosing of wild elephants by the *mela-shikar* method. But in Thailand's tourist resorts, trained elephant babies have to perform their tricks for tourists and drink beer with them. These baby-shows are so attractive that in the last decade a lucrative baby market evolved, fed by illegal capture of neonate and infant wild elephants not only in Thailand itself but also in Lao PDR, Vietnam and Cambodia (Baker & Kashio, 2002). In South Asian cities there always lived a small population of captive elephants, which was in great demand for marriage processions, social functions and occasionally, in political rallies. During markets and festivals some medicants often appeared with a captive elephant and made a handsome living through begging (Bist *et al.*, 2002). Today large numbers of begging elephants belong to the picture of South Asian cities and holiday resorts.

In intensive keeping systems elephants hardly reproduce due to several reasons. In many cases owners have no interest in breeding their elephants. The most obvious biological reason is the unbalanced social structure of these populations. In Kerala's temples there are practically only bulls. The riding elephants in Jaipur, at present consist of 78 females and 6 bulls (Mar, 2002). The begging street elephants in Thailand's cities are mainly older females, while babies are used as cheap entertainment in bars. In Sri Lanka, intensively kept elephants have a balanced sex ratio, but most of them are older than 45 years. It must also be expected that many intensively kept elephants are neither physically nor socially in a position to reproduce as mentioned above.

Intensively kept elephants are more and more held in large groups, which creates problems of food supply, maintaining hygiene in stands and, in due course, health. Intensively kept elephants suffer more under health problems as a result of the unnatural keeping system than elephants living in forest camps under more natural conditions and looked after by more experienced people. Accordingly, elephant health camps run by numerous national and international NGOs have to be established mainly in the cities (for overview: Baker & Kashio, 2002).

From the point of view of animal welfare and conservation, the number of intensively held elephants must be kept as low as possible and it makes no sense to increase their populations with animals from well breeding populations such as jungle villages or the Pinnawela Elephant Orphanage, from where recently several animals have been donated to temples and private owners. It is paramount for the future of elephant conservation that many of the elephant facilities in urban areas are replaced by more adequate keeping systems. But to reach this goal elephant owners and managers should know more about elephants than they do at present.

## Upgrade the knowledge on elephants

According to an international workshop on captive elephants, under the auspices of FAO in 2001, the traditional skill of mahouts is satisfactory only in Myanmar. In Thailand, Cambodia and Vietnam, it still exists but is declining rapidly. In Malaysia and Indonesia (Sumatra) traditional mahoutship was lost and recently reactivated by trainers from Thailand, where the traditional and modern knowledge on elephant management is taught in a mahout training school. In Kerala several mahout training courses have taken place in the last years. But Bangladesh, India, Nepal and Sri Lanka have asked officially for training facilities for their mahouts, and 7 out of 11 range countries require training of veterinarians (Baker & Kashio, 2002). Training of veterinarians takes place already on a regular base in several regions of India, in Thailand and Sri Lanka.

In the context of this chapter it is pointed out once more that conservation orientated keeping of elephants must first of all be based on the ecological and behavioural demands of the species under natural conditions. Many years of experience reveal that most of the people concerned in one or the other way with captive elephants have little knowledge on wild elephants. Accordingly, they run the risk to consider the 'abnormal' as the 'normal' as pointed out by Jane Goodall in *Chimpanzee Handlers in Show Business*. How fast even highly motivated academics can become accustomed to grievances in intensive elephant keeping could be observed on more than 100 students in veterinary medicine and biology, participating in our research projects in Sri Lanka between 1997 to 1999: they were able to work on wild elephants in the Uda Walawe National Park or on captive ones in the Pinnawela Elephant Orphanage and in Kandy. It was never a problem to shift students from a camp with captive elephants to the national park. But it was always problematic to have them work on captive elephants after they had spent only 3 to 4 weeks in close contact with wild ones, since they immediately recognised differences in appearance, daily activities or behaviour and criticised the often poor body conditions, the wounds, or the stereotypies. But students working for one or two months with captive elephants became accustomed to such characteristics quite fast. The abnormal became the normal to them.

Weaving is considered by ethologists as a stereotypy, hence as an abnormal behaviour (Section 5.4). For western people associated with circus as well as many people associated with zoo it is normal that their elephants weave. Another even more primitive superstition concerns the treatment of recalcitrant bulls. "A good beating is still the best remedy to treat them", was heard from South Indian elephant veterinarians at the recent Trichur meeting on the management of captive elephants in 2002. Some European zoos are of the same opinion and spare no expense flying in American experts to subdue young bulls by means of beating and electric goad (Haufellner *et al.*, 2003). People using such cruel methods live under the fiction that man must always be at the top in the rank order of captive elephants, unaware of the facts that captive bulls pay back at the latest when they are in musth, i.e. when they temporarily reach the position of absolute dominance and selectively attack the men that dominate them when they are not in musth. Wise mahouts leave their bulls in musth alone and their helpers look after the animal until musth has ceased.

In many respects captive elephants are considered as domesticated animals and treated accordingly, as can be pointed out. For example in management of reproduction, where the fiction often prevails that bringing a normally cycling female to a bull would be the simple key to success. Such practices work in cattle, horses or rabbits but rarely in captive elephants. In socially highly organised wild animals like elephants, mating, parturition and raising of offsprings depend on female choice, social environment and life history. 138 successful matings in European zoos occurred only when the bull was taller than the female (see Section 5.1). Shifting females from their home places to a bull kept in a different establishment often fails since the translocated female has to first establish her social rank in the new group (Garaï, 1992) and when brought back to her original establishment she has to re-establish her rank, a process, which often induces the foetus to be resorbed or aborted (Kurt & Mar, 1996). In Nepal, where females are often translocated after mating, 9 (36%) out of 25 neonates were either stillborn, too weak to survive or killed by the mother. However, in the Pinnawela Elephant Orphanage, where mothers and offsprings grow up in more or less close contact, 8 orphaned and 3 captive born females parturiated 22 times between 1983 to 2003 and all neonates survived the first years and only one mother, Sharmi, did not accept her offspring. Sharmi was socially not integrated in the Pinnawela herd (Section 5.3).

Life history is paramount for reproductive success. After reaching the third year young elephants are encouraged by older females to care for neonates and infants. This facilitates the duties of mothers and enables maturing females to learn vital maternal skills before becoming mothers (Section 7.2). The importance of an undisturbed genesis of maternal behaviour can be shown from experiences in European zoos. Of 63 reproducing females 29 were weaned from their mothers early in their life, i.e. at the latest in their third year. 34 were separated at the earliest in their fourth year or not at all. On reaching maturity the early weaned group gave birth to 51 offsprings. Only 33.3% of these offsprings were accepted by their mothers, and 67.7% were stillborn, killed or rejected. In the second group with no or late weaning 80.3% newly born were accepted by their mothers and only 19.7% were stillborn, killed or not accepted. This comparison is only one of the many similar observations to show the importance of a basic knowledge on natural behaviour for a successful management of captive elephants.

## Alternative keeping systems: Elephant Parks and Transit Homes

2000 years ago the *Arthashastra* proposed the establishment of *Mrgavana*, a fenced park for (tamed) elephants and other wild animals. This idea should be adopted again. Large areas of at least a quarter of a km² should be fenced and furnished with adequate water sources, shady places etc. to become a home for orphaned and otherwise problematic (e.g. dangerous, surplus) elephants. Elephants kept in such elephant parks have to be fed and regularly checked by veterinarians. But, otherwise contacts between men and elephants should be reduced to a minimum. The Pinnawela Elephant Orphanage may become a good example of such an elephant park after renovation although the area is relatively, small. In Spain a large park is currently being established for surplus zoo born Asian elephant bulls. There are already two large parks in the country for problematic African elephants (Cabarceno and Reserva Natural el Castillo de las Guardas). South Africa has tremendous experience on elephants kept in fenced range areas and it would be worthwhile to learn from South African experts (e.g. Garaï, 1997, 2001, 2002).

It is believed that well-maintained elephant parks, orphanages, and transit homes will be the future establishments to keep Asian elephants in captivity, since they allow the keeping of elephants free of chains and in social groups in appropriately furnished paddocks. This allows the elephants to express their natural behaviour patterns, and provides them with exercise and socialisation opportunities. Elephants kept in such a way convey more understanding of the species and its requirement to the visitor and insight into the necessity to protect the species than 'dancing' elephants or elephants that are forced to ride bicycles, paint canvas, play football, play the guitar or a mouth-organ.

# 7

# Base Lines and Proposals

## 7.1 Genesis of Social Behaviour

Most of the captive Asian elephants stem from wild populations. Certain capturing methods can be rather unselective like *kheddas* (the wild animals are driven in groups into an enclosure), or highly selective like *mela-shikar* (noosing wild elephants from the back of a hunting elephant) and immobilization. Out of 3073 wild elephants captured in *khedda* operations in Myanmar 8.6% were neonates and infants, 23% juveniles, 10.8% sub-adults, 30.4% adults upto an estimated age of 30 and 27.2% older adults. Selective methods like *mela-shikar* and immobilisation refrain from capturing neonates as well as older adults, which are either too weak, or too reluctant, for taming, training and later for work. Of 298 elephants noosed and 714 immobilized in Myanmar 21.8% and 14.7% respectively were between 4 to 5-year-old, 42.6%  and 39.8% were juveniles, 15.8% and 14.7% were sub-adults, 17.3% and 19.4% were young adults and only 2.6% and 11.4% were older than 30 years. Some of these old adults had to be selectively captured because they were a threat to human lives and crops (March, 2002).

Out of 438 Asian elephants shipped to western zoos 28.5% were in their first year of life, 22.8% in their second, 13.4% in their third and 9.8% in their fourth year and only 25.3% were older (Haufellner *et al.*, 1993, 1997, 1999), i.e. about 50% were neonates and 25% infants, when they were separated from their families and confronted with a life in captivity. More or less the same applies to the population at Pinnawela (Section 3.1.). Since capturing operations have been stopped in most of the range countries, but increasingly more wild herds are disturbed to the extent that neonates and infants have been lost and later brought to orphanages like Pinnawela. It must be expected that in the future, dwindling captive populations will be restocked mainly with very young elephants, like it is the case presently in Sri Lanka.

As already pointed out several times in this volume, elephants growing up without their families, run the risk of retarded body growth (Section 4.1), and of unnatural behaviour in many respects (e.g. Sections 4.2, 5.3, 5.4). For a species-specific management of captive elephants, it is therefore paramount to know the genesis of social behaviour in wild populations. The following pages, summarise some findings from research on 61 elephants in the Ruhuna National Park in 1968–1969 by F. Kurt and in the Uda Walawe National Park in 1998–1999 (Kurt, 2001). The comments made do not involve research on chemical signals (e.g. Rassmussen & Krishnamurthy, 2000) and bioseismic signals by high-amplitude, low frequency vocalisations (e.g. Langbauer, 2000, Hart *et al.*, 2002). The Ruhuna population lived in Block I and belonged to one clan. A clan is considered as the number of closely associated elephants with synchronized movements over time and space. The socio-ecological organisation of a clan follows a pattern of fusion and fission, i.e. most of the times the clan members disperse into several groups but for longer movements they join up into a herd. Clans are controlled and led by matriarchs, i.e. adult females with at least one already reproducing female offspring. The population studied in the Uda Walawe National Park in 1998–1999 (Kurt, 2001, Heine *et al.*, 2002) consisted of 6 clans, each with 22 to 53 members.

In Asian elephants the birth of a new member is a spectacular social event, which is announced by loud rumbling of the group which concentrates around the place of birth. The mother is assisted by other females who help to clean the neonate. The newcomer is touched and inspected by all members of the group. And in the year to come it will never be left alone but always looked after by older group members who try to stand as close as possible to it, as can be illustrated by a the Maximum Spanning Tree of nearest neighbours (Fig.7.1.1). Such protection and close contact is necessary, since in the early weeks and months some of the most simple behaviour patterns have to be formed.

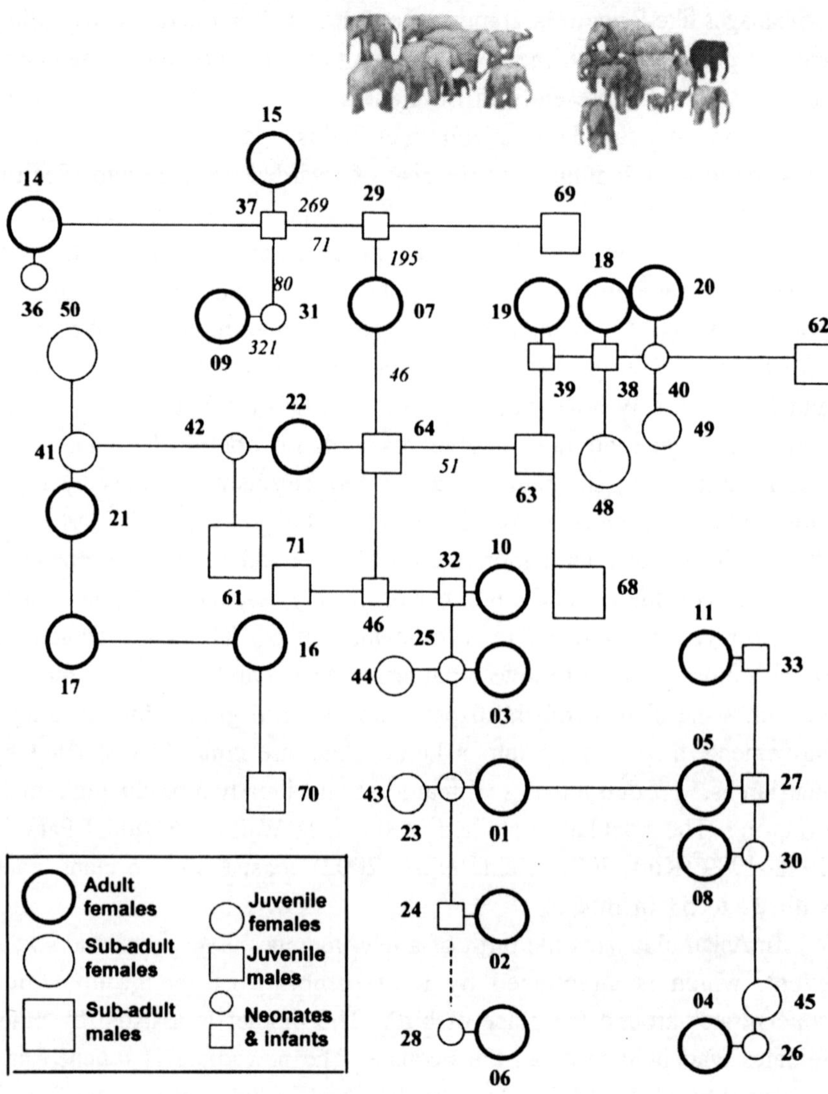





**Fig. 7.1.1** Maximum Spanning Tree of nearest neighbours in the Yala Herd (Ruhuna National Park, Block I) in 1968–69. The closer two individuals are, the more often they have been seen as nearest neighbours of each other.

## Some early behaviour patterns and activity patterns

Upto the age of about 3 months, neonates sometimes follow large moving objects, in rare cases even buffaloes, bullock carts or landrovers. Later they follow only specific, obviously known members of the group. Although young elephants suckle until the age of 4 years or even more, marked differences appear with age (Fig. 7.1.2). Regular suckling, i.e. 10 to 14 suckling bouts of 25 to 40 seconds per hour, occurs only until the age of about 3 months, later the suckling rhythm slows down to one event in 2 to 3 hours, and suckling bouts become shorter (15 to 20 seconds). The mother often ends suckling with a forward step, moves a few steps and with low rumbling sounds and loud ear flapping invites the neonate to follow her. Finally she waits, allowing another short suckling. After the age of about 2 years suckling still occurs but only exceptionally. By this time the mother is generally pregnant again.

Young elephants always suckle with their mouth. At a young age they also drink or pick up small objects with their mouth. At the age of about one week they try to use their trunk but they can only begin to use it effectively at the age of about one month. Upto the age of about 5 months, neonates increasingly take small objects into their mouth, chew them but rarely swallow them. They regularly eat faeces of conspecifics. This behaviour most probably is important to incorporate necessary micro-organisms to the digestive system. They also increasingly manipulate objects and already at the age of 2 months, they try to prepare food which is done more or less inefficiently. During this time they are allowed to snatch prepared food from older elephants. Efficient food preparation begins at the age of about 30 months (see Section 4.2).

All neonates and infants of the same mother-offspring group synchronise their activities, they play together, move together, and sleep together, often in a bunch lying over each other. The genesis of common behaviour patterns such as using the trunk, taking up objects or sleeping will take weeks and months (Fig. 7.1.2), and accordingly, neonate and infant elephants have a different activity pattern than older ones. Offsprings upto the age of 3 years sleep and play a lot, but spend little time in searching and preparing food and feeding, as they are suckled or offered prepared food. Adults, however, spend most time in searching and

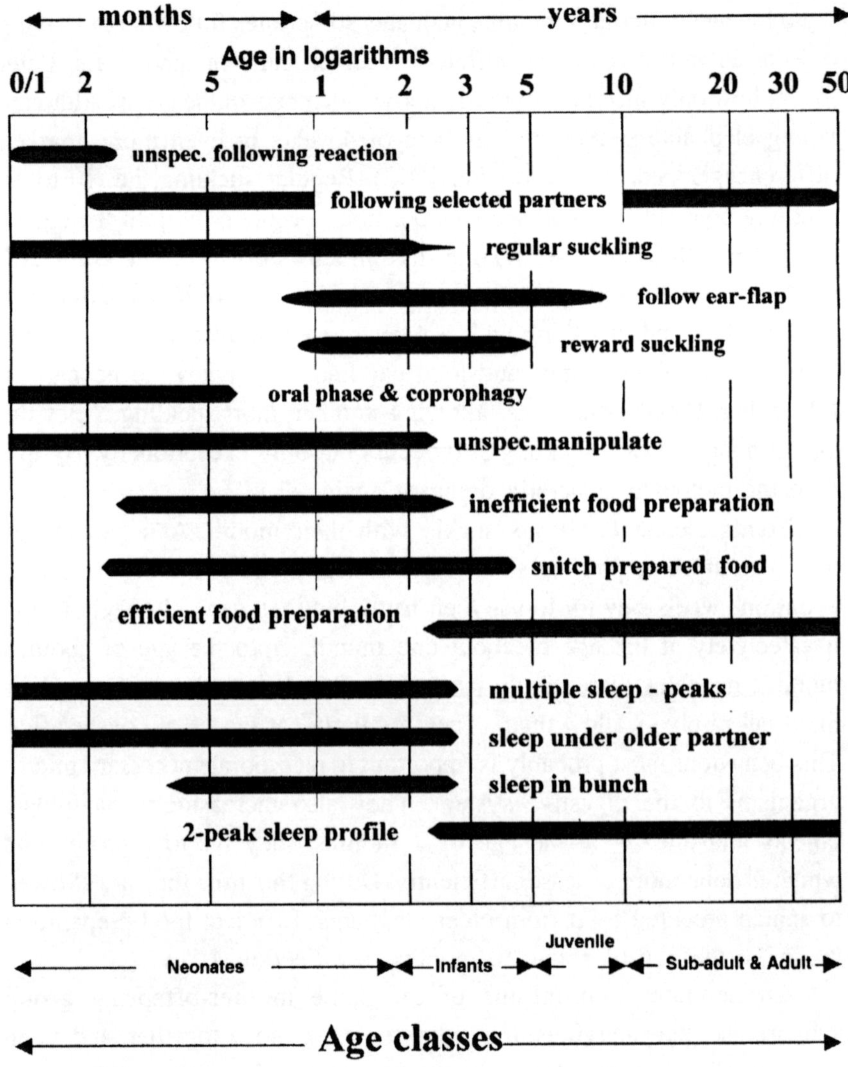


**Fig. 7.1.2** Genesis of some behaviour patterns in young wild Asian elephants in the Ruhuna N.P.. The black lines mark the periods when the respective behavioural patterns are used.

preparing food and feeding. They rest during the hottest hours and never play, their social behaviour is reduced to minimum.

For the Ruhuna population, activity patterns of different social classes can be summarised as follows, for the period between 6 a.m. to 6. p.m.: Neonates and infants spend 24% of the time in suckling and feeding; juveniles prepare food and feed for 43% of the time, sub-adult females during 47% of the time and adult females during 42% of the time. The time spent for bathing and drinking is similar in all the classes and encloses 16% to 21%, but neonates and infants spend about 17% in social behaviour and playing. In juveniles, these activities take place for 10% of the time, in sub-adult females for 3% of the time and in adult females for 10% of the time. The rest of the time is used for sleeping and resting (Kurt, 1992).

Elephants depend on a rather inefficient digestive system (see Section 4.1). Furthermore, sub-adult and adult females are either lactating or pregnant and have to economise their energy, which can be done by several means. They can reduce aggressive behaviour and choose the best possible feeding grounds and utilise the resources by using the shortest routes between the places for feeding, drinking, bathing and resting. This can be achieved only with leading positions in the dominance hierarchy and under the assumption that at least one member of the group has a profound knowledge of the home range and the whereabouts of optimal resources at a given time of the year. Furthermore, they can reduce mothering behaviour to minimum, if they can take for granted that the vulnerable neonates and infants are cared for by members of the group that are under less physical stress. They can stay in groups, where all sub-adult and adult females follow a more or less synchronised rhythm of reproduction (Kurt, 1992). Such a complex social organisation can only be achieved and maintained by an adequate genesis of social behaviour.

It can be expected that protective and maternal behaviour towards the rare and vulnerable neonates and infants has evolved to optimise the reproductive success of the individual. Sexual as well as dominance and submissive behaviour has developed to optimise partner choice and to establish and maintain hierarchies. Behaviour patterns related to social organisation in space and time have been formed in such a way so as to bond certain social units, to avoid contacts with certain other units and to use ecological resources in an optimal way.

## Care giving and allomothering behaviour

Here, protective and allomothering behaviour is discussed (Fig.7.1.3). At the age of about 3 months neonates start to play with siblings, i.e. they touch or run after each other. Lying down by a young animal triggers play behaviour in other youngsters and they frequently climb on top of each other. Infant and young juveniles are irresistibly attracted to neonates they play with, frequently and gently. The first patterns of allomothering behaviour start at the age of 3.5 to 4 years, i.e. at about the time when the mother gives birth to another offspring. They follow the neonate in order to look after it and bring it back to the herd, if necessary. This gives time to the mother to eat. They gently touch and inspect the neonate and probably indicate their interest in the neonate. This in turn gives the neonate a sense of security. They hold the neonate either by its leg or trunk and prevent it from running away. They place their trunks over the back of the neonate and pull it back, thus preventing it from running away, or they even lead it back to the mother. Furthermore, they often touch the mouth and genitals of neonates with their trunk tip. This behaviour starts at the age of 4.5 years, but it is most frequently displayed by mothers, and most probably provides olfactory clues to the state of health of the neonate.

At the age between 5 to 10 years the most care giving behaviour is that of juvenile females. This includes looking after the neonate when it is recumbent and sleeping and is most vulnerable. Another typical form of care giving behaviour includes helping the neonate and infant up steep banks, over large logs etc., or otherwise helping infants that are stuck in mud or in deep water. Later, they clean the neonate from mud, sand or leaves.

Infants frequently attempt suckling the teats of their non lactating allomothers, especially if these are older and larger, and they will do so particularly after a stressful situation. Shortly before the age of 10 years, females allow such comfort to infants while suckling. Normally mothers stand for suckling and put one leg forward, so the infant can reach the

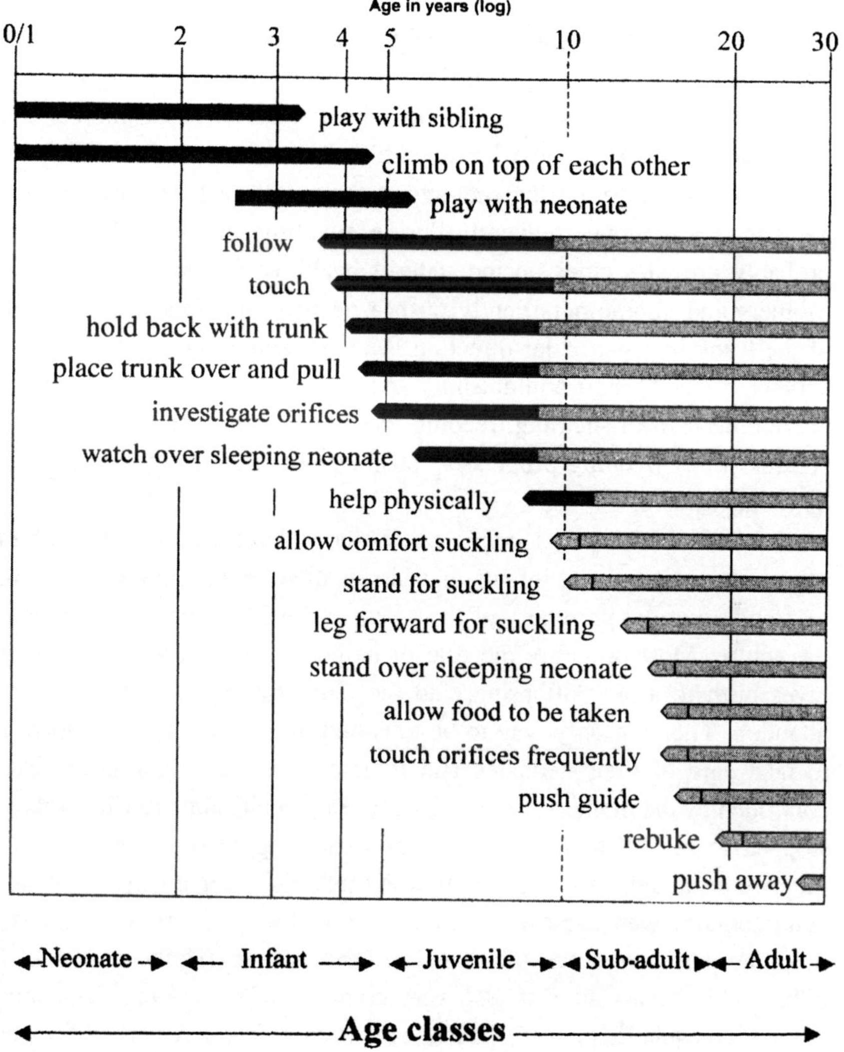



**Fig. 7.1.3** Genesis of protective and nursing behaviour in wild Asian elephants in the Ruhuna N.P.. The black lines mark the periods when the respective behavioural patterns are used.

teat for suckling. This is a behaviour that has to be learned by the first-time mother. This behaviour can sometimes be seen in nulliparous older allomothers, who probably learnt it through observation. Standing over a recumbent sleeping infant is typically done by care giving mothers or allomothers, thus protecting the infant from danger and also from sunshine.

Young elephants are allowed to take food from their mother and allomothers, thus learning what is edible. Typically a mother will touch her infant frequently during suckling at the mouth or face. Inspection of genitals is also done frequently during any time of the day and most probably provides clues on the state of health of the neonate or infant. Mothers and allomothers gently push a neonate or infant with the help of the trunk in a particular direction. Mothers and allomothers that have another offspring start admonishing and rebuking the first (older) infant, preventing it from suckling, feeding too close or even getting too close. As this (first) weaning progresses, pushing the infant away can become more aggressive.

Juvenile females and males can be considered as 'helpers', which play with neonates and infants and hinder them from leaving the group by standing at the periphery of the group and leading the babies back to the centre. Most probably the role of helper is formed, when a mother gives birth to a new offspring, and the older sibling loses its mother's attention. Then the only way to be tolerated in the vicinity of mothers is to take care of their neonates and infants. This behaviour is however controlled by the mothers: for instance, when juvenile bulls in Uda Walawe stopped to play with infants or even became aggressive towards them they were immediately attacked by the mothers. Older juvenile and sub-adult females were observed while offering food to certain infants by dropping particular small food parts in front of them (Heine *et al.*, 2001, 2002). This behaviour was also very common in reproducing elephants in the Pinnawela Elephant Orphanage (Section 4.2), where certain females allowed others than their own offsprings to suckle (Section 5.2). In wild populations such allosuckling was seen only in rare instances.

## Social organisation over time and space

In the Uda Walawe National Park the average daily distance covered by mother-offspring groups was 4.5 ± 1.9 km with extremes of 1 km and

9 km, respectively. Groups with at least one newborn offspring showed daily routes of 1 to 5 km with a mean value of 3.3 ± 1.1 km (n = 23). Groups without neonates daily covered a distance of 1 to 9 km with a mean value of 5.6 ± 1.8 km (n = 26). During their daily movements groups tended to 'navigate' from one water hole to the next. When groups of two different clans came close to each other, both changed direction in more or less sharp angles of movements and returned towards the centre of the actual home range. When resources were exhausted rapidly due to man made grass fires in Uda Walawe, old matriarchs left their groups, met with matriarchs of other clans and explored larger areas on their own. After they returned, they led their social units to new feeding grounds (Reimers *et al.*, 2002). Due to their age, old matriarchs have followed the seasonal movements of their herds many times in their lives and must have experienced rare irregularities in the availability of resources due to exceptional weather conditions (droughts, floods), fires, or forest clearings and have learnt from older conspecifics how to cope with them. Otherwise, we could not explain the fact that elephant herds suddenly appear at water holes or salt licks, where they are not seen on a regular yearly basis (Kurt, 1993; Sukumar, 2003).

Leaving the family group is the last event in the genesis of behaviour patterns concerned with social organisation over time and space (Fig. 7.1.4). The first behaviour pattern elephants have to learn is, following the mother, whom they recognise optically and acoustically at the age of about one month. Already in one-year-old neonates the mother ends suckling bouts in 4 out of 5 instances by moving away from the offspring with loud ear flapping and rumbling. The neonate follows and gets rewarded with the opportunity to continue suckling. In most cases, the neonate is followed by another neonate or infant. The third to follow the initial forward movement will be the mother or allomother of this second young one. Older offsprings are no longer suckled but rewarded with prepared food. As is known from some experiments with 3–4-year-old captive born infants in Mudumalai (Tamil Nadu), they recognise the rumbling sounds of their mothers at a distance of 1.5 to 2 km (Kurt, 1992).

Rumbling by mothers and allomothers leads to concentration of the groups. It occurs for example, when groups of two different clans come close to each other, when a parturition takes place, or when a potential

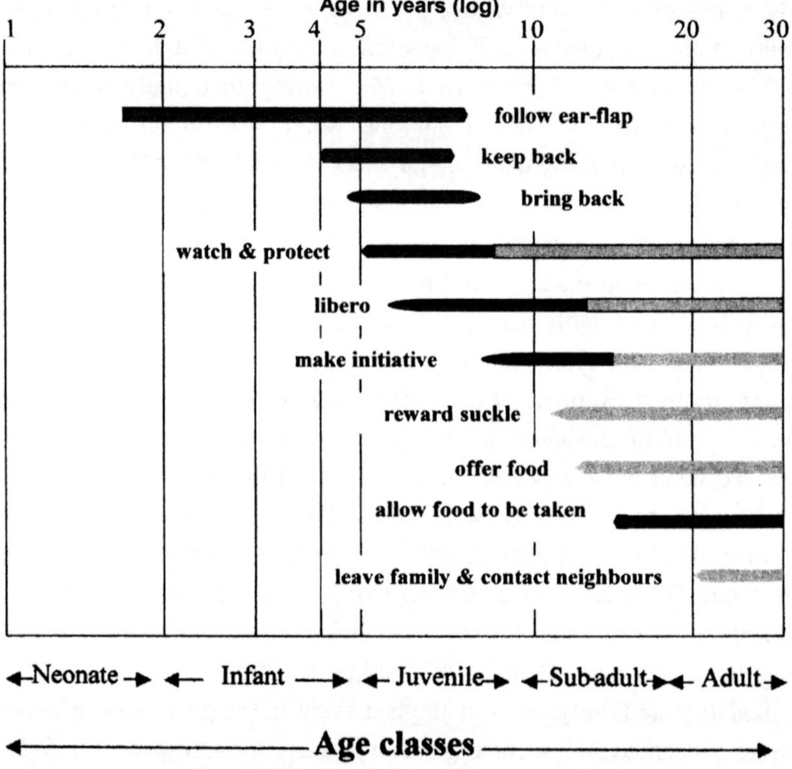


**Fig. 7.1.4**  Genesis of behaviour for group movements in the Ruhuna N.P. The black lines mark the periods when the respective behavioural patterns are used.

danger is expected and the fusion of group members is arranged in a typical 'hedge hog' formation, i.e. the oldest and strongest members face the potential danger while the vulnerable neonates and infants stand behind them protected and kept in line by juvenile and sub-adult helpers at the rear (Fig. 7.1.5). Fusion of group members is an initial step to shift from one activity centre to another. Feeding, drinking, bathing, wallowing or resting places are connected with a well maintained system of paths, which are left only in very rare occasions. The possibility to move from one activity centre to another one is therefore restricted. Sub-adult or younger adult females often start to initiate shifts by marching away from the group in a certain direction. If this is in agreement with the intention

## Spatial organisation of groups

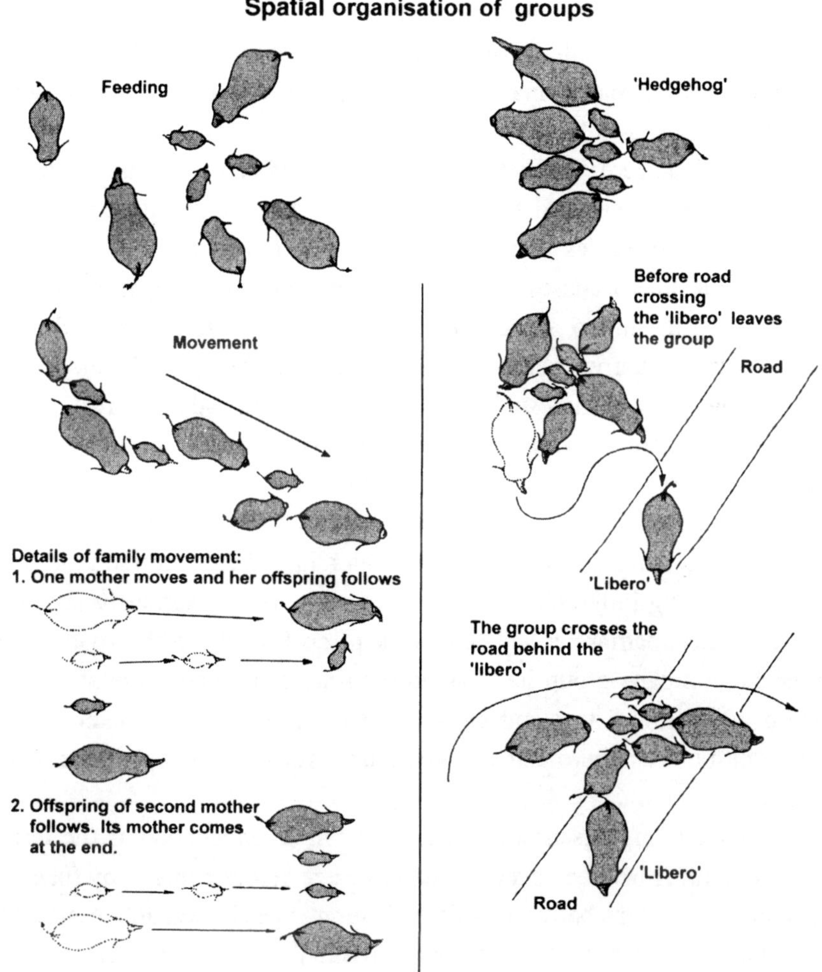



**Fig. 7.1.5**  Some strategies in the spatial organisation of a wild group of Asian elephants

of the martriarch, she will follow with the rest. If not, she starts her own initial march in the desired direction but later allows a younger female to head the column.

When a mother-offspring group is confronted with an insecure situation its members concentrate in the 'hedge hog' formation, the older members rumble and occasionally, beat their trunks on the ground with a hollow thud. Matriarchs keep younger group members in line with their trunks and occasionally, even feed them by putting a trunk full of grass or branches in their mouths. However, one specific member, often an older juvenile or a sub-adult male and occasionally a sub-adult or an adult female is allowed to leave the formation and approaches the area, where the potential danger is suspected, alone. This individual was termed 'libero'. The potential danger can be a carcass of a deer or buffalo or a water hole, where crocodiles can be expected or a road with regular tourist traffic. This libero chases away crocodiles and when the group has to cross roads, the libero attracts the attention of tourists and their drivers and wards off tourist vehicles or even attacks them, while the group crosses the road unmolested at a nearby spot. In Uda Walawe, many instances of herd movements being seriously blocked by 10 or more vehicles were observed. One of the most efficient liberos at the time was an almost sub-adult young male. After mock charging the tourist vehicles, he wanted to join a part of the group, which had already crossed the road and was resting in the shade of a nearby tree, he was seriously attacked by two older matriarchs and sent back to do his job on the road. Only after the complete group had passed the road and reached the shade the libero was allowed to stand close to them. The same bull later served two adult bulls as libero and attacked tourist vehicles when the older ones waited to cross.

The social organisation of wild bulls can be understood as the result of play-fighting between peers of the same age before puberty, by forming special relationships between bulls of different age classes, and afterwards by establishing a time niche for an individual musth period of absolute dominance of the adult male concerned, as well as by olfactory and acoustical marking, and by maintenance of a strict dispersal pattern in space and time. Hence, the social organisation of bulls depend on well established relationships between its members as well as on a fundamental knowledge of their own range. In South India and Sri Lanka translocated wild bulls often did not establish themselves in the new male society and showed a high rate of leaving the new environment and walking back to their former ranges displaying high homing capacities (for details see Sukumar, 2003).

## Dominance, submission and sexual behaviour

An adult bull in musth is absolutely dominant and mother-offspring groups are more or less permanently followed by a musth bull, who keeps other bulls at a distance from the herd. Aggressive encounters between a growing up juvenile or sub-adult males and other members of the family group are very rare. Even if such encounters occur between infants or juveniles in the group and a cheeky young male who has already smelled the freedom of being a male, the molested young group member immediately seeks closeness of its mother, allomother or even the matriarch. The young male will respect the more dominant female and refrain from further attacks, even if the youngest shows threat behaviour.

Furthermore, bulls follow a different ecological strategy to female – offspring groups,which are conservative in their choice of habitat and therefore, often live in ecologically completely different areas. Bulls often live in areas like swamps, steep hills or other higher risk areas. Fights between bulls are rare occasions and encounters normally end with threats and submission. Adult females already dominate other members of their group by their impressive body size. They control the movement of their groups over time and space. They initiate specific social roles and, if necessary, physically compel, and later reward, by using strategies such as reward-suckling, offering food or allowing access to food and shelter. Aggressive encounters between group members of different age and sex are rare and end after some pushing and shoving by the older, and the submissive retreat of the younger one.

In rare direct encounters between two matriarchs one touches the teats, the armpits and the tip of the trunk of the other and places the tip of the trunk into the mouth of the partner. Possibly these patterns are ritualised forms of suckling, offering food and being fed by others. The mutual 'placing the tip of trunk in the mouth' means more than only a test of smell and taste, but signals friendly intentions (Garaï, 1997). It often occurs between high ranking females. In seriously aggressive encounters, dominant females fight like bulls. They push each other with their foreheads and ventral trunk bases, try to wound the opponent with sharp (since often broken) tushes and try to bite its lower lip with their molars. If the loser suddenly turns around submissively and presents the hind quarter to the biting opponent, it can happen that a part of the tail is bitten off or at least broken. In Uda Walawe, where man-made fires lead to dense concentrations of different clans, one found 2 matriarchs with tails bitten off and several ones with broken ones. In Ruhuna, where ecological conditions never led to extreme concentrations such incidents never occurred (Kurt, 2001).

Adult females are dominant over juvenile as well as sub-adult bulls. They chase them away either when they are in oestrus themselves or when they try to approach a younger female of their group to test and mount, or even to mate with it. But a close relationship between a certain adult and certain younger bulls may render social advantages also for the younger bulls, as shown in the following example: an adult female in oestrus rejected very aggressively the mounting attempts of a young sub-adult male until an adult bull attacked her seriously. In due course, the female allowed the sub-adult male to mount, which was immediately followed by mating with the adult bull. Obviously, there was a special relationship between the adult bull and the newcomer to the reproducing scene, and it would be interesting to know, whether maybe, the father had helped his son. According to the observations, successful mating is always carried out by adult bulls that are either shortly before or at the beginning of their musth period. Female choice in favour of relative old males seems to be adaptive in a long-lived species like the Asian elephant. However, females in oestrus gradually attract upto 7 juvenile males which follow their group. When mating with a dominant bull finally takes place, pandemonium (Moss, 1988) starts and young bulls, long before their own puberty, mount infant and juvenile females with an erected penis (Kurt, 1975, 1992, 2001). Erection of the penis occurs already in the first days of life and takes place for instance while urinating or suckling. And, by the way, the erected penis is one of the best criteria to detect the sex of neonate elephants.

## A comparison between social behaviour in wild and captive elephants

In wild elephants parturition takes place for the first time at the age of 8 to 14 years, but males have to wait for successful reproduction until they are able to establish a time niche for their own musth periods, which may happen after they pass the age of 25 years. Females older than 20 to 25 years gradually become leading matriarchs of a clan and their dominance, social and ecological experiences increase with age. Speaking of the genesis of social behaviour and the acquisition of social roles, it can be summarised that a young mother had the opportunity to learn how to enjoy a playful well protected childhood being cared for by older members of the group. Later she learned under a 'tight trunk' by older females how to care for neonates and infants and that good social behaviour is rewarded with mother milk or, well prepared food. In all these years she had several opportunities to watch parturitions and met all 20 to 60 members of her clan, since mother-offspring groups occasionally exchange their members. She had also met numerous juvenile, sub-adult and adult males, who had contacted her for a short while, when the temporarily dominant musth bull was not close by.

Furthermore, a young mother had, for 8 years in her life, followed the seasonal movement of the clan and acquired the knowledge of the whereabouts of feeding grounds, waterholes, wallows, resting places and the system of paths connecting them. Probably, she had already met with an exceptional drought, which happens in Ruhuna and Uda Walawe about every 8 to 10 years, and followed her clan to a far off ecological exile until resources in their home range recovered. Matriarchs had many more chances to increase their social and socio-ecological knowledge. They had actively helped their daughters deliver their babies. They knew the matriarchs of neighbouring clans from occasional, often aggressive encounters, as well as how to handle exceptional weather conditions by retreating into rarely used ecological sub-optimal exiles until conditions improve.

In males the genesis of social behaviour follows more or less that of females upto the age of 4 to 5 years (when his mother has a new baby) which means: a wild bull knows since childhood, and remembers life-long very well, that neonates and infants are important fragile members of the community and, being gentle to them is rewarded by mothers, allomother and matriarchs. However, infant and juvenile males are much more explorative than females. They often try to sneak away from the group on their own. Dangerous encounters like being bitten by a cobra or slipping down a crevice is typical for young males in Sri Lanka (Kurt, 1992).

When mating takes place in their family groups, young males of 4 to 5 years imitate the mating bull and mount infant and juvenile females. As they grow up, bulls gradually leave their family groups, they learn through play-fighting to weigh out carefully between attack and retreat, they learn to establish a position in the rank order, and follow older bulls.

In many captive populations social behaviour seems to be present only in rudimentary form and its genesis is often hindered by lack of adequate social partners, space, species-specific ecological niches, and the permanent interference by owners and keepers. Captive born timber elephants have the best chance to grow up in a more natural social environment, since they are faced with a heterogenous, regularly reproducing society including even wild-living conspecifics. In contrast to this, intensively kept temple and circus elephants have to spend their life in chains and they end up as more or less insane, asocial, permanently stereotyping zoombies – as living dead. Some western zoos still keep their elephants more or less like circuses do. Others have improved and have tried to find species-specific solutions. Nevertheless, it must be stressed that even best zoos cannot completely imitate natural conditions, neither concerning the number and the diversity of social partners, nor the space or the environmental diversity or stimulation.

At the end of 2005 elephant groups in 109 European zoos have a mean size of 4.1 members (minimum 1, maximum 17) (European Elephant Group, 2003), but the mean size of temporary groups in Uda Walawe enclose 12 members (minimum 2, maximum 61) (Kurt, 2001) and each elephant has social contacts with many more conspecifics belonging to his or neighbouring clans.

Enclosures in 109 European zoos have a mean size of 0.8 ha (minimum 0.015 ha, maximum 20 ha). In wild elephants the temporary home range used for a few weeks measures between 1100 ha to 1800 ha, and the yearly home ranges between 4000 ha to 6000 ha. Of course, wild elephants are actually using only certain parts of their ranges, such as feeding grounds, baths, wallows or dry and shady resting places. But many zoo enclosures neither have a species-specific architecture, nor provide even the basic social and ecological environment.

Most ethological studies on captive elephants stem from zoos and circuses and reveal that a number of behavioural patterns like partner choice, spatial orientation or fighting, are potentially still present, although living conditions are often far from optimal. Like wild conspecifics, captive females select partners taller than they are for mating and some of them care for young ones and even nurse them without having had own offspring before. Like wild conspecifics the elephants from the Swiss National Circus show extreme capacities in spatial orientation in their yearly range. The circus elephants remember for instance, their way from the railway station to the place where the circus builds up its tents in 60 Swiss towns (Kurt & Knie, 1980). Zoo and circus elephants are involved in serious accidents with humans at the same age, when their wild conspecifics increase their aggressive behaviour. In 121 serious or deadly accidents in Europe and North America bulls were involved in 26%, although the bull fraction is hardly 10% of the captive population (European Elephant Group, 2002, 2003). 62.5% of the 40 recorded accidents happened with young adult bulls, i.e. those establishing their individual musth period, and 35.0% with sub-adult bulls, when first signs of increased aggression appear. In 81 accidents with females, 64% occurred at an age, when wild females establish and maintain their roles as matriarchs.

Humans are either attacked like conspecifics, i.e. with forefeet kicks, trunk beats and biting, or like large predators (e.g. crocodiles), i.e. taken with the trunk and either thrown to the ground or beaten against an abutment. It is still an unanswered question, whether mahouts, keepers and trainers are considered as socially high ranking conspecifics or not. At a first glance, the fact that bulls in musth selectively attack their mahouts, may advocate this opinion. However, wild bulls who had experienced being shot at in paddy fields by humans, will later, when in musth, attack all humans (e.g. Kurt, 1992). It has long been known that female elephants and the neonates they care for, form close bonds to the exclusion of other individuals. The suggestion has been made frequently that ungulates, elephants included, rapidly form such bonds through a process similar or identical to imprinting as initially described in birds by Konrad Lorenz (1937). In elephants this seems not to be the case. Their neonates do not recognize their mother immediately after parturition and after an instant process of learning at first sight, as it is the case in geese or certain species of bovids and deer (Kurt, 1991). In African elephants the mother is individually recognised after a relative slow learning process at the age of one month, and allomothers and other group members after 3 to 4 months (for overview see Lundberg *et al.*, 2001). Furthermore, close bonds are not established only at young ages (Garaï, 1992).

However, once established, close bonds between elephants or elephants and humans, seem to remain life-long (e.g. Kock, 1994). Even captive elephants recognise bond partners after year-long separation, by testing chemical signals in the urine, as experiments by Rasmussen and Krishnamurthy (2000) with timber elephants in Tamil Nadu revealed. Plasticity, in a long period of social and socio-ecological learning, is paramount in a long-lived mammal living in a social and ecological environment, where situations can change suddenly and profoundly. In captive elephants, which have been weaned at a very young age and kept out of the reach of conspecifics with adequate experience, the lack of knowledge can be fatal. Their hereditary behavioural equipment does not even provide them with information such as how to efficiently prepare their food at a time when food intake is most important for fast body growth (Section 4.2). Furthermore, without the learning opportunity they lack the knowledge of how the offsprings or breeding partners look like and how to behave towards them. They react to such incidents either with flight or aggression. Captive elephants that grew up in an inadequate social environment and never learned how to care for neonates show a high tendency towards infanticide, as already pointed out. Young inexperienced African elephants, translocated into reserves without older resident conspecifics, may consider rhinos or buffaloes as conspecifics. Furthermore, many captive bulls are unwilling to mate but become aggressive when confronted with oestrus females.

There are at least two explanations for this fact: In many cases it can be assumed that a captive bull has never had the opportunity to watch mating behaviour at a young age and practice mounting with young females and learn accordingly. Moreover, elephant trainers try to keep the bulls docile through beating (amongst other areas, the penis, the penis sheath and the perineum), it is also a common practice with intensively kept bulls in Asia. Elephant bulls, maltreated in this way, usually refuse to mate (Kurt, 1992, 1995; Schmidt *et al.*, 1992). For example: out of 10 older bulls (15 years) in European zoos, which had never lived in a circus, 7 mated successfully. 2 former circus bulls, younger than 15 years (when they began to show strong aggression towards man), were moved to a zoo where they mated, successfully. Out of 6 bulls, which were kept in circuses until the age of 15 years and most likely subjected to severe beating, 5 refused to mate. The remaining bull, Siam from Circus Knie and later in the zoo of Paris Vincennes successfully fathered 13 offsprings (Kurt, 1995).

Our examples make it clear, that in elephants a normal genesis of social behaviour can only take place in a social intact group with mother, allomothers and peers of the same age, as it is possible with timber elephants in Southern India and Myanmar and, to a certain degree in the Pinnawela Elephant Orphanage. Modern zoos (e.g. Emmen, Hamburg, Rotterdam, Zurich) have learnt this lesson. Now they allow growing up bulls to be present when mating takes place and allow the whole group to be present, unchained when a female parturiates. Accidents have never happened so far.

## 7.2  The Importance of the 'Mother Figure' in Elephant Society

All mammals are dependent on their mothers for a certain period of time during which socialisation takes place, this being the important life period when the young brain develops, and the young individual learns how to behave within its own complex society and how to cope with unpredictable events and stress.

While there are species-specific differences, all mammals share the same generalized 'emotional brain' (LeDoux, 1996) that includes the prefrontal cortex, cingulate cortex, amygdala, insula, hypothalamus, brainstem and associated physiological psycho-physiological and behavioural traits (Berridge, 2003; Bradshaw *et al.*, 2005; Bradshaw, 2005; Bradshaw *et al.*, in prep). Elephants have been shown to have cognitive abilities and sharing emotions and behavioural traits similar to humans. Behavioural traits such as fear conditioning, attachment, social bonding, pain, aggression, anxiety and extinction learning are generalised features of the brain shared by all mammals that include elephants (Panksepp, 1998; Berridge, 2003; Bradshaw & Schore, 2005 in review).

Both field observations and experimental studies document that all mammals including elephants are extremely sensitive to alterations in early life. Caretaker-infant interactions not only impart socio-ecological knowledge to the young animal, but also shape emotional and cognitive capacities (Schore, 2005). These early patternings form the basic template of stress and affect regulation of behaviour throughout the animal's lifetime. Disruption to these offspring-mother (and allomothers in the case of elephants) interactions can lead to impaired ability to regulate stress and brain dysfunction (Schore, 1994). Profound disruption to the attachment bonding process, such as maternal separation, deprivation, or trauma, can produce psycho-biological and neuro-chemical dysregulation in the developing brain (Schore, 1994; Meaney, 2001), leading to abnormal neurogenesis. The absence of compensatory social structures, such as an older female, can impede recovery (Bradshaw, 2005, Bradshaw *et al.*, 2005, Schore, 2005).

There is considerable variation between species as to the time mother and infant spend in each other's immediate proximity. Weaning is often the time when spatial and temporal distances increase between mother and offspring. In both the African and Asian elephants weaning is generally at the time of birth of a new sibling (around the age of 3.5 – 4 years), however, the bond between the mother and the older offspring remains strong, and is probably among the strongest in any animal society, to last a lifetime between mother and daughters. The role of mother in elephant society, and the bonds between family members have been described for both species (African: Douglas–Hamilton, 1975; Moss & Poole, 1983; Moss, 1988). This chapter looks in more detail at the consequences of separation, or total loss of the mother, to the offspring or orphan.

## The roles of the mother, allomother and matriarch

Female mammals characteristically show care giving behaviour towards their offsprings. The term given to this type of care giving behaviour is 'maternal behaviour'. The degree and type of maternal behaviour differs between the different orders. Care giving behaviour by others than the mother has been termed allomothering behaviour, and is seen most frequently in animal groups characterized by a high degree of kinship among its members (Riedman, 1982). Both the African and Asian elephants show a high degree of both maternal and allomothering behaviour, as well as the formation of life-long bonds between female kin. Maternal behaviour includes suckling, nurturing, protection from danger, helping physically when necessary (e.g. push the calf up a steep bank), providing shelter, passing on traditions (e.g. long range seasonal movement patterns). But it also includes the provision of possibilities for the socialisation and development processes, learning opportunities not only for satisfying physical requirements but also for acquiring the necessary mechanisms to function as a member of the specific society, learning of strategies for copying with varying environmental variables.

Allomothers can provide most of the above, even including suckling (Garaï, 1997) and form a vital function in elephant society (Lee, 1987). However, allomothers are generally very young and still have much to learn and therefore, a mother figure is still required. Allomothers are nevertheless, highly important to the survival chances of an infant (Lee, 1987), and one may presume also in the socialisation and learning processes.

The matriarch of an elephant herd has additional responsibilities over and above the maternal ones. The matriarch is responsible for and protects the whole herd, makes decisions as to daily and seasonal movements, feeding and drinking places, social contacts, etc. One of the many important functions of the matriarch is to pass on to the next generation her social and environmental knowledge, which she has acquired partly through experience and partly through her mother and an older matriarch. The accumulated knowledge of an older matriarch has been shown to enhance productivity of a group. In a broader sense the matriarch is the mother figure of the whole herd. The ability to be a matriarch depends on a variety of factors, such as age, experience, willingness, individual disposition and possibly the social position within a group (Garaï, 1997).

## The orphans at Pinnawela

The Pinnawela Elephant Orphanage of Sri Lanka harbours captive born elephants, which grow up in their family units, as well as orphans stemming from wild populations (Chapters 4 and 5). The orphans were considered either to be socially integrated into the herd or not, depending on their degree of association with other group members, their possibly aberrant or different behaviour and their degree of apathy, which resulted in asocial behaviour patterns (Section 5.3). At Pinnawela all neonates, infants and juveniles born at the orphanage and who had their mothers present selected her as preferred partner (Fig. 7.2.1). The same applies for wild elephants of both species. In the Ruhuna National Park the choice of the nearest neighbour for wild young elephants was found to be as follows: Neonates and infants of both sexes have an adult female as nearest neighbour. Juvenile males (5–10 years) start to seek the company of similar aged play mates, but still remain near an adult female. Juvenile females of that age show a distinct interest in infants (this is the age when allomothering begins), but an adult female is never far and usually the second most frequent nearest neighbour. Even the sub-adult females still preferred the company of an adult female, although this is normally the age of first conception. Males increasingly tend to prefer other male company. This tendency increases with age until adult bulls over the age of 15 years leave the family unit.

The first and second choice of nearest neighbour of wild African elephants in the Mashatu Game reserve, Botswana, based on similarity values from Garaï 1997 were found to be as follows (Table 7.2.1): Neonates and infants of both sexes prefer the company of an adult female at all times, this corresponds to the data from the Ruhuna population. Juvenile females, as in Asian elephants, prefer adult females, but start seeking the company of neonates and infants. This tendency increases in sub-adult females, which prefer being near juveniles, but they stay in the vicinity of an adult female. This again corresponds to the Ruhunu data. Neonate males show an interest towards same aged peers, corresponds to their more playful nature. Juvenile and sub-adult males prefer same aged or older male company, as they slowly move away from the older females and increase their distance from the breeding herd.

The young elephants born at the Pinnawela orphanage all had their respective mothers as preferred partners (see Section 5.3). The choice of partners of the different age/sex groups was analysed. If one compares the Pinnawela-born choice to that of the orphans, there is a marked difference. Only the neonate orphaned female Sandali managed to stay near an adult female. The infant males were either near an adult female

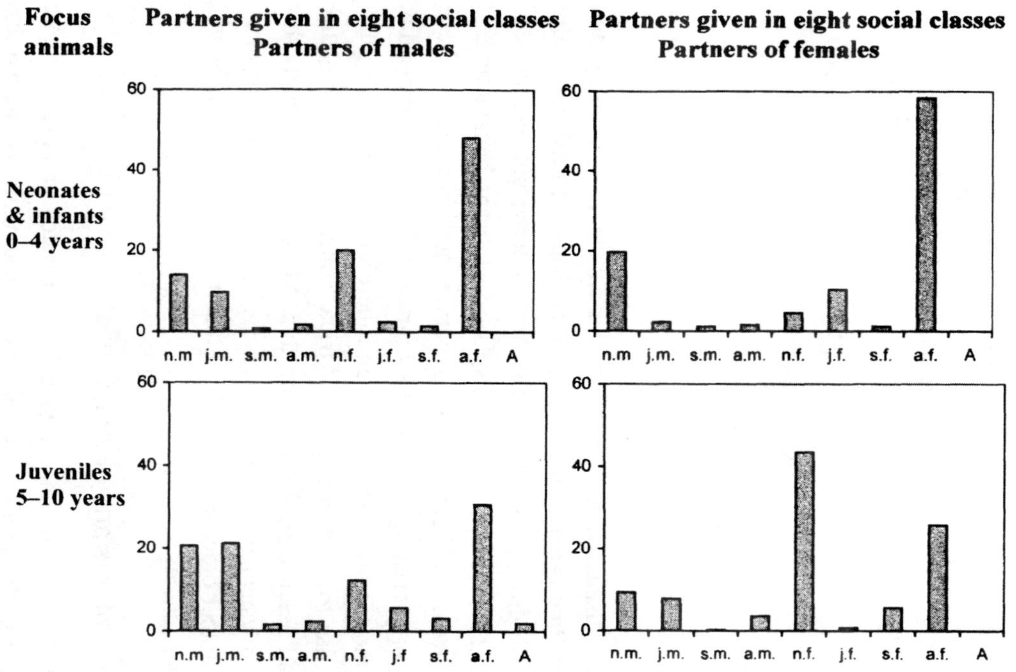

**Fig. 7.2.1** Preferred partner (next neighbour) expressed in percentage of the total of all preferred partners in wild neonates, infants and juveniles. Partner classes: n.m.: neonate and infant males; j.m.: juvenile males; s.m.: sub-adult males; a.m.: adult males; n.f.: neonate and infant females; j.f.: juvenile females; s.f.: sub-adult females, a.f.: adult females; A: alone. Data collected by F.Kurt in the Ruhuna N.P. in 1967/68 on 61 elephants (4303 scans).

**Table 7.2.1**  First and second choice of preferred nearest neighbour for young wild African elephants at Mashatu Game Reserve, Botswana.

|  | First choice | Second choice |
|---|---|---|
| Neonate f. (0–2 y) | Adult female | Adult female |
| Infant f. (2–4 y) | Adult female | Adult female |
| Juvenile f. (5–9 y) | Adult female | Infant |
| Sub-adult f. (10–14y) | Juvenile | Adult female |
| Neonate males | Adult female | Neonate |
| Infant males | Adult female | Adult female |
| Juvenile males | Young adult male | Adult female |
| Sub-adult males | Young adult male | Juvenile male |

or same aged peers, and all other young orphans were with other similar aged peers.

As has been discussed at length, in the chapters on ecology and behaviour of the Pinnawela elephants, many orphans suffered physically and psychologically from their status. Many were underweight, had retarded body growth and/or were ill. Some displayed aberrant behaviour and apathy, they showed a different sleeping pattern and they did not have the same skill for preparing food as the Pinnawela-born offsprings did, indicating that the learning opportunity and therefore, the mother figure, is lacking. Similar to wild orphans in Africa, some take on premature roles, but not all seem capable of doing this. Most of the orphans that showed severe physical problems and retarded growth did not form an attachment to another older female, although some of the females showed allomothering tendencies.

## Bonds between infants and older juveniles

Allomothers of elephant infants are usually their older siblings. As these grow up together in the family it would seem reasonable to assume that the bonds between allomother and young are strong. Whether an allomother is capable of adopting a young, if the mother dies, will depend on her age, experience and disposition, and also whether another older female in the group is present or not. Adoptions may be affected by grandmothers or older females who do not have their own offsprings, but is not a must in elephant society. As the investment of a mother in her young one is very high in elephants, an adult female is not in the position of looking after more than her own offspring, unless she has the help of other females (Garaï, 1997). In Pinnawela some young females showed great allomothering tendencies towards the neonates and not towards orphans. For example 10 years old Soma in the year 1997 was a lonely elephant, but in 1998 she did not move from 20-month-old Isuru's side. Sama II, the elephant with a severed front foot, did likewise, did not move from one-year-old Arjuna's side. Each of these calves had several allomothers.

The same behaviour was observed in wild African elephant orphaned juvenile groups. In both species more than one allomother and a mother or 'mother figure' looked after a young, and many of the other orphans were left to fend for themselves. Some of these orphans may form bonds with each other, others do not. In several cases in South Africa where juveniles (survivors of culling operations) were initially released together onto a reserve, and where family units were subsequently introduced to the reserve, the juveniles did not always join up with the respective family unit. In some cases the juvenile groups had formed their own 'family' with a leader. Not all groups however, were capable of doing this, and where the matriarch of the family unit allowed them to stay nearby, they did. Whether the juveniles join up with a family herd, probably depends on various factors, such as the leading ability of the juvenile in charge of that group, but also whether the matriarch of the family unit will accept them. Two examples may illustrate that bonds between the juveniles are not as strong as might be expected, and that releasing juveniles without adult is not always successful.

In 1998 a controversial capture of 30 wild juvenile elephants (aged 2–7 years) in Botswana and their translocation to a training facility in South Africa took place. Here, after being subjected to harsh training methods for about an year, 14 of the juveniles were reintroduced to the wild area. This introduction was not properly planned. The elephants were placed in an enclosure and released after a few days into an unknown environment, with no adult female, nor much chance of bonding with each other. They were not monitored, but left to their own devise. Although the project was subsequently kept very secret, the death rate allegedly seemed to be around 30%, and on one occasion a young elephant was seen by tourists running after vehicles.

In 1998 and 2000, in Sri Lanka 9 juveniles (aged 4 to 6 years) were released from the elephant Transit Home into Uda Walawe National Park. Some of them were collared and monitored by Muynudeen Mohamed (2002). Of the first batch of 5 orphans released in 1998, one elephant probably died. One male and one female joined separate herds and had their collars removed after 3 years. Another male joined a herd but had no collar so could not be monitored further. In the second batch released in 2000 one female probably died. The remaining 2 males and 2 females initially joined up with wild herds, but after a couple of years, 4 collared juveniles had left the herds and joined up to form a separate group, with a 6-year-old female leading them. In general, the Uda Walawe orphaned elephants were seen to be more nervous, less playful and displayed greater lethargy than the same aged wild juveniles. These behaviour patterns mirror in all respects observations from South Africa (see Garaï, 1997).

Releasing orphaned juveniles without the care of adults may result in further problems when they grow up, as illustrated by an example from South Africa. One of the problems experienced was the high killing rate of rhino by the elephants, which had reached puberty (Slotow *et al.*, 2000). Both in Pilanesberg National Park and Hluhluwe Imfolozi Park over 30 rhinos respectively, were killed by elephants, which had been orphaned in their youth and had experienced the culling of their families. Similar incidents were reported from private reserves, where young elephants, generally after reaching puberty, had killed rhinos as well as buffaloes. All incidents happened in reserves, which had introduced orphaned elephants from culling operations. These elephants experienced trauma early in their life, grew up without the social structure of the family, without their mother to guide and teach them and without the social hierarchy and presence of older bulls, so necessary to maintain order amongst the male population. Abnormal behaviour patterns such as intraspecific and interspecific dysregulated aggression, correlate with and are indicative of symptoms of early trauma, which can impede the development of the right frontal lobe of the brain, and cause dysfunction and disorders throughout life (Schore, 1994; 2001;2002; Bradshaw *et al.*, 2005; Bradshaw, 2005).

## Stress behaviour in orphaned elephants

Effects of sustained or severe stress cannot be underestimated. Stress can be a long-term inhibitor of breeding, growth and development as the adrenocorticol system is affected (Hattingh, 1986; Haemisch, 1990; Sachser & Lick, 1991). Chronic stress can induce disturbed behaviour such as stereotypies or injurious activities (Wiepkema, 1987). Stress can be induced by social as well as environmental factors. In South Africa the death rate among juvenile elephants translocated before 1994 was 40 out of 226 elephants (17.7%) (Garaï *et al.*, 2004), nearly all of which were very young and translocated without their family. One of the main causes of death was allocated to stress in combination with cold weather or malnutrition. Often these young elephants were not able to find adequate food in their new unknown surroundings. Out of a group of 12 young juveniles more or less 2 years of age, which were translocated and released on their own onto a reserve in KwaZulu Natal, 8 died. The death of 4 neonate orphans at Pinnawela (Section 3.2) could also have been a result of stress. Recent neurobiological research has shown that stress responses are first cultivated in early developmental interactions (Schore, 1994, 2002, 2005). When, as in the case of orphans, trauma and deprivation occur during infancy or adolescence, long-lasting changes in neuroplasticity, behaviour, and psychophysiology can occur in mammals whose early environment is dominated by social interactions. Further, in the absence of compensatory social structures (e.g., allomothers and matriarch), trauma is not buffered and dysfunction likely to persist (Bradshaw *et al.*, 2005). Orphaned elephants are often prematurely weaned which means that not only have they suffered the lack of social learning and normal brain development but also nutritional deprivation, that threatens physical and mental survival (Bradshaw and Schore, in review).

## The effect of an adult female on stress behaviour and bonding in juveniles

In a study done in South Africa on 4 groups of juveniles that were in a holding facility, stress related behaviours were defined. The 4 groups were kept in separate but diagonally adjacent enclosures, so that they could see and communicate with each other through the respective wooden fence poles (Group A: 4 females 6 to 7-year-old, Group B: 3 males, 5 females 4 to 5-year-old, Group C: 5 males 2.5 to 3.5-year-old, Group D: 5 males 18 month– 2-year-old). After some months an adult female was introduced into the enclosure of group C. Stress behaviour was measured before and after the arrival of this female. Although only one of the groups were in direct contact with the adult female, all juvenile groups significantly calmed down after her arrival (Garaï, 1994, 1997).

Another group of 6, 8 to 11-year-old African elephant juveniles (4 females, 2 males) had been translocated from the Kruger National Park to the same reserve and were kept together in a 2 ha large paddock for 6 months. One of the females appeared to be the leader. The elephants were released together, but the supposed leader and another female immediately split off from the other 4, and went to a different area of the reserve, although they were no other elephants there. The remaining 4 stayed together. The following year 15 more juveniles were released into the reserve, but this time together with a young adult female. All elephants stayed together and even the initial group of 4 joined up with them (but not the 2 loners). Three years later, 2 family units had acquired several adult females. It was only after 4 years, did the 2 females that had gone off separately, joined the herd. This example illustrates firstly, that an adult female has a strong cohesive effect on the juvenile elephants and secondly, that this does not necessarily imply that all juveniles will join up, but that there is individual behaviour.

Before the adult female was introduced to the 15 juveniles, which were in the holding pens for 4 months before being released into the reserve, the juveniles had formed bonds amongst themselves. After their release with the adult female, most of these bonds changed and all juveniles tried to stay near the adult. In fact, there was a great change in who was nearest neighbour to whom (Garaï, 1997). The members of original group B had been in the holding pen together for several months, before they were released with others and the adult female. The young female 'Du' had been the dominant elephant and leader of her little group of 8 juveniles in the holding facility. Female 'Mo' had been at the bottom of the hierarchy. Once they joined up with the others in the reserve however, 'Du' was relegated to the periphery of the herd together with female 'Mo' and some of the smaller males. Similarly, original group C was totally split after the release, 2 young males were near the adult female, the others were pushed out of the group to form a small sub-group, together with 'Du' and 'Mo' from group B, which sometimes split off on their own. Again, this illustrates that the bonding effect between juveniles is not as strong as expected and can be disrupted and overruled by the presence of an adult female, showing that juvenile bonds are forged more by temporary necessity than by adaptive forces.

## Other stress related effects

Stress can lead to a reduction of complexity of exploratory behaviour and to a loss in variability of behaviours in general. Play behaviour is a so-called low priority behaviour, meaning that it is not one of the survival behaviours such as food search or defence behaviour. Play behaviour occurs when no other more important behaviour predominates and when the animal is psychologically 'in the mood' for play, i.e. not stressed or frightened. Play among neonate wild elephants can typically be seen at any time of the day, when the animals are not sleeping. In two different groups of translocated juvenile African elephants it was shown that play behaviour occurred at a significant level only two years after their translocation (Garaï, 1997). In the Pinnawela elephants play behaviour was seen mainly between young males and between neonate Pinnawela-born elephants. Similarly, a group of Rhesus monkeys stopped all play after being translocated to a new area. Play is a social behaviour and needs a partner, therefore, the more social animals will show play behaviour and mainly males, or young females and neonates.

In the Pinnawela elephants, lack of play and social contact behaviour, as well as lack of exploratory behaviour was seen in those elephants that were sick, physically underdeveloped or otherwise displayed apathetic behaviour. These were nearly all orphans (see Section 5.3). The one exception of the Pinnawela-born female elephants, which showed similar behaviour patterns as the socially not integrated orphans, i.e. apathy, little social contact, no play, very little exploration, was the female Amali who was about 3 to 4-year-old. She was the daughter of Mathali, who was constantly chained. Amali had an older sister, Sama I, but she did not take care of Amali, who was left to herself, as her mother could not get to her. Early deprivation of maternal care, due to Mathali being constantly chained, must have had a profound neurological effect on Amali, so that she lacked social abilities altogether. It was very sad to see Amali, who had a mother and sister, so totally left alone and psychological deficient due to mismanagement and obvious lack of understanding of elephant sociality.

In the Ude Walewa study (Mohamed, 2002) the released orphans that joined a herd, displayed greater lethargy than their wild peers. Mohamed observed that they showed more fear of humans and tended to stay near the adult females rather than move with same aged conspecifics. This shows that orphans without bond to a mother or older female are insecure. As was seen in the African groups in the enclosure when the adult female was introduced, even the elephants not in direct contact to her clamed down. The security factor supplied by a mother figure cannot be substituted by other same aged elephants, although the group will provide more security than an elephant on its own.

## Elephants need a mother

All mammals are dependent on their mothers for a species-specific time period, ranging from a few weeks to a few years. What then is the necessary time for an elephant to be with its mother? Elephants are long-lived animals with a relatively long maturation period, meaning they have much to learn during their young lives. This also indicates that they need the time to develop the brain structures necessary to cope with social and environmental challenges. Elephants are known to suckle for a relatively long period, over and above immediate metabolic requirements (e.g. comfort suckling). This behaviour must be evolutionary, adaptive and vital for socialisation processes as well as learning and acquiring stress response mechanisms. As has been shown in previous chapters, young elephants without a mother or mother substitute suffer psychologically and physically. It was also shown that release of juveniles on their own does not necessarily result in integration into a herd, on the contrary, most will remain separate, and therefore lack the guidance and knowledge and learning opportunities provided by an adult female (Garaï, 1997; Mohamed, 2002). This manifests itself in lesser ability of food preparation, increased stereotyped and other aberrant behaviour patterns, higher stress levels, less play behaviour and less exploratory behaviour. As discussed in this book, it is evident that young elephants will always seek the company of an adult female first, and that this preference only decreases in females when they become sexually mature, and even then, they tend to stay near an adult female. Its also discussed that the introduction of an adult female to a group of juveniles has a significant effect on calming them down. Its also suggested that bonds between juveniles are not always strong and when an adult female is available she will be given preference.

Elephant males tend to be more independent from an early age and independence increases as the bull grows up. Bulls will generally leave the maternal group when they become sexually mature, with great individual and regional variation. Until such time they rely on the maternal herd for protection, security and guidance. During this period they learn how to cope and function within a male society, an important factor later in life, when they join bull herds.

Bonds between daughters and mothers appear to stay for a lifetime. Given that the emotional brain is similar to humans, one must conclude that separation must be stressful at any time to female elephants. Young adult females with offsprings usually have their own mother and grandmother in the herd, the oldest usually being the matriarch. Therefore, the whole group relies on her as 'mother of the herd', leader, protector, security factor, knowledge carrier and decision-maker. The ability to lead a group or small herd by a young female was shown by a 5-year-old in an African orphan group female and by a 5-year-old female in Uda Walawe (Mohamed, 2002), however this does not imply that all elephant females can take over this responsibility. In fact, it has been shown that same age females in other African orphan groups were not capable of this role (Garaï, 1997).

The mother figure, be it an individual mother or the 'group mother' has a very important function within elephant society. The importance of the knowledge of an older matriarch has been discussed by McComb *et al.*, 2001. The authors suggest that the presence of an older matriarch, who has the greater accumulated knowledge, enhances the productivity of the group. In this volume, it is argued that orphans without mothers lack considerable opportunities for learning, socialisation processes and show a lack of behavioural complexity, which in turn affects their health status. It was discussed that orphans that cannot attach themselves to or form a bond with another elephant show great psychological and physical deficiencies. Interestingly, an example at Pinnawela, where the 8 to 9-year-old female Menika, showed signs of stress when housed separately from her mother and siblings, although she could socialise with these during the day, shows that proximity to the mother at all times, or at least the possibility for this, is important.

## 7.3 The Management of Elephants in Captivity with Special Reference to the Elephant Transit Home and the Pinnawela Elephant Orphanage

The status of wild elephants in Sri Lanka has been estimated to be 3000 to 3500 (overview in de Silva, 1998; Kurt, 2001). The number of traditionally kept elephants in the country is about 180. These figures have serious implications for conservation and management of wild and captive elephants, respectively. If the Sri Lankan elephant is to survive in this country and if the tame elephants needed for temples, processions and tourist facilities are not to deplete the wild populations any further, three vital programmes should be in place: (1) Implication of a conservation minded management of all captive elephants. (2) A rehabilitation programme, and (3) a breeding programme. Guidelines for a better management of captive elephants have been recently published (Kurt & Mar, 2003). Rehabilitation is currently being done in Uda Walawe National Park, and the Pinnawela Elephant Orphanage is at present the only source for captive elephants in temples and private ownership. However, these programmes need to be optimised. The following pages discuss some salient points on how to manage optimal conditions. First, some aspects of reintroduction are given, then the captive elephant situation is discussed and finally general recommendations both for reintroduction and for keeping of young elephants are given, with specific reference to the Pinnawela orphanage.

### Baselines for the management of captive elephants

The following reasons can be advocated for the AsESG and other conservation agencies to become involved in the conservation orientated management of captive Asian elephants:

1   The relative high number of Asian elephants presently living in captivity, which is estimated to enclose 25 to 33% of the total population of *Elephas maximus* living today.

2   The importance of traditional knowledge of elephants and biodiversity of the natural elephant habitat by local, often tribal elephant people. Regional members of conservation agencies need to identify individuals, who keep traditional medical and other knowledge of elephants (capturing, taming etc.), and document their knowledge before it is lost. The employment of these traditional specialists in jungle or forest-based captive elephant establishments can be paramount for conservation, applied research and local politics. In collaboration with local NGOs, vocational training for mahouts' families should be set up to supplement the mahout's income (e.g. integrated farming, weaving traditional fabrics etc.); this would hinder emigration in urban areas.

3   A database on species-specific ecological and behavioural characteristics which are paramount to changing unsuitable keeping systems and improving living conditions for captive elephants should be set up. Furthermore, the formation of databases and studbooks by regions should be encouraged. The facilities of AsESG/IUCN or any other conservation agency to monitor the demography of captive elephants, can only be achieved after a proper registration of captive elephants, owners and elephant holding facilities has been established.

4   The international importance of the AsESG and other international conservation agencies must be used to set up an international veterinary consultant group to give free advice and medical care in the range countries. Each range country should sign a memorandum of understanding to share elephant database. Appropriate software should be created to enable each country to maintain computerized registration database and to exchange data with neighbouring countries.

5   The AsESG and other conservation agencies should initiate a network that could lobby the respective governments of elephant range countries to improve existing national laws in order to enforce at least the Prevention of Cruelty to Animals Act to protect misuse or abuse of captive elephants.

6   Governmental and non-governmental conservation agencies have to fight for relevant elephant keeping facilities, where elephants are kept under modern welfare guidelines but also under the considerations of conservation, such as captive propagation of genetic and behavioural diversities, conservation minded education and promotion. This goal is best reached by the points discussed further.

7   Maintenance of jungle based establishments and, where selective logging had to be stopped, local people and captive elephants working in these stations should be employed in research, eco-tourism and education.

8   In urban areas, zoological gardens should be improved so that they reach the standards defined by the International Zoo Conservation Strategy.

9   For elephants living in establishments for religious and other traditional ceremonies (intensive keeping systems), conservation agencies should encourage the improvement of elephant welfare and, if necessary, restrict or reduce the size of these populations.

Some of the most deplorable sights were seen in so-called tourist facilities (Section 3.2 and 3.3). In this modern day and age, with the available knowledge we have on the intelligence of elephants, it should be a thing of the past to see adult elephants chained for their entire life with unattended wounds on their legs and body. Keeping these elephants unchained in paddocks during their free time does not prevent training or even riding, provided the appropriate training methods are used. Veterinary care must be a pre-condition of any facility. Forced bathing to satisfy tourists should be eliminated. Riding of elephants which have wounds or that are otherwise sick must be prevented. In this respect, the tourists themselves could contribute substantially by refusing to ride on a sick or wounded elephant. Breeding programmes should also be a priority, so that the young animals can be integrated into the training programme from a young age and in the presence of their mothers. Capture of wild elephants for training purposes, whether within Sri Lanka or from other countries must stop.

10  Conservation agencies should find ways to update the scientific knowledge on elephants of range officers, managers and

representatives of animal welfare organisations. The life of elephants in their natural habitat must be the model for conservation orientated keeping systems.

11  All conservation and welfare agencies concerned with captive elephants have to find alternative keeping systems (e.g. orphanages, transit homes), where the animals can be kept under near-natural conditions, where genetic and behavioural diversity is maintained and, if suitable, where a release into their natural habitat is possible.

## Rehabilitation of young elephants to the wild areas

In Nepal, a total of 17 captive elephants were released into the wild areas in 1904, most probably due to lack of work. Nothing became known of their future. On the Interview Island (Andamans) the P.C. Ray Timber Company abandoned their 40 working elephants in 1962 when the company went bankrupt. Thirty years later, the population had increased to about 70 animals. In 1962 another herd of timber elephants was abandoned in the Chirpur Forest in the North Andamans which presumably travelled through mangrove creeks and finally ended in Diglipur Forest Division, where 30 years later at least 8 feral elephants still survived (Sivaganesan & Kumar, 1995). In recent times, one experiment with releasing 6 captive adult females and one 2-year-old captive bull into Thailand's Doi Phameung Wildlife Sanctuary produced promising results (Sukumar, 2003). In Myanmar it is easier to release captive elephants into their old home ranges ('soft release') than into unfamiliar ranges ('hard release'), and third or fourth generation young captive elephants are said to be the easiest to be reintroduced into the wild (Mar in Pimmanrojnagool & Wanghongsa, 2002).

Successful reintroductions of adult trained female zoo elephants was done in South Africa by Randall Moore (1989), who spent many months teaching the elephants how to survive. Reintroduction of orphans from the Elephant Transition Home into Uda Walawe National Park (Jayawardena, 2002; Mohamed, 2002) appear to be partially successful, although the time since rehabilitation is too short to know the long-term implications on the life of the elephants. Translocations of orphans in situ (i.e. from wild to wild situation) used to be carried out on a large-scale in the 80s and 90s in South Africa. This practice has been stopped due to the many problems that may arise, and the fact that technology to move the entire family group has been available since 1994 (see Section 7.2).

The effect that the enclosure can have on stress behaviour of elephants was shown in a study on translocated juvenile African elephants (Garaï, 1994, 1997), that were kept in an enclosure for several months prior to being released onto the new reserve. It appeared that the design of the enclosure played an important part in increasing or decreasing the stress levels in newly translocated wild African elephants. The more open the design, so that the elephant could see what was happening on the outside of their enclosure, the less stress they experienced. It was also shown that older juveniles of 5 to 7-year-old remain more nervous for longer than very young ones in a new enclosure. This does not mean that the initial stress to the young infants is not severe. In fact, they are more prone to dying of stress when very young. Older juveniles probably have had the opportunity to develop strategies to cope with stressful situations. As was discussed in Section 7.2 the introduction of an adult female was instrumental in reducing stress related behaviour in the juveniles.

Very young mothers giving birth under the age of 10 years in African elephants in Amboseli, Kenya, only had a 50% probability of their calf surviving the first year of life. For females between 10 to 15 years calf survival in the first year increased to 76%, which increases to 78.6% in 15-20-year-olds and upto 90% in adult females (Moss, 2001). This shows that young mothers are inexperienced and will require the help of older females, as well as the protection of the group. As was shown in a study at Amboseli (Lee, 1987) calf survival rate increases with the number of allomothers present. It is therefore reasonable to assume, that young females released without the family group (as would be the case in a rehabilitation programme of orphans), will have an even higher rate of calf mortality as they will have no experienced matriarch to help them, nor the protection and help of other herd members.

In the Uda Walawe reintroduction (Jayawardene *et al.*, 2002; Mohamed, 2002), the young elephants eventually joined a herd, but did not appear to form strong bonds with herd members, as they were seen to change herds. Similarly, in some instances in South African private reserves, where young orphaned juveniles were introduced, and later family units introduced to the same reserve, the two groups rarely joined them (EMOA unpublished database). Whether a young individual, or several, are allowed to join a female-offspring herd, seems to depend on the individual character of the matriarch of that family. Some matriarchs accept them, some don't. But even 2 orphaned groups in the same reserve, but introduced in separate years may not necessarily join them, as was observed in a private reserve with 2 groups of 3 elephants, 4 to 5-year-old and 5 to 7-year-old, respectively. These 2 units remained separate for many years (Garaï, 1997). An interesting fact is that the matriarch is highly likely to accept an orphan into her group if she had been orphaned herself in younger years, and a few reintroductions to such a group were successful, despite much initial scepticism (EMOA records unpublished).

One particular good example is a group of African elephants, orphaned in their young years, that were moved to a second private game reserve, due to repeatedly breaking out of their original reserve. After about one year, a lone 10-year-old female was introduced to this reserve, and not only did the other elephants accepted her, they even went to the enclosure to fetch her after she had been offloaded from the transport vehicle, although they had never seen each other before (EMOA, 2001). In contrast, a young orphaned, hand-raised male, which was to be reintroduced to a wild family group (not his natal group) on a game reserve at the age of 3 years, was repeatedly pushed out, despite all human efforts, and eventually became petrified of elephants. This male was later successfully introduced to a herd of trained elephants, where he started thriving.

These facts together with those mentioned in Section 7.2 emphasise the danger of reintroducing young elephants to the wild without their mothers or other older females or family members. If one wants to optimise the population growth rate in terms of conserving an endangered species, as well as reduce stress to the individual animals concerned, which from a welfare point should be the goal, then it is vital that young elephants that are to be reintroduced to the natural environment are given the opportunity to bond with other, especially older female elephants.

Any young animal is quick to lose its fear of humans and to bond with the humans, who provide food and comfort. There is a plethora of anecdotal evidence that a once tame or habituated animal, especially one with such a long-lasting memory as the elephant, will remain fearless of humans. This is a very dangerous attribute in view of the great human-elephant conflicts that many Asian and African countries are faced with. Therefore, reintroduction programmes should, among other vital factors, ensure that the young elephants are in an appropriate group with older females and that they are not given the chance to bridge the natural gap of fear between them and humans.

## Elephants in Captivity

Keeping of such strong and large animals in adequate conditions, and preventing them from breaking out or injuring people or themselves, is indeed a challenge. Despite this, modern zoos are moving towards a new way of keeping elephants and facilities are being built to allow the keeping of elephants chain-free and in social groups in appropriately furnished paddocks. This allows the elephants to express their natural behaviour patterns, such as preparing food, skin care, social interactions, and provides them with exercise and socialisation and learning opportunities. The most salient factors inhibiting breeding in captivity are social and psychological constraints as well as inadequate management. Animals are not genetically adapted to confinement. Adaptations will be at the phenotypic level allowing an animal to cope with the environment. This coping behaviour is a response to an aversive situation (Wechlser, 1995). Learning to cope with stress is an important part of development. However, if an animal cannot escape or remove the aversive stimulus, this can either lead to frustration, aggression or abnormal behaviour patterns, or alternatively lead to apathy. The genetic make up of an animal allows it to adapt to specific conditions for a while, but if the genetically programmed behaviour patterns are suppressed altogether due to constant chaining or confinement, this will lead to abnormal behaviour patterns, or the animal 'gives up' and is apathetic, a condition that was seen in some of the young elephants at Pinnawela Elephant Orphanage (Section 5.3).

There has been a marked increase in interest shown in enrichment programmes for captive animals, including elephants (e.g. Markovitz,

1997). Environmental enrichment programmes serve to prevent boredom and to enhance normal behaviour patterns, which are essential for a normal physical and psychological development. Enrichment programmes should have a more holistic approach than just preventing boredom and, as Müri (in prep) suggests, enrichment and training programmes have to focus on allowing the animal to realise their own species-specific adaptive and life history strategies, such as foraging strategies, social life as well as resting and body care behaviours. Animals characteristically develop abnormal behaviour patterns in systems, which restrict their movements and offer a stimulus poor environment (Wechsler, 1992). This means the animal must be allowed to utilise its genetically programmed and ontogenetically adapted behaviour patterns, even in a new environment.

Not much has been said about the importance of the development of learning in captivity. As the environment is less complex and varied, there is less necessity for the brain to develop varied responses. Experiments have shown that rats reared in a more enriched environment have a higher cerebral cortex weight and more glial cells (Carlstead, 1996). They exhibit higher motor activity and more exploration. A complex environment may also enhance the learning ability to respond to a novel situation. Learning to cope with novel or aversive situations reduces stress. If an animal is less capable of adapting to novel situations it may result in long-term stress or apathy. Stress activates the adrenocortical system. Chronic stress such as separation, loss of attachment, close confinement may result in chronic elevations of adrenal hormones or produce adrenal hypersensitivity to ACTH. Chronic stress has a wide range of negative effects, such as loss of calcium from the bones, inhibition of growth and reproduction, development of ulcers and inhibition of the immune system (Carlstead, 1996).

Social learning is also a vital part of growing up. Young females without the help of experienced mothers in the herd will not know how to look after their calves. In zoos it is well-known that young primiparous females often attempt to kill their new born calf, not quite knowing what to do with it and they have to be taught how to suckle their calves (see Section 7.1). Management has to frequently intervene and prevent the mother killing her baby (see example of Sharmi at Pinnawela as discussed further). In the wild, experienced mothers take over this function. In the previous chapters various problems with captive orphans, such as health problems and growth deficiencies, stereotypy, inadequate feeding methods or insufficient nutrition, social problems and methods of restraint, all of which can severely affect a healthy physiological and psychological development of the young animals have been discussed. The following pages have some general recommendations for (1) keeping of wild orphans and rehabilitating them and (2) keeping elephants in a zoo like facility with special reference to the Pinnawela Elephant Orphanage.

## Recommendations for keeping of wild orphans and rehabilitation

1. Orphans that are to be used in a rehabilitation programme should be kept in a group in a large enclosure. The size will depend on the number of elephants and available space, however an area about one ha is recommended (100m x 100m) for a group of upto 10 young animals. This is large enough for the elephants to move around in and escape from aggressive conspecifics if necessary, but not too large to allow easy access and observation of the animals. It is also advised that the enclosure be built in an area containing natural trees. These will provide protection from sun or rain, they can be used by the elephants to scratch against and if the branches are low enough the elephants can feed on them and play with the branches. If there are no trees available, or the elephant have destroyed them, then a covered area must be constructed. The enclosure can be divided into two or more sections, so that cleaning of one is possible while the elephants are in another section, or for the seclusion of a sick animal. A remote control system for opening and closing gates between the compartments prevents direct contact between elephants and keepers.

2. Shade is often underestimated. Experiments have shown that 80% of heat loss in elephants is through radiation to cooler objects (Spinage, 1994). Hence, standing in the shade is important.

   It is necessary that certain features be installed, if not naturally present:

3. A source of freshwater must be provided, so that elephants can drink at will and play with water, this keeps them occupied and enhances social contacts, as most play behaviour occurs in and around water.

4. They must have the opportunity to make mud-wallows. Mud is a vital form of protecting the skin against parasites and sunburn. Elephant skin needs to be kept damp and flexible. This happens mainly through diffusion within the body. Diffusion increases with higher skin temperature and raised humidity. Experiments have shown that a young 1.2 tonne Asian elephant can lose about 2.5 litres of water every hour (Benedikt, 1936). Mud has proven to be about eight times more efficient in preventing evaporation of water from the surface (Spinage, 1994), therefore throwing mud onto the body is a much better cooling method than spraying with water.

5. Sand must be available for 'dusting', which is an important skin care behaviour. Very young infants may not be able to control their trunk sufficiently to throw either mud or sand onto their body. In this case, management should do this for the animals. Sand also helps prevent sunburn and overheating.

6. Feeding troughs can be provided, but should not be near the water, as the food will become wet and foul. Each elephant should have a separate trough. If older elephants bully smaller ones and prevent them from feeding, this must be monitored and appropriate measures should be taken accordingly. One suggestion is to place the troughs outside the enclosure, so the elephants must reach through the bars for feeding. This allows control over the amount of feed and decreases bullying possibilities.

7. Branches must be given *ad hoc* at all times to keep the elephants occupied. It also allows them to utilise several behaviour patterns such as food preparation or tool use and skin care. Branches also help relieve boredom, which will otherwise ensue in such a stimulus poor captive environment.

8. Large logs should be provided for scratching or playing.

9. The elephants should never be chained. The fence of the enclosure must be robust enough to contain the elephants, but should always be constructed in such a manner that the elephants can see what is happening outside. This is important to reduce stress and allow the animal to see any approaching human or animal. It also provides the elephants with stimulus from outside and allows them to familiarise themselves with the environment in which they will be released.

Additionally, they can easily be observed from a distance by the keepers. Electric wires have proven highly successful in containing elephants even within a lightly constructed fence. These wires must be fixed to the fence with brackets, so that the live and earth wires are held by the same bracket, but at least 10 cm away from the fence. Solar energy can be used for the energizers. Maintenance of these electric systems is of paramount importance.

10. Handling of the elephants must be kept to minimum, but this will of course depend on the age (whether the elephant is weaned) and on its health state (i.e. does not need veterinary care).

11. All efforts should be made to introduce at least one or two older females to the orphans, preferably also an orphan. If necessary a female from another orphanage could be used. Of course the ideal would be an experienced mother to teach the young orphans. The animals will need time to bond with each other. Probably a few months, but this will depend on the behaviour of the elephants and should be monitored.

12. The enclosure must be cleaned out daily and all remains of dung and old food removed, as this is a focus for parasites. Concrete floors are to be avoided unless these are cleaned daily and ample provisions of hay or grass are provided for bedding to prevent sores induced by the hard ground when the elephants lie down.

13. When the time for release is considered appropriate the group including any older females, can just be released together. It is advised to collar one or two of the elephants to facilitate monitoring. If there is an adult female with the group, the elephants will most likely remain together. Release of juveniles alone, as has been experienced in South Africa and Sri Lanka, does not guarantee that they remain together, in fact it is highly likely that they will split up.

14. Monitoring of the released group should be implemented in all cases.

## Recommendations for Pinnawela Elephant Orphanage

Much of the above also applies to a facility for keeping of trained elephants. The area should be large enough for the animals to move around, shelter must be available, water for drinking and mud wallowing must be available, hygiene must be a daily concern and chains should not be necessary for keeping the animal at any time of the day or night. The elephants should only be chained for training or treatment purposes. If several infants have to be bottle fed, they could be placed separately into small pens during feeding time, or chained only for the duration of feeding, so that they learn to associate chains with something positive. This will later facilitate training. Keeping the elephants chain-free allows them to choose their social partners and will reduce stereotypies. In a captive situation veterinary care must be a priority and a veterinarian should be available at all times and check the animals daily.

Much of the following recommendations will hold true for any similar facility. The interrelationships between the factors influencing elephant well-being are depicted in the diagram (Fig. 7.3.1). The initial important factors, such as stress through capture and the psychological impact of the loss of their family, already places the orphans in a vulnerable position on arrival at Pinnawela. These factors are acerbated by inadequate medical care, too much direct sun rays, inadequate feeding programme as well as chaining and missing social partners. In turn, these factors have a negative impact on the immune system, growth and development and behaviour, which will influence social integration and breeding success and well-being of the animals.

In order to reverse this negative flow several management changes should be implemented (Fig.7.3.2). Proper medical care is paramount as is proper feeding and daily management programme as suggested below.

1. In a large orphanage such as at Pinnawela, there will be elephants of all ages, some with offsprings and some adult bulls, and therefore the elephants must be kept in separate enclosures during the night for their own safety.

2. The existing family units should be supported by allowing all female family members to be together at all times, and the bulls should not be separated from these before they reach the age of 5 years, when they have made the transition from infancy to becoming a more independent juvenile (see Section 5.3). This means that these related elephants must also be housed together at night-time. Non-kin females that display allomothering behaviour towards a particular calf, should also be integrated into these family groups. There could be two or three family groups thus formed, one for example under the leadership

## Present situation of elephant management at Pinnawela

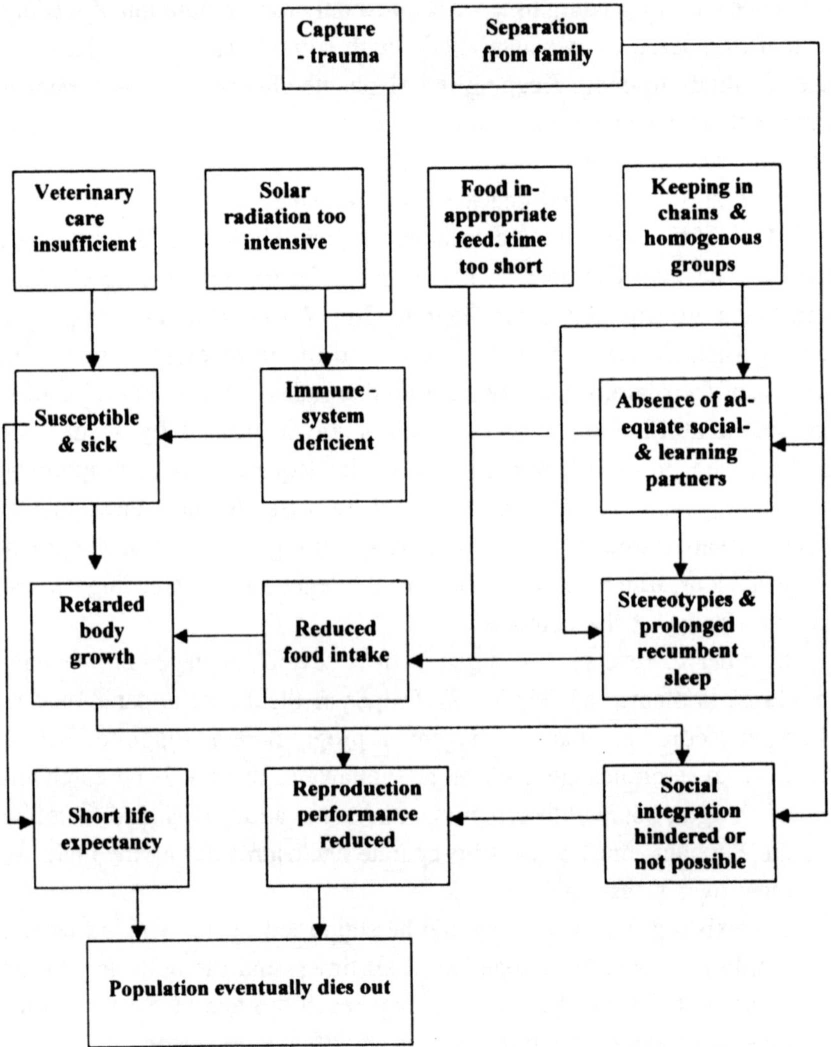


**Fig. 7.3.1** Flow chart on causes and effects of inadequate keeping conditions of elephants at the Pinnawela Elephant Orphanage. Arrows indicate the direction of effect.

**Proposal for an improved elephant management at Pinnawela**

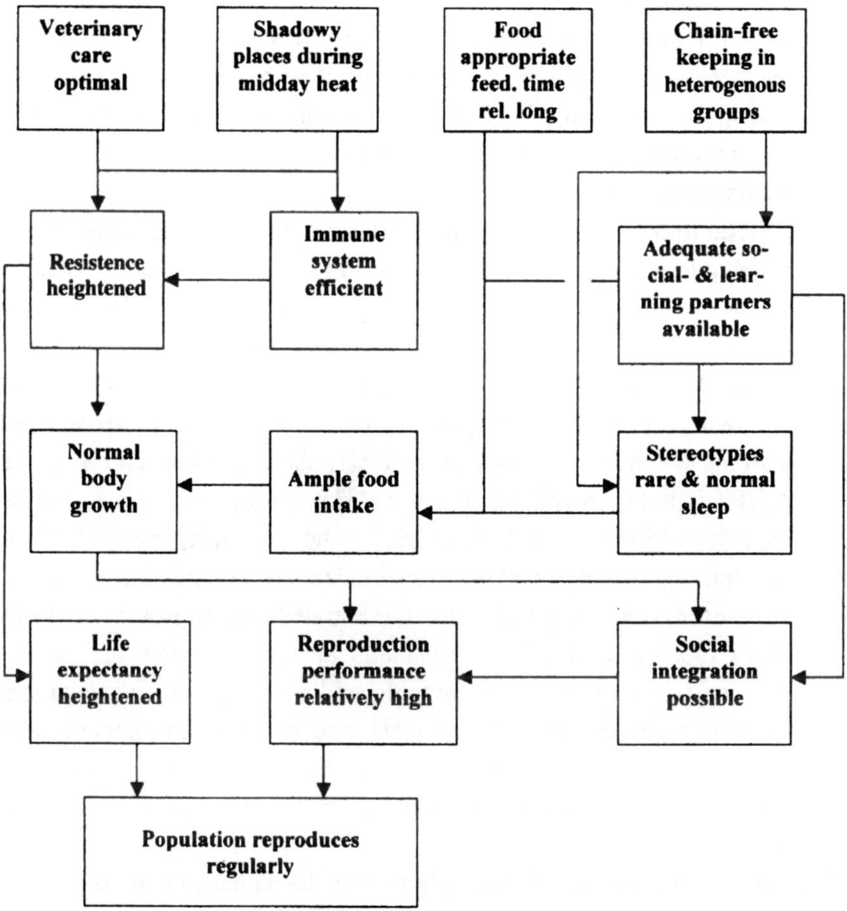


**Fig. 7.3.2**   Flow chart on causes and effects of an improved management of elephants at the Pinnawela Elephant Orphanage. Arrows indicate the direction of effect of certain conditions.

of Mathali, together with her offspring including the females displaying allomothering behaviour towards the calves. Another group could be under the leadership of Kumari, together with her offspring and their respective allomothers. A third group would be possible under Komali. The older female Anusha and the younger mothers Anuradhika and Rejina, together with their offsprings could be integrated into any of the three groups depending on their social preferences. Those females that spend much time in each other's company should be allowed to stay together at any time including the night, as they have formed bonds with each other.

The importance of integrating other young orphaned females into existing mother herds becomes apparent in the case of Sharmi. Sharmi was estimated to have been born in 1990 and was about 7 to 8-year-old at the time of the study. She was chained under a palm tree at night, alone. She had very little contact with any other elephant and not with any mother-calf group. This means she grew up without learning about being a mother. In 2001, she had a calf and attacked it. This is not surprising, as it is a well-known fact that zoo elephants attempt to kill their newborn calves, if they are inexperienced. What is surprising is that Pinnawela management ignored this fact! It would have been vital for Sharmi to be integrated into a group and learn about looking after calves, especially at such a young age. In the absence of an older female helping and teaching her, management must be prepared for such incidents, and be ready to intervene and teach the young mother how to suckle her baby. This illustrates the importance of having older females in the same group as young orphans.

3. Possibilities for social integrations can be enhanced by providing better bathing and mudding facilities. Wild elephants have shown that even asocial elephants will display some amount of social behaviour at the water hole, which seems to stimulate certain behaviour patterns such as play, tool use and body care. At Pinnawela a short rain shower provided the elephants with puddles and endless opportunity of playing, rolling in the mud, spraying themselves with mud and water and generally displaying behaviour otherwise not seen in the river, as the river bed is stony and uncomfortable for the elephants

and the only mud available for skin care is at the opposite river bank, where they are not allowed to go to. But mud is highly important to prevent sunburn, keeps the skin humid and supple and free of parasites. Therefore, a muddy water hole in the existing depression in the open area at the orphanage could be turned into a mud-wallow and bath. It would provide the elephants with opportunity for better skin care and socialisation and, resulting from this, better social integration. Bonding between elephants is important to induce security. Play provides the males with opportunities to measure and improve their strength and acquire a better position in the hierarchy. Play also provides females with opportunities to acquire the necessary social skills for later part of life. Social partners are often learning partners and thus can help the orphans learn skills, they would otherwise not learn without their mothers. Social integrations will allow the elephants to cope with the trauma of the capture and loss of family more readily. This in turn will help them overcome physical difficulties. Social integration will also allow young females to learn how to look after young calves. The role of the allomother is very important and must be promoted.

4. The open area could be divided into two fenced paddocks, with a path for visitors between them, allowing the visitors to get close to the elephants, but safely protected by a fence. The fence would allow the elephants to be kept free of chains. The current method of a mahout constantly calling the elephant and herding them together to stand in one spot is not necessary. In fact this method is detrimental to the elephants and boring for the public. Any modern zoo will try to allow the natural behaviour patterns of the animals, as has been stated before. Having to stand in a bunch, not allowed to do anything except to sway their trunks, and usually in full sun, must be agonising to such a social, explorative and inquisitive and highly intelligent animal.

5. Shade must be provided in form of one or more roofed constructions, so the elephants can choose whether to stand in the sun or shade. It was shown in  the previous chapters how intensive rays of the sun affect their eyes, skin and health.

6. Food should be provided frequently during the day in the form of large branches or freshly cut tree stems. This provides the elephants with opportunities to prepare their food in a natural manner. This will also provide the slow eaters with a chance of obtaining sufficient food. The elephant digestive system is rigged to process food continuously and not only once a day. Food preparation enhances natural behaviour patterns and helps fight boredom. It is also much more interesting and fun for visitors to watch elephants actively doing something than just standing in a group doing nothing surrounded by mahouts, who keep shouting at them.

7. The elephants should not be herded to the river more than once a day, and then only to bathe for a short while. Standing in the sun and water for hours is detrimental to their skin (sun burn), eyes (damaged by the glare) and feet (which get soaked and soft from the water) (see Section 3.3).

8. The elephants should also be free from chains at night and housed according to the respective groups. For this, there should be several fenced paddocks, each with a roofed housing section for protection from the elements. This roofed housing facility should preferably have a slightly elevated floor that can be easily cleaned from faeces and urine each morning, and that prevents the elephants from lying down in wet mud during the rains. Single chaining to a tree or tree stump should be prohibited. The elephants should not be made to stand in piles and piles of their own faeces. The urine drenched ground will cause sores on their soles and allow the germs, which proliferate in these piles, to enter the skin. Single chaining also prevents elephants from social contacts, which are important, even during the night.

9. Bulls should be housed individually and in their own paddocks, but should be allowed to socialise with the females during the day occasionally. This will promote breeding, as has been done at Pinnawela.

10. The paddock fence should be a robust steel construction so that the elephants cannot break out and can be enhanced by electric wires. To provide the visitor better viewing, the paths between the paddocks could be slightly elevated platforms. The Pinnawela tradition is to bring certain elephants close to the public, this can still be done with a couple of the more friendly young elephants.

11. Husbandry requirements include:
    - Routine checks by a veterinarian as to the health of the individual animals, the hygiene of the facility and the food.
    - Checks on the physical condition and development of the elephants (i.e. regular measurements and weight control).
    - Improved food supply for the orphaned calves (milk formula and mineral supplements).
    - Improved food supply for all the elephants (the currently provided palm fronds are expensive and not particularly species appropriate). Grass would be a more natural food source, which grows along the Maha Oya river.

# Appendix I

1.1  Proportion of maknas among sub-adult and adult males (Pma) and sex ratio (number of adult and sub-adult males per 100 adult and sub-adult females, Sm) in wild populations of Asian elephant as found in extensive field studies (I), extensive censuses (E) and non-selective capturing operations (C ). Sample sizes (n) are given as the number of adult and sub-adult males.

| Region | Period | Method | n | Pma | Sm |
|---|---|---|---|---|---|
| *Southern India* | | | | | |
| 1.  Karnataka | 1875 | C | 16 | 18.8 | – |
| 2.  Tamil Nadu | 1926–1980 | C | 217 | 1.8 | – |
| 3.  Travancore – Cochin | 1939–1943 | C | 34 | 0.0 | – |
| 4.  Travancore – Cochin | 1946–1951 | C | 31 | 0.0 | – |
| 5.  Karnataka | 1968–1974 | C | 55 | 5.5 | – |
| 6.  Satyamangalam – Chamaraj | 1981–1983 | I | 24 | 8.3 | 22.0 |
| 7.  Kerala (Ke) | 1983 | E | 641 | 32.3 | 37.8 |
| 8.  Ke: Southern Forest Circle | 1983 | E | 174 | 30.5 | 54.4 |
| 9.  Ke: High Range Circle | 1983 | E | 55 | 30.9 | 20.1 |
| 10.  Ke: Central Circle | 1983 | E | 54 | 37.0 | 36.2 |
| 11.  Ke: Northern Circle | 1983 | E | 176 | 9.1 | 51.6 |
| 12.  Ke: Periyar (a) | 1983 | E | 119 | 67.2 | 21.8 |
| 13.  Ke: Periyar (b) | 1983 | E | 92 | 57.6 | 16.9 |
| 14.  Ke: Thekkady | 1983 | E | 55 | 61.8 | 18.1 |
| 15.  Ke: Vallakaduwa | 1983 | E | 37 | 51.4 | 15.4 |
| 16.  Ke: Periyar | 1987 – 1990 | E | 54 | 57.6 | 16.9 |
| *North eastern India* | | | | | |
| 17.  Chittagong | 1874 | C | 25 | 8.0 | – |

| | | | | | | |
|---|---|---|---|---|---|---|
| 18. | Assam | 1937–1946 | C | 606 | 46.2 | – |
| 19. | Assam | 1947–1950 | C | 707 | 50.8 | – |
| 20. | Arunachal Pradesh | 1961–1977 | C | 1117 | 59.4 | – |
| 21. | Assam | 1981 | C | 67 | 46.3 | – |
| *Sri Lanka* | | | | | | |
| 22. | Yala | 1867 | E | 31 | 77.4 | 59.5 |
| 23. | Yala | 1967–1968 | I | 13 | 84.5 | 65.0 |
| 24. | Yala | 1990 | E | - | 83.8 | 67.0 |
| 25. | Yala | 1993 | E | 12 | 89.8 | 26.3 |
| 26. | Gal Oya | 1967 | E | 13 | 84.6 | 34.0 |
| 27. | Gal Oya | 1967–1968 | I | 31 | 97.2 | 37.8 |
| 28. | Lahugala | 1967 | E | 49 | 89.8 | 41.0 |
| 29. | Lahugala | 1968 | E | 25 | 92.0 | 40.0 |
| 30. | Gal Oya & Lahugala | 1993 | E | 51 | 91.0 | – |
| 31. | Mahaweli | 1945 | E | - | 100.0 | – |
| 32. | Mahaweli | 1968 | E | 27 | 100.0 | – |
| 33. | Mahaweli | 1970–1985 | E | - | 100.0 | – |
| 34. | Mahaweli | 1977–1986 | I | 229 | 100.0 | 56.1 |
| 35. | Mahaweli | 1993 | E | 147 | 98.7 | – |
| 36. | Sri Lanka | 1967–1968 | E | 190 | 92.1 | – |
| 37. | Sri Lanka | 1993 | E | – | 92.7 | 71.6 |

Sources of data: 1,17 – Sanderson (1907); 2 to 6, 21 – Sukumar (1992); 7 to 12 – Surendran (1992); 13 to 15 – Karror (1992); 16 – Chandran (1990); 18, 19, 31 – Deraniyagala (1955); 20 – Chowdhury (1992); 22, 26, 28, 32, 36 – Kurt (unpublished); 23 – Kurt (1974); 24. Dissanayake (1992); 30, 37 Hendavitharana *et al.*, (1994).

1.2 Proportion of maknas among sub-adult and adult males (Pma) and sex ratio (number of adult and sub-adult males per 100 adult and sub-adult females, Sm) in intensively (IC) or extensively (EC) kept populations of Asian elephants. Sample sizes (n) are given as the number of adult and sub-adult males. In chi$^2$ - tests the respective findings were compared to the maximum natural makna-tusker ratio in wild populations of 65 males $_{a+sa}$ : 100 females $_{a+sa}$ (Ps), to a maximum natural makna-tusker ratio of 10% for the Asian mainland (Pm), and an average tusker ratio of 7% for Sri Lanka (Pm).

| Region | Period | System | n | Pma | Sm | Ps | Pm |
|---|---|---|---|---|---|---|---|
| *Kerala* | | | | | | | |
| 1. Private owners | 1979–1989 | IC | 140 | 2.1 | – | – | <0.01 |
| 2. Private owners | 1993 | IC | 25 | 0.0 | 625.0 | <0.001 | <0.2 |
| 3. Forest Dept. | 1993 | EC | 25 | 0.0 | 208.3 | <0.001 | <0.2 |
| *Tamil Nadu* | | | | | | | |
| 4. Forest Dept. | 1978–1979 | EC | 48 | 2.1 | 129.7 | <0.02 | 0.05 |
| *Myanmar* | | | | | | | |
| 5. Forest Dept. | 1950 | EC | 200 | 10.0 | – | – | 1.0 |
| 6. Forest Dept. | 1981–1991 | EC | 582 | – | 60.2 | <0.1 | – |
| 7. Forest Dept. | 1993 | EC | 23 | 43.5 | – | – | <0.001 |
| *Sri Lanka* | | | | | | | |
| 8. Processions | 1940–1945 | IC | 32 | 69.0 | 229.0 | <0.001 | <0.001 |
| 9. Total population | 1940–1945 | IC | 364 | 89.0 | 119.0 | <0.001 | <0.002 |
| 10. Processions | 1967 | IC | 33 | 81.8 | 80.5 | <0.5 | <0.02 |
| 11. Processions | 1991 | IC | 68 | – | 73.9 | <0.5 | – |
| 12. Processions | 1992 | IC | 43 | 81.4 | 57.2 | <0.5 | >0.5 |

Sources of data: 1 – Panicker (1990); 2,4,7,10 – Kurt (unpublished); 3 – Chandran (1990); 5,8,9 – Dernainyagala (1955); 6 – Khyne U Mar (1992); 11 – Ilangakoon (1993); 3 – Rantnasooriya & Fernando (1992).

1.3 Number of captive elephants (Nc), estimated size of wild populations (Nw), and relative size of captive populations (Nc : Nw) in different regions of Asia during different periods. Captive breeding (CB) can be absent (–), rare (+/–) or regular (+). Unselective capturing operations (CU) are *khedda* (ke) and pitfalls (pe) . Selective capturing operations are hand-noosing (hs), *mela-shikar* (ms) and decoy (de). * Animlas in private ownership have been found as orphans.

| Region / reign | Period | Nc | Nw | Nc:Nw | CB | CU | | CS | | |
| --- | --- | --- | --- | --- | --- | --- | --- | --- | --- | --- |
| | | | | | | ke | pe | hn | ms | de |
| *Indian peninsula* | | | | | | | | | | |
| 1. Mauriya | 330 BC | 9000 | 375000 | 0.024 | – | +/– | – | – | + | + |
| 2. Delhi Sultanate | 1340 | 3000 | 375000 | 0.008 | – | – | – | – | – | – |
| 3. Vijayanagar | 1565 | 1000 | 125000 | 0.008 | – | + | + | | | |
| 4. Moghul | 1650 | 30000 | 375000 | 0.080 | +/– | – | – | – | +/– | – |
| 5. South India | 1990 | 300 | 7150 | 0.042 | +/– | + | + | – | – | – |
| *Southeast Asia* | | | | | | | | | | |
| 6. Northeast India 1990 | | 1600 | 12135 | 0.132 | + | – | – | – | – | – |
| 7. Vietnam | 1990 | 600 | 2000 | 0.333 | ? | +/– | – | – | + | – |
| 8. Laos | 1990 | 1332 | 3000 | 0.444 | + | – | – | – | + | – |
| 9. Thailand | 1990 | 5300 | 2000 | 2.650 | + | – | – | – | + | + |
| 10. Myanmar | 1992 | 4500 | 6500 | 0.692 | + | – | – | – | + | – |
| *Sri Lanka* | | | | | | | | | | |
| 11. United kingd. | 1680 – 1696 | 2500 | 12000 | 0.208 | – | – | – | + | – | + |
| 12. Private owners | 1945 – 1950 | 670 | 2000 | 0.335 | – | +/ | – | + | – | – |
| 13. Private owners* | 1967 – 1979 | 532 | 2500 | 0.021 | – | – | – | – | – | – |
| 14. Private owners* | 1992 | 480 | 2000 | 0.240 | – | – | – | – | – | – |

Sources of data: 1 – Trautmann (1982); 2 – Digby (1971); 3 – Lahiri – Choudhuri (1991); 4 – Tavernier (1984); 5 to 9 – Santiapillai & Jackson (1990); 10 – Khyne U Mar (1992); 11 – Baldaeus (1960), Pieris (1920), McKay (1973); 12 – Deraniyagala (1955); 13 – Jayasinghe & Jainudeen (1970), McKay (1973); 14 – Ratnasooriya & Fernando (1992), Santiapillai (1993).

1.4  Initial population size and mortality schedules used in the simulation of elephant populations. Data are derived from Table 2.1.1

| Region | Initial | 300 BC– | Rate of mortality (in %) | | | | | |
|---|---|---|---|---|---|---|---|---|
| Classes | size of population | 600 AD | 600–1250 | 1250–1600 | 1600–1800 | 1800–1800 | 1900–1950 | 1950–2000 |
| *Mahaweli* | | | | | | | | |
| 1. Females | 360 | 5.5 | 5.5 | 5.5 | 5.5 | 5.5 | 5.5 | 5.5 |
| 2. Tuskers | 130 | 10.0 | 10.0 | 10.0 | 15.0 | 15.0 | 8.5 | 8.0 |
| 3. Maknas | 10 | 7.5 | 7.5 | 7.5 | 12.0 | 15.0 | 7.0 | 7.0 |
| *Yala* | | | | | | | | |
| 1. Females | 360 | 5.5 | 5.5 | 5.5 | 5.0 | 5.0 | 5.0 | 5.0 |
| 2. Tuskers | 280 | 9.0 | 10.0 | 10.0 | 15.0 | 15.0 | 8.5 | 7.0 |
| 3. Maknas | 20 | 7.0 | 7.5 | 7.5 | 15.0 | 15.0 | 7.0 | 7.0 |
| *South India* | | | | | | | | |
| 1. Females | 4000 | 5.5 | 5.5 | 5.5 | 5.5 | 5.5 | 5.5 | 5.5 |
| 2. Tuskers | 1920 | 8.0 | 8.0 | 8.0 | 8.0 | 8.0 | 8.0 | 8.0 |
| 3. Maknas | 80 | 7.0 | 7.0 | 7.0 | 7.0 | 7.0 | 7.0 | 7.0 |

# Appendix II

2.1 Accuracy, standard deviation and figures for minimum and maximum figures (in %) of 3 methods to estimate body weight by means of shoulder height (S) and chest girth (T)

| | Males> 6 years<br>n = 21 | Females> 6 years<br>n = 34 | Juveniles:s 6 years<br>n = 9 |
|---|---|---|---|
| Males: W = 0.915*S*P/10000<br>Females: W = $0.982*S*T^2/10000$<br>Juv.: W = S*P/10000 | 0.0 ± 5.08 %<br>min: -7.8 %<br>max: +7.6 % | 0.0 ± 4.00 %<br>min : - 7.7 %<br>max: + 7.8 % | 0.0 + 4.26 %<br>min : -5.8 %<br>max: + 6.0 % |
| W = [0.053*(S+22.39)]3<br>(Kurt and Nettasinghe 1968) | 6.23 ± 13.80 %<br>min: -19.96 %<br>max: + 45.59 % | -2.10 ± 9.94 %<br>min: - 22.50 %<br>max: + 21.41 % | -1.98 + 8.16 %<br>min : - 20.84 %<br>max: + 7.59 % |
| W = [0.035*(T+60.60)]3<br>(Kurt and Nettasinghe 1968) | 6.66 ± 6.90 %<br>min : - 6.62 %<br>max: + 20.88% | 4.66 ± 5.83 %<br>min :- 6.37 %<br>max:+ 17.62 % | 17.15 + 7.28 %<br>min: + 7.0 %<br>max: + 29.0% |
| Fem.: W = [(0.06*S) - 0.335]3<br>(Sukumar *et al.* 1988) | | -0.63 ± 10.21 %<br>min.: - 22.81 %<br>max.: + 25.26 % | |

2.2  Checklist to assess body weights for captive male elephants older than 6 years. Figures are given in 1000 kg.

| Girth (in cm) ↓ | ← (Shoulder height (in cm) → | | | | | | | | | | | |
| --- | --- | --- | --- | --- | --- | --- | --- | --- | --- | --- | --- | --- |
| | 171–180 | 181–190 | 191–200 | 201–210 | 211–220 | 221–230 | 231–240 | 241–250 | 251–270 | 261–270 | 271–280 | 281–290 |
| 221–230 | 0.81 | 0.86 | | | | | | | | | | |
| 231–240 | 0.88 | 0.94 | | | | | | | | | | |
| 241–250 | 0.96 | 1.02 | 1.08 | | | | | | | | | |
| 251–260 | 1.04 | 1.10 | 1.16 | | | | | | | | | |
| 261–270 | | 1.19 | 1.25 | 1.32 | | | | | | | | |
| 271–280 | | 1.28 | 1.35 | 1.42 | | | | | | | | |
| 281–290 | | | 1.45 | 1.52 | 1.59 | | | | | | | |
| 291–300 | | | 1.55 | 1.63 | 1.71 | 1.79 | | | | | | |
| 301–310 | | | | 1.75 | 1.83 | 1.92 | 2.00 | | | | | |
| 311–320 | | | | 1.86 | 1.95 | 2.04 | 2.13 | 2.22 | | | | |
| 321–330 | | | | | 2.08 | 2.18 | 2.27 | 2.37 | | | | |
| 331–340 | | | | | 2.21 | 2.31 | 2.41 | 2.52 | 2.62 | | | |
| 341–350 | | | | | | 2.45 | 2.56 | 2.67 | 2.77 | | | |
| 351–360 | | | | | | 2.60 | 2.71 | 2.83 | 2.94 | 3.05 | | |
| 361–370 | | | | | | | 2.87 | 2.99 | 3.11 | 3.23 | | |
| 371–380 | | | | | | | 3.02 | 3.15 | 3.28 | 3.41 | 3.54 | 3.67 |
| 381–390 | | | | | | | | | 3.46 | 3.59 | 3.73 | 3.87 |
| 391–400 | | | | | | | | | 3.64 | 3.78 | 3.93 | 4.07 |
| 401–410 | | | | | | | | | | 3.98 | 4.13 | 4.28 |
| 411–420 | | | | | | | | | | 4.18 | 4.33 | 4.49 |
| 421–430 | | | | | | | | | | 4.38 | 4.55 | 4.71 |

2.3  Checklist to assess body weight of captive female elephants older than 6 years. Figures are given in 1000 kg.

| Girth | ← (Shoulder height (in cm) → | | | | | | | | | | |
| (in cm) ↓ | 141–150 | 151–150 | 161–170 | 171–180 | 181–190 | 191–200 | 201–210 | 211–220 | 221–230 | 231–240 | 241–250 | 251–260 |
| --- | --- | --- | --- | --- | --- | --- | --- | --- | --- | --- | --- | --- |
| 201–210 | 0.60 | 0.64 | | | | | | | | | | |
| 211–220 | 0.66 | 0.70 | 0.75 | | | | | | | | | |
| 221–230 | 0.72 | 0.77 | 0.82 | 0.87 | | | | | | | | |
| 231–240 | | 0.84 | 0.90 | 0.95 | 1.00 | | | | | | | |
| 241–250 | | 0.91 | 0.97 | 1.03 | 1.09 | 1.15 | | | | | | |
| 251–260 | | | 1.05 | 1.12 | 1.18 | 1.25 | | | | | | |
| 261–270 | | | | 1.21 | 1.28 | 1.35 | 1.41 | | | | | |
| 271–280 | | | | 1.30 | 1.37 | 1.45 | 1.52 | 1.60 | | | | |
| 281–290 | | | | | 1.48 | 1.55 | 1.64 | 1.72 | 1.80 | | | |
| 291–300 | | | | | | 1.67 | 1.75 | 1.84 | 1.92 | 2.01 | | |
| 301–310 | | | | | | 1.78 | 1.87 | 1.96 | 2.06 | 2.15 | | |
| 311–320 | | | | | | | 2.00 | 2.10 | 2.19 | 2.29 | 2,39 | |
| 321–330 | | | | | | | 2.13 | 2.23 | 2.33 | 2.44 | 2.54 | |
| 331–340 | | | | | | | 2.26 | 2.37 | 2.48 | 2.59 | 2.70 | |
| 341–350 | | | | | | | | 2.51 | 2.63 | 2.75 | 2.86 | 2.98 |
| 351–360 | | | | | | | | 2.66 | 2.79 | 2.91 | 3.03 | 3.16 |
| 361–370 | | | | | | | | | 2.94 | 3.07 | 3.21 | 3.34 |
| 371–380 | | | | | | | | | | 3.24 | 3.38 | 3.52 |
| 381–390 | | | | | | | | | | | 3.57 | 3.71 |
| 391–400 | | | | | | | | | | | 3.75 | 3.91 |
| 401–410 | | | | | | | | | | | 3.95 | 4.11 |

2.4 Checklist to assess body weight of captive juvenile elephants younger than 6 years. Figures are given in  kg.

| Girth | ← Shoulder height (in cm) → | | | | | | |
|---|---|---|---|---|---|---|---|
| (in cm) ↓ | 121–130 | 131–140 | 141–150 | 151–160 | 161–170 | 171–180 | 181–190 |
| 151–160 | 300 | | | | | | |
| 161–170 | 340 | | | | | | |
| 171–180 | 380 | 410 | | | | | |
| 181–190 | | 460 | 500 | | | | |
| 191–200 | | 510 | 550 | 590 | | | |
| 201–210 | | | 610 | 650 | | | |
| 211–220 | | | | 720 | 770 | | |
| 221–230 | | | | 790 | 840 | 890 | |
| 231–240 | | | | 860 | 910 | 970 | |
| 241–250 | | | | | 990 | 1050 | 1110 |
| 251–260 | | | | | 1070 | 1140 | 1200 |
| 261–270 | | | | | | 1230 | 1300 |
| 271–280 | | | | | | | 1400 |

2.5 Comparison of different I–Types with different spinal configurations. The difference between all males and all females, and the difference between 'ethas' and 'aliyas' are significant ($chi^2 = 15.21$; $p < 0.01$, $chi^2 = 15.21$; $p < 0.001$). Data from Godagama (1996) were included.

| | 'Flat' back (%) | 'Broken' back (%) | 'Sloping' back (%) | Total (%) |
|---|---|---|---|---|
| *Males* | | | | |
| 1. Ethas | 6 (22.2) | 9 (33.3) | 12 (44.4) | 27 (99.9) |
| 2. Aliyas | 18 (17.3) | 74 (71.2) | 12 (11.5) | 104 (100.0) |
| 3. Pussas | 4 (26.7) | 7 (46.7) | 4 (26.7) | 15 (100.1) |
| 4. All males | 28 (19.2) | 90 (61.6) | 28 (19.2) | 146 (100.0) |
| *Females* | | | | |
| 5. Ethinnas | 28 (36.4) | 44 (57.1) | 4 (6.5) | 77 (100.0) |
| 6. Alidenas | 28 (34.1) | 47 (57.3) | 5 (8.5) | 82 (99.9) |
| 7. All females | 56 (35.2) | 91 (57.2) | 12 (7.5) | 159 (99.9) |

2.6 Extent of depigmentation on trunk, ears, temples, and shoulders of different I-types in Sri Lankan elephants. The amount of depigmentation for each area has been ranked in classes 0 – 4 (see text for an explanation). The minimum extent of depigmentation for 1 animal is 0, the maximum 16. Significant differences were found between 'ethinnas' and 'alidenas' in all 3 age groups investigated. (Mann-Whitney – U, $p < 0.05$):

| | | Age groups in years | | | | | | | |
|---|---|---|---|---|---|---|---|---|---|
| | Type | A: 15 – 24 | | | B: 25 – 40 | | | C ≥ 40 | | |
| | | Mean | SD | n | Mean | SD | n | Mean | SD | n |
| I. | Ethas | 4.75 | 1.71 | 4 | 7.00 | 2.92 | 5 | 9.67 | 1.63 | 6 |
| 2 | Aliyas | 4.72 | 2.10 | 11 | 9.64 | 2.15 | 22 | 12.30 | 3.20 | 4 |
| 3. | Pussas | 6.00 | – | 1 | 10.00 | 5.29 | 3 | 12.00 | 1.00 | 3 |
| 4. | All males | 4.81 | 1.90 | 16 | 9.29 | 2.71 | 30 | 11.00 | 2.45 | 13 |
| 5. | Ethinnas | 8.71 | 2.93 | 7 | 10.00 | 3.12 | 9 | 10.50 | 3.18 | 12 |
| 6. | Alidenas | 4.42 | 1.83 | 14 | 7.53 | 2.59 | 15 | 7.33 | 3.20 | 9 |
| 7. | All females | 5.86 | 3.00 | 21 | 8.45 | 2.92 | 24 | 9.14 | 3.50 | 21 |

2.7 Degree of accentuation of eyes and openings of the temporal gland in Sri Lankan elephants older than 20 years. The degree has been estimated for each animal and could range from 0 to 4 (see text for an explanation). In males eyes are significantly more accentuated by depigmentation than in females ($p = 0.002$). Between 'ethas' and 'aliyas' the difference in depigmentation is hardly significant ($p = 0.042$), but 'pussas' have a significantly more prominent depigmentation around the eyes than 'ethas' ($p = 0.013$). No significant differences in depigmentation of the opening of the temporal gland was found between males and females. But in 'ethas' the extent of depigmentation is significantly smaller than in 'aliyas' ($p = 0.031$) and 'pussas' ($p = 0.029$) (Mann – Withney – U).

| Type | Degree of accentuation of eyes | | | | | | Degree of accentuation of openings of temporal gland | | | | | |
|---|---|---|---|---|---|---|---|---|---|---|---|---|
| | 0 | 1 | 2 | 3 | 4 | n | 0 | 1 | 2 | 3 | 4 | n |
| 1. Ethas | 6 | 1 | 1 | 2 | – | 10 | 2 | 2 | 3 | 2 | 1 | 10 |
| 2. Aliyas | 3 | 6 | 5 | 7 | 2 | 23 | – | 1 | 8 | 9 | 5 | 23 |
| 3. Pussas | – | 1 | 1 | 2 | 1 | 5 | – | – | – | 4 | 1 | 5 |
| 4. All males | 9 | 8 | 7 | 11 | 3 | 38 | 2 | 3 | 11 | 15 | 7 | 38 |
| 5. Ethinnas | 6 | 10 | 2 | – | 1 | 19 | – | 1 | 9 | 5 | 4 | 19 |
| 6. Alidenas | 14 | 6 | 3 | 2 | 1 | 26 | – | 5 | 13 | 6 | 2 | 26 |
| 7. All females | 20 | 16 | 5 | 2 | 2 | 45 | – | 6 | 22 | 11 | – | 45 |

2.8  Mean of ages and body weight of elephants belonging to 4 different categories with respect to the form of upper ear edges (see text for an explanation). In the correlations with age 4 significant differences were found among ear edge types. In the correlation with weight only type 3 and type 4 in females were not significantly different.

| Form of | Age (yrs) | | | | Weight (kg) | | | |
|---|---|---|---|---|---|---|---|---|
| upper ear edge | Mean | SD | n | p | Mean | SD | n | p |
| *Males* | | | | | | | | |
| 1 | 6.10 | 2.23 | 10 | – | 986.7 | 191.5 | 10 | – |
| 2 | 16.11 | 11.53 | 19 | 1.12:0.023 | 2040.3 | 983.5 | 19 | 1.12:0.005 |
| 3 | 23.22 | 14.36 | 18 | 2.13:0.132 | 2610.0 | 811.2 | 18 | 2.13:0.050 |
| 4 | 32.73 | 13.78 | 30 | 3.14:0.025 | 3310.5 | 832.5 | 27 | 3.14:0.011 |
| *Females* | | | | | | | | |
| 1 | 12.53 | 8.48 | 15 | – | 1340.7 | 540.5 | 15 | – |
| 2 | 15.88 | 10.07 | 17 | 1.12:0.374 | 1819.1 | 604.3 | 15 | 1.2:0.022 |
| 3 | 26.19 | 16.27 | 26 | 2.13:0.028 | 2346.1 | 706.8 | 26 | 2.13:0.024 |
| 4 | 37.23 | 14.49 | 35 | 3.14:0.006 | 2454.2 | 531.4 | 35 | 3.14:0.516 |

2.9  Density of hair per cm² in 30 elephants of different size classes. 14 regions of the body have been distinguished. n: number of individuals. Data stem from Pinnawela.

| Region | Size classes | | | | | |
| --- | --- | --- | --- | --- | --- | --- |
| | 1&2<br>n=3 | 3&4<br>n=6 | 5&6<br>n = 10 | 7&8<br>n=8 | 9<br>n=3 | Total<br>n=30 |
| Dorsal regions | | | | | | |
| Front | 2.7 ± 1.4 | 7.2 ± 6.5 | 3.8 ± 2.2 | 5.2 ± 2.6 | 6.5 ± 7.5 | 5.0 ± 4.0 |
| Shoulders | 3.2 ± 3.8 | 4.8 ± 5.6 | 2.8 ± 1.2 | 2.9 ± 1.2 | 6.5 ± 7.5 | 3.6 ± 3.5 |
| Back | 7.3 ± 6.6 | 5.7 ± 5.0 | 3.8 ± 2.2 | 5.1 ± 4.4 | 5.3 ± 3.8 | 5.0 ± 4.0 |
| Pelvis | 4.8 ± 2.3 | 2.7 ± 1.3 | 1.3 ± 0.8 | 2.4 ± 2.3 | 3.2 ± 3.8 | 2.4 ± 2.1 |
| **Total** | | | 4.0 ± 3.6 (n = 120) | | | |
| Ventral regions | | | | | | |
| Chin | 4.0 ± 3.3 | 3.8 ± 2.1 | 3.7 ± 1.6 | 3.6 ± 2.7 | 3.5 ± 0.0 | 3.7 ± 2.0 |
| Chest | 1.0 ± 0.0 | 2.3 ± 1.4 | 1.3 ± 0.8 | 1.0 ± 0.0 | 1.0 ± 0.0 | 1.3 ± 0.9 |
| Belly | 1.8 ± 1.4 | 2.7 ± 1.3 | 2.0 ± 1.3 | 1.3 ± 0.9 | 2.7 ± 1.4 | 2.0 ± 1.2 |
| **Total** | | | 2.3 ± 1.7 (n = 90) | | | |
| Lateral regions | | | | | | |
| Edge of eyes | 1.0 ± 0.0 | 1.0 ± 0.0 | 1.3 ± 0.8 | 1.0 ± 0.0 | 1.2 ± 0.6 | 1.2 ± 0.6 |
| Base of trunk | 1.0 ± 0.0 | 1.3 ± 1.2 | 1.0 ± 0.0 | 2.3 ± 2.4 | 1.5 ± 1.5 | 1.5 ± 1.5 |
| Trunk | 1.0 ± 0.0 | 1.8 ± 1.3 | 1.5 ± 1.1 | 1.0 ± 0.0 | 1.0 ± 0.0 | 1.3 ± 0.9 |
| Cheeks | 1.0 ± 0.0 | 1.8 ± 1.3 | 1.5 ± 1.1 | 1.0 ± 0.0. | 1.0 ± 0.0 | 1.4 ± 0.9 |
| Upper arm | 1.0 ± 0.0 | 1.8 ± 1.3 | 1.8 ± 1.2 | 2.1 ± 2.3 | 4.0 ± 3.3 | 2.0 ± 1.8 |
| Flanks of belly | 2.7 ± 1.4 | 3.3 ± 2.4 | 2.0 ± 1.3 | 2.6 ± 1.3 | 3.5 ± 0.0 | 2.6 ± 1.5 |
| Thigh | 1.8 ± 1.4 | 2.3 ± 1.4 | 1.5 ± 1.1 | 1.3 ± 0.8 | 1.8 ± 1.4 | 1.7 ± 1.1 |
| **Total** | | | 1.7 ± 1.3 (n = 210) | | | |

2.10 Length of hair (cm) in 30 elephants of different size classes. 14 different body regions have been distinguished. n: number of individuals. Data stem from Pinnawela.

| Region | Size classes | | | | | |
| --- | --- | --- | --- | --- | --- | --- |
| | **1 & 2**<br>**n = 3** | **3 & 4**<br>**n = 6** | **5 & 6**<br>**n = 10** | **7 & 8**<br>**n = 8** | **9**<br>**n = 3** | **zusammen**<br>**n = 30** |
| Dorsal regions | | | | | | |
| Front | 6.8 ± 2.0 | 6.3 ± 1.9 | 6.0 ± 2.0 | 5.4 ± 1.6 | 6.8 ± 2.0 | 6.1 ± 1.9 |
| Shoulder | 4.8 ± 3.0 | 4.3 ± 2.2 | 3.0 ± 1.3 | 4.1 ± 2.6 | 4.8 ± 3.0 | 3.9 ± 2.2 |
| Back | 8.0 ± 0.0 | 5.3 ± 2.3 | 4.6 ± 1.4 | 3.6 ± 1.3 | 6.0 ± 3.5 | 4.9 ± 2.1 |
| Pelvis | 5.7 ± 2.0 | 5.3 ± 2.3 | 3.5 ± 1.3 | 3.5 ± 1.3 | 6.0 ± 3.5 | 4.7 ± 2.2 |
| **Total** | | | 4.9 ± 2.2 (n = 120) | | | |
| Ventral regions | | | | | | |
| Chin | 5.7 ± 2.0 | 5.7 ± 1.8 | 4.9 ± 2.8 | 5.9 ± 2.4 | 6.8 ± 2.0 | 5.6 ± 2.3 |
| Chest | 5.7 ± 2.0 | 4.8 ± 2.7 | 4.4 ± 1.7 | 3.6 ± 1.3 | 2.8 ± 1.4 | 4.2 ± 1.9 |
| Belly | 5.7 ± 2.0 | 4.7 ± 1.9 | 3.0 ± 1.3 | 3.7 ± 2.1 | 5.7 ± 2.0 | 4.1 ± 2.0 |
| **Total** | | | 4.6 ± 2.2 (n = 90) | | | |
| Lateral regions | | | | | | |
| Edges of eyes | 2.0 ± 0.0 | 2.0 ± 0.0 | 2.0 ± 0.0 | 2.0 ± 0.0 | 1.4 | 2.1 ± 0.5 |
| Base of trunk | 2.0 ± 0.0 | 1.7 ± 0.8 | 2.0 ± 0.0 | 2.1 ± 1.2 | 4. ± 3.5 | 2.2 ± 1.3 |
| Trunk | 2.0 ± 0.0 | 2.8 ± 1.3 | 2.8 ± 1.2 | 2.9 ± 1.3 | 2.8 ± 1.4 | 2.8 ± 1.2 |
| Cheeks | 4.8 ± 3.0 | 2.0 ± 0.0 | 3.0 ± 1.3 | 2.6 ± 1.2 | 4.8 ± 3.0 | 3.1 ± 1.7 |
| Upper arm | 2.8 ± 1.4 | 3.0 ± 2.5 | 3.1 ± 2.0 | 2.9 ± 1.3 | 2.8 ± 1.4 | 3.0 ± 1.7 |
| Belly flanks | 2.8 ± 1.4 | 4.1 ± 1.0 | 2.8 ± 1.2 | 2.9 ± 1.3 | 4.5 ± 0.0 | 3.3 ± 1.3 |
| Thigh | 3.7 ± 1.4 | 2.4 ± 1.0 | 2.5 ± 1.1 | 3.3 ± 1.3 | 2.8 ± 1.4 | 2.8 ± 1.2 |
| **Total** | | | 2.7 ± 1.4 (n = 210) | | | |

# References


Abul'l – Fazl' Alami, *Ain-i-Akbari* (H. Blochmann, D.V. Phillot, Transl.), 1927, Asiatic Society of Bengal, Calcutta.

Alahakoon, J. & Santiapillai, C., 'Elephant: unwitting victims of Sri Lanka's civil war', *Gajah,* 1997, 12: 46–47.

Alcock, J., 'The evolution of the use of tools by feeding animals', *Evolution,* 1972, 26: 464–473.

Ananthasubramaniam, C. R., 'Some aspects of elephant nutrition', In: *The Asian Elephant: Ecology, Biology, Diseases, and Management* (E.G. Silas, M. Krishna, P.G. Nair & G. Nirmalan, eds.), Lumière Printing Works, Trichur, 1992, 86–90.

Ashraf, N.V.K., Choudhury, B., Mainkar, K. & Barman, R., *Elephant Health Camp*, Wildlife Trust of India, 2002.

Aung, U.T. & Nyut, U.T., 'The care and management of domesticated elephants in Myanmar', In: *Giants on our Hands. Proc. Int. Workshop on the Domesticated Asian Elephant* (I. Baker, M. Kashio, eds.): FAO, Bangkok, 2002, 89–102.

Baker, C.M.A. & Manwell, C., 'Man and elephant'. 'The 'dare theory' of domestication and the origin of the breeds', 1983, *Z.Tierzücht./Züchtungsbiol.* 100: 55–75.

Baker, I. & Kashio, M., *Giants on our Hands. Proc. Int. Workshop on the Domesticated Asian Elephant*, FAO, Bangkok, 2002.

Baldaeus, P., *A True and Exact Description of the Great Island of Ceylon (1672)*, Saman Press, Maharagama, 1960.

Balke, J.M.E., Barker, I. K., Hackenberger, M. K., Mcmanamon, R. & Boever, W.J., 'Reproductive anatomy of three nulliparous female Asian elephants: the development of artificial insemination techniques', *Zoo Biology*, 1988: 7: 99–113.

Barua, P. and Bist, S.S., 'Cruelty to elephants – A legal and practical view', *Zoo's Print*, 1996 11(6): 47–51.

Baskaran, N., Balasubramanian, M., Swaminathan, S., & Desai, A. 'Home ranges of elephants in the Nilgiri Biosphere Reserve, South India', *A Week with Elephants*. J.C. Daniel & H. Datye (eds.), 1995, Bombay Natural History Society, 296–313.

Bazé, W., *Just Elephants*, Elek Books, London, 1955.

Beck, B. B., *Animal Tool*, Garland STPM Press, New York, 1980.

Becker, R.-B., *Werkzeuggebrauch im Tierreich*, Wissenschaftliche Verlagsgesellschaft, Stuttgart,1993.

Beissinger, S. R. & Westphal, M. I., 'On the use of demographic models of population viability in endangered species management', *Journal of Wildlife Management*, 1998, 62: 821–841.

Benedict, F.G. *The Physiology of the Elephant*, 1936, Carnegie Institution, Washington D.C.

Berridge, K.C., 'Comparing the motional brains of humans and other animals'. In R.J. Davidson, K.R. Scherer, and H. Hill Goldsmith (eds.), *Handbook of Affective Sciences*, Oxford University Press, 2003.

Bist, S.S., 'Standards and norms for elephant owners, draft for comment', *Zoo's Print*, 1996, 11 (6): 52–54.

Bist, S.S., Cheeran, J.V., Choudhury, S., Barua, P. & Mishra, M.K., 'The domesticated Asian elephant in India'. In: *Giants on our hands. Proc. Int. Workshop on the Domesticated Asian Elephant* (I. Baker, M. Kashio, eds.), 2002, 129–148. FAO, Bangkok.

Boner, A. & Sarma. R., *New Lights on the Sun Temple of Konarka*, 1972, Varanasi.

Bongso, T.A. Perera, B.M.O.A. & Motha, M.X.L, 'Estimation of shoulder height from forefoot circumference in the Asian elephant (*Elephas maximus*)', *Ceylon Journal of Science (Biological Science)*, 1981, 14 (1&2): 79–82.

Boorer, M. K., 'Some aspects of streotyped patterns of movement exhibited by zoo animals', *International Zoo Yearbook*, 1972, 12: 164–168.

Bradshaw, G.A., 'Not by bread alone', *Society and Animals*, 2004, 12(2) 143–158.

Bradshaw, G. A., 'Elephant trauma and recovery from human violence to liberation ecopsychology' Unpublished dissertation, Pacifica Graduate Institute, USA, 2005.

Bradshaw, G.A., Schore, J.L. Brown, J.H. Poole, & C.J. Moss. 'Elephant breakdown', *Nature*, 2005, 433 : 807.

Brannian, J.D., Griffin, F., Papkoff, H. & Terranova, P. F., 'Short and long phases of progesterone secretion during the oestrus cycle of the African elephant' (*Loxodonta africana*), *Journal of Fertility and Reproduction*, 1988, 84: 357–365.

Brohier, R.L., *Seeing Ceylon*, Lake House Press, Colombo, 1965.

Brown, J. L., Citino, S.B., Bush, M., Lehnhardt & Phillips, L.G., 'Cyclic patterns of luteinizing hormone, follicle–stimulating hormone, inhibin, and progesterone secretion in the Asian elephant (*Elephas maximus*)', *Journal of Zoo and Wildlife Medicine*, 1991, 22: 49–77.

Brown, J. L. & Lehnhardt, J., 'Serum and urinary hormones during pregnancy and the peri–and postpartum period in an Asian elephant (*Elephas maximus*)', *Zoo Biology*, 1995, 14: 555–564.

Bubenik, A.B., *Das Geweih*, Paul Parey, Hamburg & Berlin, 1966.

Burne, E.C., 'A record of gestation periods and growth of trained Indian elephant calves in the Southern Shan States, Burma', *Proceedings of the Zoological Society*, London, 1942, 42: 27–29.

Buss. I.O. 1990: *Elephant Life: Fifteen Years of High Population Density*, Iowa State University Press, Ames, 1990.

Buss, I. O., Rasmussen, L.E. & Smuts, G. L., 'The role of stress and individual recognition in the function of the African elephant's temporal gland', *Mammalia*, 1976, 40: 437–451.

Byrne, R.W. & Byrne, J.M.E., 'Variability and standardization in the complex leaf – gathering task of mountain gorillas (*Gorilla gorilla berengei*)', *American Journal of Primatology*, 1993, 31: 241–261.

Byrne, R.W. & Russon, A.E., 'Learning by immitation: a hierarchical approach', *Behavioural and Brain Sciences*, 1998, 21 (5): 667–684.

Carlstead, K., 'Effects of captivity on the behaviour of wild mammals'. In: *Wild Mammals in Captivity: Principles and Techniques* (D.G. Kleinman, M.E. Lang, K.V. Thompson & S. Lumpkin, eds.), University of Chicago Press, Chigaco & London, 1996.

Carlstead, K., Seidensticker, J. & Baldwin, R., 'Environmental enrichment for zoo bears', *Zoo Biology*, 1991, 10: 13–16.

Chandler; D.P., *A History of Cambodia*, Silkworm Books, Chiang Mai, 1993.

Chandran, P.M., 'Population dynamics of elephants in Periyar Tiger Reserve'. In: C.K.Karunakaran (ed.), *The Proceedings of the Elephant Symposium*, Kerala Books and Publication Society, Kakkanad, Kochi, 1990.

Chandrasekharan, P., Radhakrishnan Nair, K. N. M. & Prabhakaran, 'Some observations on musth in captive elephants in Kerala (India)'. In: E.G. Silas, M. Krishna, P.G. Nair & G. Nirmalan, (eds.), *The Asian Elephant: Ecology, Biology, Diseases, and Management*, Lumière Printing Works, Trichur, 1992, 71–74.

Chevalier – Skolnikoff, S., 'Spontaneous tool use and sensorimotor intelligence in *Cebus* compared with other monkeys and apes'. *Journal of Behaviour and Brain Sciences*, 1989, 12: 561–627.

Chevalier – Skolnikoff, S., 'Tool use by wild cebus monkeys at Santa Rosa National Park, Costa Rica', *Primates*, 1990, 31: 375–383.

Chevalier – Skolnikoff, S., Galdikas, B.M.F. & Skolnikoff, A.Z., 'The adaptive significance of higher intelligence in wild orang–utans: a preliminary report', *Journal of Human Evolution*, 1982, 11:: 49–64.

Chevalier – Skolnikoff, S. & Liska, 'Tool use by wild and captive elephants', *Animal Behaviour*, 1993, 46 : 209–216.

Chowdhury; K.D., 'Conservation of Asiatic elephant with special reference to Arunachal Pradesh (India)'. In: E.G. Silas, M. Krishna, P.G. Nair & G. Nirmalan, (eds.), *The Asian Elephant: Ecology, Biology, Diseases, and Management*, 1992, Lumière Printing Works, Trichur, 132–136.

Clemens, E. T. & Maloy, G.M., 'The digestive physiology of three East African herbivores: the elephant, rhinoceros and hippopotamus', *Journal of Zoology*, London, 1982, 198: 141–156.

Clubb, R. & Mason. G., *A Review of the Welfare of Zoo Elephants in Europe*, University of Oxford, Oxford, 2001.

Cooper, K.A., Harder, J.D., Clawson, D.H., Fredrick, D.L., Lodge, G.A., Peachy, H.C., Spellmire, T.J. & Winstel, D.P., 'Serum testosterone and musth in captive male African and Asian elephants', *Zoo Biology*, 1990, 9: 297–306.

Cordiner, J., *A Description of Ceylon*, Tisara Press, Dehiwala, 1983.

Craig., D., 'Appetites and aversions as constituents of instincts', *Bulletin of Biology.*, 1918, 34: 91–107.

Cronin, G.M.,Wiepkema, P.R. & Hofstede, G.J., 'The development of stereotypies in tethered sows'. In: *Proceedings of the International Congress on Applied Ethology in Farm Animals* (J. Unshelm, G. van Putten & K. Zeeb, eds.), Darmstadt. KTBL, 1984, 97–100.

Cronin, G.M., Wiepkema, P.R. & Van  Ree, J. M., 'Endogenous opioids are involved in abnormal stereotyped behaviours of tethered sows', *Neuropeptides*, 1985, 6: 527–530.

Cuong, T.V., Lien, T.T. & Giao, P.M.: 'The present status and management of domesticated Asian elephants in Vietnam'. In: *Giants on our Hands. Proc. Int. Workshop on the Domesticated Asian Elephant* (I. Baker, M. Kashio, eds.), FAO, Bangkok, 2002, 111–128.

Daniel, J. C., 'The Asian elephant: future prospects', 1993, *Gajah* 11: 2–5.

Dany, C., Weiler, H., Tong, K. & Han, S., 'The status, distribution and management of domesticated elephants in Cambodia'. In: *Giants on our Hands. Proc. Int. Workshop on the domesticated Asian elephant* (I. Baker, M. Kashio, eds.), 2002, FAO, Bangkok 177–188.

Dallaire, A., 'Rest behavior', *Veterinary Clinic of North America (Equine Practice)*, 1986, 2: 591–607.

Dantzer, R., Behavioral, physiological, and functional aspects of stereotyped behavior: a review and a re-interpretation, *Journal of Animal Science*, 1986, 62: 1776–1786.

Darwin, C., *The Descent of Man and Selection in Relation to Sex. Vol. I.*, Murray, London, 1871.

Dastig, B., Birth and reproduction rate in a herd of captive Asian elephants at the Pinnawela Elephant Orphanage (Sri Lanka). In: *Research Update on Elephants and Rhinos* (H.M. Schwammer, T.J. Foose, M.Fouraker, D. Olson, eds.), Schüling Verlag, Münster, 2002, 19–23.

Dastig, B., Kurt, F., Petzold, S., Rastelli, J., Schmelz, J., Tragauer, V. & Zacha C, Geburt und Jugendentwicklung von Asiatischen Elefanten – Beobachtungen aus der Pinnawela Elephant Orphanage. *Zeitschrift des Kölner Zoo*, 1999, 42(2): 83–91.

Dehnhardt, G., Friese, C. & Sachser, N., Sensivity of the trunk of Asian elephant for texture differences of actively touched objects. *Proceedings of the International Symposium on Physiology and Ecology of Wild and Zoo Animals*, 1996 Supplement. 2: 37–39.

Demment, M.W. & Van Soest, P.J., A nutritional explanation for body-size patterns of ruminant and nonruminant herbivores. *American Naturalist*, 1985, 125: 641–672.

Deraniyagala, P.E.P. *Some Extinct Elephants, their Relatives and the two Living Species*, Government Press, Colombo, 1955.

De Silva, K.M., *A History of Sri Lanka*, Oxford University Press, Delhi, 1991.

De Silva, M., Status and conservation of the elephant (*Elephas maximus*) and the allevation of man-elephant conflict in Sri Lanka, *Gajah* 19: Special Issue, 1998.

De Silva, M. & Atapattu, N. *Allevation of wild elephant – human conflict and conservation of elephant in the north – western region of Sri Lanka*, Department of Wildlife Conservation. Colombo, 1997.

De Silva, M., Jayaratne, B.V.R. & De Silva, P.K., A study of elephant movements in Ruhuna National Park, Sri Lanka. *Journal of Asian Natural History*, 1997, 2(2): 183–198.

De Vos, F.H., A pertinent account and detailed description of the character of elephants by Cornelis Taay. *R.A.S. (Ceylon Branch)*, 1872, 15: 176–200.

Dhakal, F.R., 'The challenge of managing domesticated elephants in Nepal'. In: *Giants on our Hands. Proc. Int. Workshop on the Domesticated Asian Elephant* (I. Baker, M. Kashio, eds.), FAO, Bangkok, 2002, 129–148..

Digby, S., 'War horse and elephant in the Delhi Sultanate, a study of military supplies'. *Orient Monographs,* Oxford, 1971, 5–100.

Dissanayake, S.R.B., Seasonal changes in population structure, movement pattern and activity of elephants in the Ruhunu National Park–Sri Lanka. Asian Elephant Specialist Group Meeting, Bogor Indonesia, Asian Elephant Conservation Centre, Bangalore, India: 1992, 25–34.

Dissanayake, S.R.B., De Silva, M. & Santiapillai, C., 'A study of human–elephant conflict in the Accelerated Mahaweli Development Area', Sri Lanka, 1998. (unpublished manuscript)

Dittrich, L., Beitrag zur Fortpflanzung und Jugendentwicklung des Indischen Elefanten, *Elephas maximus*, in Gefangenschaft mit einer Übersicht über die Elefantengeburten in europäischen Zoos und Zirkussen, *Der Zoologische Garten (Neue Folge)* 34 (1/3), 1967, 56–92.

Dorresteyn, A., 'Forward planing and EEP management for elepants in EAZA institutions'. In: *Beiträge zur Elefantenhaltung in Europa* (H.M. Schwammer, S. de Vries, eds.), Schüling Verlag. Münster, 2001, 13–16.

Douglas – Hamilton, I. & Douglas–Hamilton, O., *Among the Elephants*, Viking Press, New York, 1975.

Drysdale, T.A. & Florkiewicz, R.F., 'Electrophoretic variation within and between the two extant elephant species (Mammalia: Probiscidea)'. *Journal of Mammology*, 1989, 70: 381–383.

Dublin, H.T., 'Cooperation and reproductive competition among female African elephants'. In: *Social Behaviour of Female Vertebrates* (ed. S. Wasser), Acad. Press, New York: 1983, 291–313.

Duke – Elder, S., *The Eye in Evolution*, 1958, Henry Kimton, London.

Edgerton, F., *The Elephant - Lore of the Hindus*, New Heaven (USA), 1931.

Eisenberg, J.F., *The Mammalian Radiation: An Analysis of Trends in Evolution, Adaptation and Behaviour*, Chicago University Press, Chicago, 1981.

Eisenberg, J.F. & Mckay, G.M., 'An annotate checklist of the recent mammals in Ceylon with keys to the species', *Ceylon Journal of Science (Biological Science)*, 1970, 8: 69–99.

Eisenberg, J.F., Mckay G.M. & Jainudeen, M.R., 'Reproductive behaviour of the Asiatic elephant (*Elephas maximus maximus* L.)'. *Behaviour*, 1971, 37: 193– 225.

Eltringham, S.K., *Elephants*, Blandford Press, Oole, U.K, 1982.

EMOA, Newsletter of the Elephant Management & Owner Association, 2001, No. 25.

European Elephant Group, *Elefanten im Zirkus*, Schüling Verlag, Münster, 2000.

European Elephant Group, *Elefanten in Zoos und Safariparks Europa*, Schüling Verlag, Münster, 2003.

Evans; G.H., *Elephants and their Diseases*, Government Printing, Rangoon, 1910.

Fernando, P. 'Keeping jumbo afloat: is translocation an answer to the human-elephant conflict?', *Sri Lanka Nature*, 1997, 1 (1): 1–12.

Fernando, P., Pfrender, M.E., Encalda, S.E. & Lande, R., 'Mitochondrial DNA variation, phylogeopraphy and population structure of the Asian elephant', *Heredity*, 2000, 84: 362–372.

Fernando, S.D.A., Jayasinghe, J.B. Panabokke, R.G., 'A study of the temporal gland of an Asiatic elephant', *Ceylon Veterinary Journal*, 1963, 11: 108–111.

Ferrier, A.J., *The Care and Management of Elephants in Burma*, 1947, Steel Brothers & Co. Ltd., London.

Fischer, M.T., Houston, E.W., O'sullivan, T., Read, B.W. & Jackson, P., 'Selected weights for ungulates and the Asian elephant *Elephas maximus* at St. Louis Zoo', *International Zoo Yearbook*, 1993, 32: 169–173.

Frädrich, H., 'Probleme der Elefantenzucht am Beispiel Thailand', *Bongo*, Berlin: 1993, 22: 23–34.

Franklin, I.R., 'Evolutionary change in small populations'. In: *Conservation Biology: An Evolutionary – Ecological Perspective*, (eds. M. E. Soulé & B. A. Wilcox). Sinauer Associates, Sunderland, Massachusetts, 1980.

Gadgil, M.; Guha, R., *This Fissured Land, an Ecological History of India*, 1992, Oxford Indian Paperbacks, Delhi, Calcutta, Chennai, Mumbai.

Gadgil, M. & Nair, P.V., 'Observations on the social behaviour of free ranging groups of tame Asiatic elephants (*Elephas maximus* Linn.)', *Proceedings of the Indian Academy of Science*, 1984, 93: 225–233.

Galdikas, B.M.F., 'Orangutan tool use at Tanjung Puting Reserve, Central Indonesian Borneo (Kalimantan Tengah)', *Journal of Human Evolution*, 1982, 11: 19–33.

Garaï, M.E., 'Special relationship between female Asian elephants (*Elephas maximus*) in zoological gardens', *Ethology*, 1992, 90:187–205.

Garaï, M.E. 'The effect of Boma design on stress – related behaviour in juvenile translocated African elephants', *Pachyderm*, 1994, 18: 55–60.

Garaï, M.E., *The Development of Social Behaviour in Translocated Juvenile African Elephant Loxodonta Africana* (Blumenbach), Ph.D. Dissertation. University of Pretoria, 1997.

Garaï, M.E., 'Are elephants suitable as game ranch animals?' In: *Proceedings of the International Wildlife Ranching Symposium*, 2001a, (H. Ebedes, B. Reilly, W. van Hoven, B. Penzhorn, eds.) South African Game Rangers' Association, Pretoria.

Garaï, M.E., 'Sozialstruktur und Sozialisation'. In: *Elefant in Menschenhand*, 2001b, (F. Kurt, ed.) Filander, Fürth, 273–286.

Garaï, M.E., 'Social behaviour of the elephants at Pinnawela Elephant Orphanage, Sri Lanka'. In: *Research Update on Elephants and Rhinos*, 2002a, (H.M. Schwammer, T.J. Foose, M.Fouraker, D. Olson, eds.) Schüling Verlag, Münster, 32–40.

Garaï, M.E., *Managing African Elephants. Guidelines for Introduction and Management of African Elephants on Game Reserves*, 2002b, Second rev. edition. EMOA, Pretoria.

Garaï, M.E. & Carr, R.D., 'Unsuccesful introduction of adult elephant bulls to confined areas in South Africa', *Pachyderm*, 2001, 31: 52–57.

Garaï, M.E., Carr, R.D., Slotow, R. & Reilly, B., 'History and population ecology of re-introduced elephants in small fenced reserves in South Africa', *Pachyderm*, 2004.

Geiger, W., 'Army and war in medieval Ceylon'. In: S.D. Saparamadu (ed.), *The Pollonaruwa Period*, Tissa Press, Dehiwala: 1973, 173–179.

Geist, V., 'On the evolution of optical signals in deer': a preliminary analysis, In: *Biology and Management of Cervidae.*, C.M. Wemmer (ed.). Smithsonian Institution Press, Washington: 1987, 235–255.

Geist, V., 'On the relation of social evolution and dispersal in ungulates during the Pleistocene, with emphasis on the Old World deer and the genus, Bison, *Quaternary Research*, 1971, 1: 238–315.

Gerbet, S., *L'éléphant de travail en Thailande: Etudes des traditions locales d'élevage, de dressage et de traitement des affections courantes*, Dissertation Faculté de Médicine à l'Univérsité de Nantes, 1994.

Ghosh, R., *Gods in Chains*, Foundation Books, Delhi, 2005.

Ginsberg, J.R. & Milner-Gulland, E.J., 'Sex-biased harvesting and population dynamics in ungulates: implications for conservation and sustainable use', *Conservation Biology*, 1994, 21: 57–166.

Ginsberg, J.R. & Rubenstein, D.I., 'Sperm competition and variation in zebra mating behaviour', *Behavioural Ecology and Sociobiology*, 1990, 26: 427–434.

Godagama, W. K., *An Ethno-Zoology of Domesticated Elephants in Sri Lanka*, Master Thesis, University of Colombo, Colombo, 1996.

Goodall, J., 'Tool-using in primates and other vertebrates'. In: *Advances in the Study of Behaviour*, (R.A. Hinde & E. Shaw, eds.) Academy Press, New York 3: 1970, 195–249.

Goodall, J., *The Chimpanzees of Gombe*, Belknap Press of Harvard University Press, Cambridge, Maasachusetts, 1986.

Gordon, J.A., 'Elephants do think', *African Wildlife*, 1966, 20: 75–79.

Gowda, K., 'Breeding Indian elephants *Elephas maximus* at Mysore Zoo', *International Zoo Yearbook.*, 1969, 121–122.

Gray, C.W. & Nettasinghe, A.P.W. 'A preliminary study on the immobilization of the Asiatic elephant (*Elephas maximus*) using Etorphine (M–99)', *Zoologica*, 1970, 55 (3): 51–54.

Guha, R. 'Radical American environmentalism and wilderness preservation: a Third World critique', *Evironmental Ethics*, 1989.

Haemisch, A., 'Coping with social conflict, and short-term changes of plasma cortisol titers in familiar and unfamiliar environments', *Physiology & Behaviour*, 1990, 47: 1265–1270.

Hall, K. R. L., 'Tool-using performances as indicators of behavioral adaptability', *Current Anthropology*, 1963, 4: 479–494.

Hamilton, W. J. III, *Life's Colour Code*, McGraw-Hill, New York, 1973.

Harshan, K.R., Chungath, J.J., Paily, L. & Ommer, P.A., 'Histology of the temporal gland of the Indian elephant (*Elephas maximus*)'. In: E.G. Silas, M. Krishna, P.G. Nair & G. Nirmalan, (eds.), *The Asian Elephant: Ecology, Biology, Diseases, and Management*, Lumière Printing Works, Trichur, 1992, 46–48.

Hart, B.L., 'Behavioural adaptation to pathogens and parasites', *Review of Neuroscience and Biobehaviour*, 1990, 14: 273–294.

Hart, B.L., 'Cognition and tool use in elephants'. In: *Reserach Update on Elephants and Rhinos*, (H.M. Schwammer, T.J. Foose, M.Fouraker, D. Olson, eds.): 41. Schüling Verlag, Münster, 2002.

Hart, B. L., & Hart, L.A., 'Fly switching by Asian elephants: tool use to control parasites', *Animal Behaviour*, 1994, : 35–45.

Hart, L.A., 'Bioseismic communication mechanisms in elephants and rhinos'. In: *Reserach Update on Elephants and Rhinos*, (H.M. Schwammer, T.J. Foose, M.Fouraker, D. Olson, eds.) Schüling Verlag, Münster, 2002, 42–46.

Hartl, G.B., Lang, G., Klein, F., & Willing, R., 'Relationships between allozymes, heterozygosity and morphological characters in red deer (*Cervus elaphus*), and the influence of selective hunting on allele frequency distribution', *Heredity*, 1991, 66: 343–350.

Hartl, G.B., Klein, F., Willing, R., Apollonio, M. & Lang, G., 'Allozymes and the genetics of antler development in red deer (*Cervus elephus*)'. *Journal of Zoology*, London, 1995a, 237: 83–100.

Hartl, G.B., Kurt, F., Hemmer, W. & Nadlinger, K., 'Electrophoretic and chromosomal variation in captive Asian elephants (*Elephas maximus*)', *Zoo Biology*, 1995, 14: 87–95.

Hartl, G.B., Kurt, F., Tiedemann, R., Gmeiner, C., Nadlinger, K., Mar, K.U. & Rübel, A., 'Population genetics and systematics of Asian elephant (*Elephas maximus*) : a study based on sequence variation at the Cyt *b* gene of PCR – amplified mitochoncrial DANN from hair bulbs', *Zeitschrift für Säugetierkunde*, 1996, 61: 285–294.

Hattingh, J., 'Physiological measurement of stress', *South African Journal of Science*, 1986, 82: 612–614.

Haufellner, A., Kurt, F., Schilfarth, J. & Schweiger, G., *Elefanten in Zoo und Zirkus. Dokumentation Teil 1: Europa*, European Elephant Group, Karl Wenschlow, München, 1993.

Haufellner, Schilfarth, J. & Schweiger, G., *Elefanten in Zoo und Zirkus. Dokumentation Teil 2: Nordamerika*, European Elephant Group, Karl Wenschlow, München, 1996.

Haufellner, A., Schilfarth, J. & Schweiger, G., *Elefanten – Dokumentation*, European Elephant Group, EOS Verlag, St. Otilien, 1999.

Haufellner, A., Schilfarth, J. & Schweiger, G., 'Haltungsbedingte Unfälle mit Elefanten in Zoos'. In: *Elefanten in Zoo und Circus*, München, 2002, 2: 3–18.

Haufellner, A., Schilfarth, J. & Schweiger, G., 'Haltungsbedingte Unfälle mit Elefanten in Zoos'. In: *Elefanten in Zoo und Circus*, 2003.

Hediger, H., *Wild Animals in Captivity*, Butterworths Scientific Publications, London, 1950.

Hediger, H., *Skizze zu einer Tierpsychologie in Zoo und Zirkus*, Gutenberg, Zürich, 1954.

Hediger, H., *Studies on the Psychology and Behaviour of captive Animals in Zoos and Circuses*, Butterworths Scientific Publications, London, 1955.

Hediger, H.,' Wie Tiere schlafen', *Medizische Kleintier Klinik*, 1969, : 20: 938–946.

Hediger, H., *Ein Leben mit Tieren*, Werd Verlag, Zürich, 1990.

Heine, N., Kurt, F. & Weihs, W., 'Soziale Rollen und die Bildung von Clans'. In: *Elefant in Menschenhand*, (F. Kurt, ed.): 111–131. Schüling, Münster, 2001.

Heine, N., Kurt, F., Pieler, E. & Weihs W., 'Social roles, family units and the formation of clans in Asian elephants of the Uda Walawe National Park (Sri Lanka)'. In: *Research Update on Elephants and Rhinos*, (H.M. Schwammer, T.J. Foose, M.Fouraker, D. Olson, eds.): 47–51. Schüling Verlag, Münster, 2002.

Herre, W. & Röhrs, M., *Haustiere – Zoologisch Gesehen*, Paul Parey, Stuttgart & New York, 1990.

Hess, D.L., Schmidt, A. M.& Schmidt, M.J., 'Reproductive cycle of the Asian elephant (*Elephas maximus*) in captivity', *Biology of Reproduction*, 1983, 28: 767–773.

Hildebrandt, T.B. & Göritz, F., 'Sonographischer Nachweis von Leiomyomen im Gentialtrakt weiblicher Elefanten'. *Verh. Ber. Erkrg. Zootiere*, 1995, 287–294.

Hildebrandt, T.B., Hermes, R., Pratt, N.C., Fritschi, G., Blottner, S., Schmitt, D.L. Ratnakorn, P.Brown, J.L., Ritschel, W. & Göritz, F., 'Ultrasonography of the urogenital tract in elephant (*Loxodonta africana* and *Elephas maximus*): an important tool for assesing female reproductive function', *Zoo Biology*, 2000, 19: 321–332.

Hinde, R.A., *Animal Behaviour*, McGraw-Hill, New York, 1960.

Hinde, R.A., 'Mother-infant separation and the nature of interindividual relationship: experiments with Rhesus monkeys', *Proceedings of the Royal Society of London. B.*, 1962: 225–233, 1977.

Hundley, G., 'The breeding of elephants in captivity', *Journal of Bombay Natural History Society*, 1922, 28: 537–538.

Hundley, G., 'Statistical record of growth in the Indian elephant (*Elephas maximus*)', *Journal of Bombay Natural History Society*, 1935, 37 (2): 487–488.

Hutajulu, B. & Janis, R., 'The care and management of domesticated elephants in Sumatra, Indonesia', In: *Giants on Our Hands. Proc. Int. Workshop on the Domesticated Asian Elephant*, (I. Baker, M. Kashio, eds.) FAO, Bangkok, 2002, 59–66.

Illangakoon, A. 'A preliminary study of the captive elephants in Sri Lanka', *Gajah*, 1993, 11: 29–42.

Ishwaran, N. *Ecological Studies on Populations of Elephants in the Gal Oya Area in Relation to Distribution, Habitat Preferences, Competition and Food Availability*, MS – thesis. University of Peradeniya, Sri Lanka, 1979.

Ishwaran, N., 'Comparative study on elephant populations in Gal Oya, Sri Lanka', *Biological Conservation*, 1981a, 21: 303–313.

Ishwaran, N., 'Elephants and woody-plant relationships in Gal Oya, Sri Lanka', *Biological Conservation*, 1981b, 26: 255–270.

Ishwaran, N., *The Ecology of the Asian Elephant (Elephas maximus L.) in Sri Lanka*, Ph D Thesis, Michigan State University, East Lansing, USA, 1984.

Ishwaran, N., 'Ecology of the Asian elephant in lowland dry zone habitats of the Mahaweli river basin, Sri Lanka'. *Journal of Tropical Ecology*, 1993, 9: 169–182.

Ishwaran, N. & Punchi Banda, A., *Conservation of Sri Lankan Elephant – Planning and Management of the Wasgamuwa–Maduru Oya–Gal Oya complex of Reserves*, Final report on project 1783 to WWF / IUCN, Gland, Switzerland, 1982.

Ishwaran, N., Santiapillai, C., Amerasinghe, F.P. & Ekanayake, U. *On the Significance and Importance of the Elephant Corridor between the Proposed Maduru Oya National Park and the Wasgomuwa Strict Nature Reserve*, University of Perandeniya, Sri Lanka, 1980.

Islam, M.A., 'The status of Bangledesh's captive elephants'. In: *Giants on our Hands. Proc. Int. Workshop on the Domesticated Asian Elephant*, (I. Baker, M. Kashio, eds.) FAO, Bangkok, 2002, 67–78.

Jaeggi, P., 'In mir ist ein Meer von Traurigkeit – Die Leiden der Tempelelefanten', *Schweizer Tierschutz*, 1993, 120 (3): 4–52.

Jaeggi, P., 'Sehen, wie viele Tränen er hat', *Schweizer Tierschutz*, 1997, 124 (4) 4–63.

Jainudeen, M.R., Eisenberg, J.F. & Tilakeratne, N., 'Oestrus cycle of the Asiatic elephant *Elephas maximus* in captivity', *Journal of Fertility and Reproduction*, 1971a, 27: 321–328.

Jainudeen, M.R., Eisenberg, J.F. & Jayasinghe, J.B., 'Semen of the Ceylon elephant *Elephas maximus*', *Journal of Fertility and Reproduction*, 1971b, 24: 213–217.

Jainudeen, M.R., Mckay, G.M. & Eisenberg, J.F., 'Observation on musth in the domesticated Asiatic elephant (*Elephas maximus*)', *Mammalia*, 1972, 36 (2): 247–261.

Janis, C., 'The evolutionary strategy of the equidae and the origins of rumen and caecal digestion', *Evolution*, 1976, 30: 757–774.

Jayasinghe, J.B. & Jainudeen, M.R., 'A census of the tame elephant population in Ceylon with reference to location and distribution', *Ceylon Journal of Science (Biological Science)*, 1970, 2: 63–68.

Jayawardena, B.A.D.S., Perera, B.V.P. & Prasad, G.A.T., 'Rehabilitation and release of orphaned elephants back into the wild in Sri Lanka', *Gajah*, 2002, 21: 87 – 88.

Jayewardene, J., 'Elephant management and conservation in the Mahaweli project areas', *Gajah*, 1993, 11: 6–15.

Jayewardene, J., *The Elephant in Sri Lanka*, The Wildlife Heritage Trust of Sri Lanka, Colombo, 1994.

Jayewardene, J., 'Elephant conservation problems in Sri Lanka's Mahaweli River Basin'. In: *A Week with Elephants* (J. C. Daniel & H. Datye, eds.), Bombay Natural History Society, Bombay: 1995, 197–216.

Jayewardene, J., 'The status of captive elephants in Sri Lanka'. In: *Proc. Workshop on Captive Elephant Management* (J.V. Cheeran, ed.), 2002a, Trichur. In print.

Jayewardene, J., 'The care and management of domesticated Asian elephants in Sri Lanka'. In: *Giants on our Hands. Proc. Int. Workshop on the Domesticated Asian Elephant* (I. Baker, M. Kashio, eds.), FAO, Bangkok, 2002b, 43–58.

Kapustin, N., Critser, J.K., Olson, D. & Malven, P.V., 'Nonluteal oestrous cycles of 3-week duration initiated by anovulatory luteinizing hormone peaks in African elephants', *Biology of Reproduction*, 1996, 55: 1147–1154.

Karoor, J. J., 'Some aspects of the ecology and behaviour of  Asian elephants in the Periyar Wildlife Reserve, Kerala (India)'. In: E.G. Silas, M. Krishna,

P.G. Nair & G. Nirmalan (eds.), *The Asian Elephant; Ecology, Biology, Diseases, and Management*, Lumière Printing Works, Trichur: 1992, 29–33.

Katugaha, H.I.E., De Silva, M. & Santiapillai, C., 'A long-term study on the dynamics of the elephant (*Elephas maximus*) in Sri Lanka', *Biological Conservation*, 1998.

Kerschbaumer, P., 'Verletzungen und Erkrankungen der Augen'. In: *Elefant in Menschenhand* (F. Kurt, ed.), Filander, Fürth, 2001, 179–184.

Kiley-Worthington, M., *Behavioural Problems of Farm Animals*, Oriel Press, London, 1977.

Kiley-Worthington, M., *Animals in Circuses and Zoos – Chiron's World?*, Little Eco-Farms Publishing, Basildon, Essex, G.B, 1990.

Knox, R., *A Historical Relation of Ceylon*, Tisara Press, Dehiwala, 1966.

Kock, K., *Elefanten-mein Leben*, Rasch & Röhring, Hamburg, 1994.

Koenig, O., *Urmotiv Auge*, München und Zürich, 1975.

Krishnamurthy, V., 'Care and management of elephant calves in captivity'. In: E.G. Silas, M. Krsihna, P.G. Nair & G. Nirmalan (eds.), *The Asian Elephant: Ecology, Biology, Diseases, and Management*, Lumière Printing Works, Trichur: 1992, 82–85.

Krishnamurthy, V., 'Reproductive pattern in captive elephants in the Tamil Nadu Forest Department, India'. In: *A Week with Elephants* (eds. J.C. Daniel & H. Datye), Bombay Natural History Society, Bombay, 1995.

Krishnamurthy, V. & Wemmer, C., 'Timber elephant management in British India (1844–1947) with special emphasis on the Madras Presidency'. In: : *A Week with Elephants* (eds. J.C. Daniel & H. Datye), Bombay Natural History Society, Bombay, 1995a.

Krishnamurthy, V. & Wemmer, C., 'Veterinary care of Asian timber elephants in India: historical accounts and current observations', *Zoo Biology* 1995b, 14: 123–133.

Kummer, H. & Goodall, J., 'Conditions of innovative behaviour in primates', *Philosophical Transactions of the Royal Society of London*, 1985, 308: 203–214.

Kummer, H. & Kurt, F., 'Social units of free-living Hamadryas baboons', *Folia Primatologica*, 1963, 1: 4–19.

Kummer, H. & Kurt, F., 'A comparison of social behavior in captive and wild hamadryas baboons'. In: *The Baboon in Medical Research* (H. Vagtborg, ed.). University of Texas Press, Austin, 1965, 1–46.

Kurt, F., 'Le sommeil des elephants'. *Mammalia*, 1960, 24: 259–272.

Kurt, F., 'Smithsonian Elephant Survey', *Loris*, 1967, 11 (2): 56–61.

Kurt, F., 'Elephant survey in Sri Lanka', *Oryx* 1968, 9: 364–365.

Kurt, F., 'Some observations on Ceylon elephants', *Loris*, 1969a, 11 (5): 238–243.

Kurt, F., Elephant Distribution Map, Ruhuna National Park, scale 1 : 31,680,4 sheets. Survey Department, Ceylon, 1969b.

Kurt, F., 'Comparison of reproduction in tame and wild elephants'. In: Elliot, H.F. (ed.), *Procs. 11th Tech. Meeting of the Commission on Ecology*, Vol. 1 IUCN, Morges: 1970a, 148–155.

Kurt, F., Final Report to WWF and IUCN on the Leuser Survey (North Sumatra), 1970b, WWF, Morges, Switzerland.

Kurt, F., Der Leuser Survey (Sumatra), Z. *Kölner Zoo*, 1970C.

Kurt, F., 'Remarks on the social structure and ecology of the Ceylon elephant in the Yala National Park'. In: V. Geist & F. Walther (eds.), *The Behaviour of Ungulates and its Relation to Management*, IUCN Publ. New Series 24, Morges, Switzerland, 1974a, 618–634.

Kurt, F., 'Asiatic elephants: shapers of the jungle', *Image Roche*, 1974b, 59: 2–13.

Kurt, F., *Wildtiere in der Kulturlandschaft*, 1977, Eugen Rentsch Verlag, Zürich.

Kurt, F., 'Socio-ecological organization and aspects of management in South Asian deer'. In: *Threatened Deer*, 1978, IUCN Publications, New Series 24: 618–634.

Kurt, F., 'Echthirsche, Muntjakhirsche, Fleckenhirsche'. In: *Grzimeks Enzyklopädie. Säugetiere Bd. V.* (B. Grzimek, ed.), Kindler Verlag, München, 1988, 201–211.

Kurt, F., *Das Reh in der Kulturlandschaft*, Paul Parey, Hamburg und Berlin, 1991a.

Kurt, F., *Die Gärtner von Eden*, Rasch und Röhring, Hamburg, 1991.

Kurt, F. *Das Elefantenbuch*, Rasch und Röhring, Hamburg, 1992.

Kurt, F., 'Zur Biologie und Geschichte der Elefantenhaltung in Südasien', *Bongo*, Berlin, 1993a, 22: 3–22.

Kurt, F., Die Erhaltung des Asiatischen Elefanten in Menschenobhut: Ein Vergleich zwischen verschiedenen Haltuingssystemen in Südasien und Europa, Z. *Kölner Zoo*: 1994a, 91–113.

Kurt, F., Elefanten in Zoo und Zirkus. In: *Mensch und Elefant. Hrsg.* Staatliches Museum für Völkerunde in München, Pinguin Verlag, Innsbruch & Umschau – Verlag, Frankfurt/Main, 1994b, 117–133.

Kurt, F., 'The preservation of Asian elephants in human care–a comparison between the different keeping systems in South Asia and Europe', *Animal Research and Development*, 1995, 41: 38–60.

Kurt, F., 'Pinnawela Elephant Orphanage–Die grösste Elefantenherde in Menschenobhut'. Berichte aus Forschung und Wissenschaft, *Schönbrunner Tiergarten Journal*, 1998a, 7(24): 1–11.

Kurt, F., Wildtiere in zerissener Welt – Beobachtungen und Gedanken über 30 Jahre Artenschutz in Indien, *Z. Kölner Zoo*, 1998b, 41: 159–195.

Kurt, F., 'Wild animals in a strife – torn world – observations and reflections on 30 years of wildlife conservation in India', *Animal Research and Development*, 1999, 50: 7–43.

Kurt, F. (ed.), *Elefant in Menschenhand*, 2001, Filander, Fürth.

Kurt, F. & Garai, M.E., 'Stereotypies in captive Asian elephants–a symptom of social isolation'. In: *Research Update on Elephants and Rhinos* (H.M. Schwammer, T.J. Foose, M.Fouraker, D. Olson, eds.), 2002, 57–63. Schüling Verlag, Münster.

Kurt, F. & Hartl, G.B., 'Asian elephants (*Elephas maximus*) in captivity – a challenge for zoo biological research'. In: *Research and Captive Propagation* (U. Ganslosser, J.K. Hodges & W. Kaumanns, eds.), 1995, Filander Verlag, Fürth: 310–326.

Kurt, F., Hartl, G.B. & Tiedemann, R. 'Tuskless bulls in Asian elephant *Elephas maximus'*, 'History and population genetics of a man-made phenomenon', *Acta Theriologica*, 1995, Supplement 3: 125–143.

Kurt, F. & Knie, L., *Tiger und Elefanten im Zirkus Knie*, 1980, Ringier Verlag, Zürich.

Kurt, F. & Kumarasinghe J. C., 'Remarks on body growths and phenotypes in Asian elephant *Elephas maximus'*, *Acta Theriologica*, 1998, Supplement 5: 135–153.

Kurt, F. & Mar, K.U. 'Neonate mortality in captive Asian elephants (*Elephas maximus*)' *Zeitschrift für Säugetierkunde*, 1996, 61: 155–164.

Kurt, F. & Mar, K.U. 'Remarks on AsESG/IUCN and the management of captive Asian elephants', *Gajah*, 2003, 22: 1–18.

Kurt, F., Mar, K.U. & Garai, M.E., 'Giants in chains: History, biology and preservation of Asian elephants in captivity'. In: *Never Forgetting: Elephants and Ethics* (C.M. Wemmer & C.A. Christen, eds.), The John Hopkins University Press (in print), 2006 .

Kurt, F. & Nettasinghe, A.P.W., 'Estimation of body weight of the Ceylon elephant (*Elephas maximus*)', *Ceylon Journal of Veterinary Sciences*, 1968, 14: 24–28.

Kurt, F. & Pucher, H. Körpermasse, 'Gewicht und Gewichtsschätzung *(Elephas maximus)*'. In: *Elefantenhaltung in Zoo und Zirkus.* K. Schaller & C. Ziegler (eds.), Zoo Münster: 1998, 13–20.

Kurt, F. & Touma, C., 'Musth in wild-living and captive Asian elephants in Sri Lanka'. In: *Research Update on Elephants and Rhinos* (H.M. Schwammer, T.J. Foose, M.Fouraker, D. Olson, eds.), Schüling Verlag, Münster, 2002, 64–69.

Lahiri – Choudhury, D.K., 'Elephants in northeast India', *WWF Monthly Report*, Gland, Switzerland, 1986, 7–17.

Lahiri – Choudhury, D.K., 'Direct count of elephants in North – East India'. In: U. Ramakrishnan, R.A. Santosh & R. Sukumar (eds.), *Censusing Elephants in Forests*, Techn. Report No. 2, Asian Elephant Conservation Centre, Bangalore: 1991, 33–45.

Lahiri – Choudhury, D.K., 'Conservation of the Asian elephant – 1971–92, an overview'. In: J.C. Daniel & H. Datye (eds.), *A Week with Elephant*, Bombay Natural History Society, 1995, 9–31.

Lair, R.C., *Gone astray – the care and management of the Asian elephant in domesticity.* FAO Regional Office for Asia and the Pacific (RAP), Thailand. RAP Publ. Nr. 1997/16, 1997.

Lamprecht, J., 'Mechnismen des Paarzusammenhaltes bei Cichliden *Tilapia mariae* Boulenger 1899 (Cichlidae, Teleostei)', *Z. Tierpsychol*, 1973, 32: 10–61.

Lang, E.M. & Eggenberger, U., 'Protokoll einer Elefantengeburt', *Zoologische Garten (New Folge)*, 1991, 1: 5–7.

Langbauer, W.R., 'Elephant communication', *Zoo Biology*, 2000, 19: 425–445.

Laurie, A. 'Das Indische Panzernashorn'. In: *Die Nashörner*, Filander Verlag, Fürth: 1997, 95–114.

Laws, R, R. M., 'Biology of African elephants', *Sci. Progr.*, 1970, 58: 251–262.

Leader - Williams, N. & Dublin, H.T., Charismatic megafauna as 'flagship species'. In: *Priorities for the Conservation of Mammalian Diversity*, Cambridge University Press, Cambridge, 2000.

LeDoux, J., *The Emotional Brain: The Mysterious Underpinnings of Emotional Life*, Touchstone Books, 1996.

Lee, P.C., 'Allomothering among African elephants', *Animal Behaviour*, 1987, 35: 278–291.

Lee, P. C. & Moss, D.J., 'Early maternal investment in male and female African elephant calves', *Journal of Behaviour, Ecology and Sociobiology*, 1986, 18 (5): 353–361.

Lekagul, B. & McNeely, J. A., 'Elephants in Thailand: Importance, status and conservation', *Tigerpaper*, 1977, 4 (3): 22–25.

Lent, P. C., 'Mother-infant relations in ungulates'. In: *The Behaviour of Ungulates and its Relation to Management*, V. Geist & F. Walter (eds.). IUCN Publications New Series, 1974, 24: 14–55.

Lincoln, G.A. & Rantansooriya, W.D., 'Testosterone secretion, musth behaviour and social dominance in captive male Asian elephants living near the equator', *Journal of Fertility and Reproduction*, 1996, 108: 107–133.

Lohanan, R., 'The elephant situation in Thailand and a plea for co-operation'. In: *Giants on our Hands. Proc. Int. Workshop on the Domesticated Asian Elephant* (I. Baker, M. Kashio, eds.). FAO, Bangkok, 2002, 231–238.

320 ✦ *The Asian Elephant in Captivity – A Field Study*

Löhlein, W., *Untersuchungen zur Verdaulichkeit von Futtermitteln beim Asiatischen Elefanten* (*Elephas maximus*), Dissertation, Ludwig – Maximilians–Universität, München, 1999.

Lorenz, K., 'Über den Begriff der Instinkthandlung', *Folia biotheoretica*, 1937, 2: 17–50.

Ludescher, B., 'Haltungssysteme'. In: *Elefant in Menschenhand* (F. Kurt, ed.), Filander, Fürth, 2001, 169–174.

Ludescher, B. 'Verletzungen und Narben'. In: *Elefant in Menschenhand* (F. Kurt, ed.), 2001, 175–178. Filander, Fürth.

Lundberg, U., Szdzuy, K. & König, I., 'Beobachtungen zur Verhaltensontogenese Afrikanischer Elefanten (*Loxodonta africana*) im Tierpark Berlin–Friedrichsfelde', *Milu*, 2001, 10: 392–408.

Lydekker, R., *Catalogue of the Ungulate Mammals in the British Museum* (N. H.), 1916, Vols. IV & V.

Mahsavangkul, S., 'Domestic elephant status and management in Thailand'. In: *A Research Update on Elephants and Rhinos* (H.M. Schwammer, T.J. Foose, M. Fouraker & D. Olson, eds.), Schüling, Münster, 2002, 71–82.

Mainka, S.A. & Lothrop, C.D., 'Reproductive and hormonal changes during the estrous cycle and pregnancy in Asian elephants (*Elephas maximus*)', *Zoo Biology*, 1990, 9: 411–419.

Mar, K.U., 'The studbook of timber elephants of Myanmar with special reference to survivorship analysis'. In: *Giants on our Hands. Proc. Int. Workshop on the Domesticated Asian Elephant* (I. Baker, M. Kashio, eds.), FAO, Bangkok, 2002, 195–212.

Mar, K.U. & Win, S., 'Working elephants for the timber extraction of Myanmar', *Draught Animal News*, 1997, 2: 6–13.

Mar, K.U., 'Development of artificial insemination in Myanmar elephants (*Elephas maximus*)', *Forestry Research Paper*, Union of Myanmar, Ministry of Forestry, Forest Department, Yangon, 1992, 1–12.

Mar, K.U.; Kurt, F., Nyut, T.U. & Mynth, A., 'Fortpflanzung bei Arbeitselefanten in Myanmar'. In: *Elefantenhaltung in Zoo und Zirkus*, K. Schaller & C. Ziegler (Eds), Zoo Munster, 1997, 26–37.

Markowitz, H., 'Environment enrichment and the conservation of behaviour'. In: *Proceedings of the Second International Conference on Behavioural Enrichment* (B. Holst, ed.), Copenhagen Zoo, 1997.

Markowitz, H., Schmidt, M., Nadal, L. & Squier, L., Do elephants ever forget? *Journal of Applied Behaviour Research*, 1975, 8: 333–335.

Martin, F., *Rüsselmotorik und Freßstrategien beim Asiatischen Elefanten* (*Elephas maximus*), 1999, Diplomarbeit, Technische Universität Berlin.

Martin, P. & Bateson, P., *Measuring Behaviour–an Introductory Guide*, 2nd Edition, Cambridge University Press, Cambridge, 1993.

Marriner, L.M. & Drickamer, L.C., 'Factors influencing stereotyped behavior of primates in a zoo', *Zoo Biology*, 1994, 13: 267–275.

Mason, G.J., 'Stereotypies: a critical review', *Animal Behaviour*, 1991, 41: 1015–1037.

Mathew, A.J., 'Management of the captive elephant population in West Bengal with special reference to the Jaldapara Wildlife Sanctuary'. In: *Workshop on Captive Elephant Management* (J.V. Cheeran, ed.), Trichur, 2002, 29.

Mawberry, B., 'Breeding Indian elephants *Elephas maximus* at Portland Zoo', *International Zoo Yearbook.*, 1965: 80 –83.

Meyer, J., *Das Altindische Buch vom Welt – und Staatsleben*, 1929, Leipzig.

McGrew, W. C., 'Intelligent use of tools by chimpanzees: ten propositions', *American Journal of Primatol.*, 1990, 20: 211–212.

Mckay, G.M., *'Behavioural and Ecology of the Asiatic Elephant in Southeastern Ceylon*, Smithsonian Contribution to Zoology, No. 125, Smithsonian Institution Press, Washington D.C, 1973, 113 pp.

McNeilly. B. & Sinha, 'Protected areas for Asian elephants', *Parks* 6 (1): 1981, 4–7.

McNeilly, A.S., Martin, R.D., Hodges, J.K. & Smuts, G.L., 'Blood concentrations of gonadotrophins, prolactin and gonadal steroids in males and non-pregnant and pregnant female African elephants (*Loxodonta africana*)', *Journal of Fertility and Reproduction*, 1983, 67: 113–120.

Meaney, M.J. 'Maternal care, gene expression, and the transmission of individual differences in stress reactivity across generations', *Annual Reviews of Neuroscience*, 2001, 24: 1161–1192.

Menon, V., *Tusker. The Story of the Asian Elephant*, Penguin, Delhi, 2002.

Meyer, J., *Das Altindische Buch von Welt-Staatsleben*, Leipzig, 1929.

Meyer – Holzapfel, M., 'Abnormal behavior in zoo animals'. In: *Abnormal Behavior in Animals* (M. W. Fox ,ed.), Philadelphia, 1968, 476–503.

Milroy, A.J., *A Short Treatise on the Management of Elephants*, Government Printer, Shillong, Assam, 1922.

Mohamed, M.R., 'Behaviour of juvenile elephants relaesed from the Elephant Transit Home to the Uda Walawe NP', *Sixth Newsletter, Biodiversity & Elephant Conservation Trust*, Sri Lanka, 2002.

Moller, A.P., 'Ejaculate quality, testes size and sperm competition in primates', *Journal of Human Evolution*, 1989, 17: 479–488.

Morgan, B.J.T., Simpson, M.J.A., Hanby, J.P. & Hall-Craggs, J., 'Visualizing interaction and sequential data in animal behaviour: theory and application of cluster–analysis methods', *Behaviour*, 1976, 56, 1–43.

Moore, R.J., *Back to Africa*, Southern Book Publ. (Pvt.) Ltd, Johannesburg, 1989.

Morris, D., 'The response of animals to a restricted environment', *Symposia Transactions of the Royal Society of London*, London, 1964, 13: 99–118.

Morris, D., 'Abnormal rituals in stress situations', *Philosophical Transactions of the Royal Society of London*, 1966, 251: 327–118.

Moss, C.J., 'Social circles', *Wildlife News*, 1981, 16 (1): 2–7.

Moss, C.J., 'Oestrus behaviour and female choice in the African elephant', *Behaviour*, 1983, 86: 167–196.

Moss, C.J., 'The demography of an African elephant (*Loxodonta africana*) population in Amboseli, Kenya', *Journal of Zoology*, 2001, 255: 145–156.

Moss, C.J. & Poole, J.H., 'Relationships and social structure of African elephants'. In: *Primate Social Relationships: An Integrated Approach* (R.A. Hinde, ed.), Blackwell Scientific Publishing: 1983, 315–325.

Moss, C.J., *Elephant Memories: Thirteen years in the Life of an Elephant Family*, William Morrow, New York, 1988.

Mueller – Dombois, D., 'Crown distortion and elephant distribution in the woody vegetations of Ruhunu National Park, Ceylon', *Ecology*, 1972, 53 (2): 208–226.

Murphy, C. & Kern, T., 'Refractive state, corneal curvature, accomodative range and ocular anatomy of the Asian elephant (*Elephas maximus*)', *Vision Research*, 1992, 32 (11): 2013–2021.

Murphy, R., 'The ruins of ancient Ceylon', *Journal of Asian Studies*, 1957, 16: 1–185.

Nair, P.G., Radhakrishnan & K. Chandrasekharan, 'Mating behaviour of the Asian elephant in captivity'. In: E.G. Silas, M. Krishna, P.G. Nair & G. Nirmalan, (eds.), *The Asian Elephant: Ecology, Biology, Diseases, and Management*, Lumière Printing Works, Trichur, 1992, 38–40.

Nettasinghe, A.P.W., *The Inter-relationship of Livestock and Elephants at Thamankaduwa Farm with Special Reference to Feeding and Environment*, MVSc thesis. University of Ceylon, Peradeniya Campus, 1973.

Niemuller, C. A. & Litrop R.M., 'Altered androstenedione to testosterone ratios and LH concentrations during musth in the captive male Asian elephant (*Elephas maximus*)', *Journal of Fertility and Reproduction*, 1991, 91: 139–146.

Niemuller, C.A., Shaw, H.J. & Hodges, J.K., 'Non-invasive monitoring of ovarian function in Asian elephants (*Elephas maximus*) by measurement of urinary 5beta – pregnanetriol', *Journal of Fertility and Reproduction*, 1993, 99: 617–625.

Norris, C.E., *Preliminary Report on the Ceylon Elephant Field Survey*, Wildlife Protection Society, Colombo, 1959.

Norris, C.E., *PART II. Preliminary Report on the Ceylon Elephant Field Survey*, Wildlife Protection Society, Colombo, 1967.

Nozawa, K. & Shotake, K, 'Genetic differentiation among local populations of Asian elephant', *Zschr. f. zool. Systematik & Evolutionsforschung*, 1990, 28: 40–47.

Ödberg, F.O., 'Abnormal behaviours: (stereotypies)'. In: *First World Congress on Ethology Applied to Zootechnics*, 1978, Garsi, Madrid.

Odum, E.P., *Fundamentals of Ecology*, W.B. Saunders Company, Philadelphia, London, Toronto.

Olivier, R.C.D., *On the Ecology of the Asian Elephant Elephas maximus Linn.: with Particular Reference to Malaya and Sri Lanka.* PhD thesis, University of Cambridge, UK.

Olsen, J.H., Chen, C.L., Boules, M.M., Morris, L.S. & Coville, B.R., 'Determination of reproductive cyclicity and pregnancy in Asian elephants (*Elephas maximus*) by rapid radioimmunoassay of serum progesterone', *Journal of Zoo and Wildlife Medicine*, 1994, 25: 349–354.

Owen – Smith, R.N., *Megaherbivores: The Influence of very Large Body Size on Ecology*, Cambridge University Press, Cambridge, 1988.

Panicker, K.C. 'Morphological pecularity of maknas and cow elephants'. In: C.K. Karunakaran (ed.), *The Proceedings of the Elephant Symposium*, Kerala Books & Publication Society, Kakkanad, Kochi: 1990, 56–57.

Panksepp, J., *Affective Neurosciences: The Foundations of Human and Animal Emotions*, Oxford University Press, 1998.

Payne, K.B., Langbauer, W.R. Jr & Thomas, E.M., 'Infrasonic calls of the Asian elephant (*Elephas maximus*)', *Behavioral Ecology and Sociobiology*, 1986, 18: 297–301.

Petters, R.H., *The Ecological Implication of Body Size*, Cambridge University Press, Cambridge, 1983.

Phillips, W.W.A., *Manual of the Mammals of Sri Lanka*, Wildlife and Nature Protection Society of Sri Lanka, Colombo, 1980.

Pieris, P.E., *Ceylon and the Portugese, 1505–1658*, 1920, American Ceylon Mission Press, Telipalai.

Pieler, E. & Kurt, F., 'Tool behaviour in Asian elephants'. In: *Research Update on Elephants and Rhinos* (H.M. Schwammer, T.J. Foose, M.Fouraker, D. Olson, eds), Schüling Verlag, Münster, 2002, 104–109.

Pieris, P.E., *Ceylon in the Portugese Era*, Tisara Press, Dehiwala, 1929.

Pimmanrojnagool, V. & Wanghongsa, S., 'A study of street wandering elephants in Bangkok and the socio-economic life of their mahouts'. In: *Giants on our Hands. Proc. Int. Workshop on the Domesticated Asian Elephant* (I. Baker, M. Kashio, eds), FAO, Bangkok, 2002, 35–42.

Plotka, E.D., Seal, U.S., Zaremba, F.R., Simmons, L.G., Teare, A., Phillips, L.G., Hinshow, K.C. & Wood, D.G., 'Ovarian function in the elephant: Luteinizing hormone and progesterone cycles in African and Asian elephants', *Biology of Reproduction*, 1988, 38: 309–314.

Poole, J.H., 'Rutting behaviour in African elephants', *Behaviour*, 1987, 102: 283–316.

Poole, J.H., 'Mate guarding, reproductive success and female choice in African elephants', *Animal Behaviour*, 1989, 37: 842–849.

Poole, J.H., 'Signals and assessment in African elephants: evidence from playback experiments', *Animal Behaviour*, 1999, 58: 185–193.

Poole, J.H., Kasman, L.H. & Lasley, B.L., 'Musth and unrinary tesosterone concentrations in the African elephant (*Loxodonta africana*)', *Journal of Fertility and Reproduction*, 1984, 70: 255–260.

Poole, J.H. & Moss, C.J., 'Musth in the African elephant (*Loxodonta africana*)', *Nature*, 1981, 292: 830–831.

Poole, T.B., Taylor, V.J., Fernando, S.B.U., Ratnasooriya, W.D., Ratnayake, A., Lincoln, G., McNeilly, A. & Manatunga, A.M.V.R., 'Social behaviour and breeding physiology of a group of captive Asian elephants', *International Zoo Yearbook*, 1997, 35: 297–310.

Portmann, A., *Die Tiergestalt*, Verlag Friedrich Reinhardt, Basel, 1948.

Rajapaksa, E.C., Mendis, G.U.S.P., Wijesinghe, C.G., Alahakoon, J., Kumarawadu, P.L. & De Silva; L.N.A., 'Treatment and management of an elephant calf with a head injury', *Workshop on Captive Elephant Management*, Trichur (in print), 2002.

Rappaport, L. & Haight, J., 'Some observations regarding allomaternal caretaking among captive Asian elephants (*Elephas maximus*)', *J. Mammalogy*, 1987, 68 (2), 438–442.

Ramsay, E.C., Lasley, B.L. & Stabenfeldt, G.H., 'Monitoring the estrous cycle of the Asian elephant (*Elephas maximus*) using urinary estrogens', *American Journal of Veterinary Research*, 1981, 42: 256–260.

Rasmussen, L.E., Buss, I.O., Hess, D.L. & Schmidt, M.J., 'Testosterone and dihydrotestosterone concentrations in elephant serum and temporal gland secretions', *Biology of Reproduction*, 1984, 30: 352–362.

Rasmussen, L.E.L., Bryce, L. & Munger, B. L., 'The sensorineuronal specialisations of the trunk (finger) of the Asian elephant, *Elephas maximus*', *The Anatomical Record*, 1996, 246: 127–134.

Rasmussen, L.E.L., Hess, D.L. & Haight, J.D., 'Chemical analysis of temporal gland secretions collected from an Asian bull elephant during a four-month musth period', *Journal of Chemical Ecology*, 1990, 16: 2167–2181.

Rasmussen, L.E.L. & Krishnamurthy, V., 'How chemical signals integrate Asian elephant society: the known and the unknown', *Zoo Biology*, 2002, 19: 405–423.

Rasmussen, L.E.L., Lee, T.D., Roelofs, W.L., Zhang, A. & Daves, G.D.Jr., 'Insect pheromone in elephants', *Nature*, 1996, 379: 684.

Rasmussen, L.E.L. & Riddle, H.S., 'Musth in teenage male Asian elephants (*Elephas maximus*): the what & the why of their chemical signals.' In: *Research Update on Elephants and Rhinos* (H.M. Schwammer, T.J. Foose, M.Fouraker, D. Olson, eds.), Schüling Verlag, Münster, 2002, 110.

Rasmussen, L.E.L. & Schulte, B.A., 'Chemical signals in the reproduction of Asian (*Elephas maximus*) and African (*Loxodonta africana*) elephants', *Animal Reproduction Science*, 1998, 53: 19–34.

Ratnasooriya, W.D., Fernando, S.B.U., & Manatunga, B.M.V.R., 'Pregnancy duration of Sri Lankan elephant (*Elephas maximus*) in captivity', *Journal of Medical Science Research*, 1991, 19: 623–624.

Ratnasooriya, W.D., Fernando, S.B.U. & Manatunga, A.M.V.R., 'Serum tesosteron levels of Sri Lankan female elephants (*Elephas maximus*)', *Journal of Medical Science Research*, 1992a, 20: 79–80.

Ratnasooriya, W.D. & Fernando, S.B.U., Pattern of toe nail distribution of Sri Lankan elephant, *Elephas maximus*, *Journal of Medical Science Research*, 1992b, 20: 221–222.

Ratnasooriya, W.D., Fernando, S.B.U. & Manatunga, B.M.V.R., 'Serum prolactin levels of Sri Lanka elephants (*Elephas maximus maximus*)', *Journal of Medical Science Research*, 1993, 21: 259–261.

Reilly, J. & Sutkatmoto, P., 'The elephant training centre at Way Kambas National Park, Sumatra: a review of its operations and recommendations for the future', *Gaja*, 2002, 21: 1–40.

Reimers, M, Schmidt, S. & Kurt, F., 'Daily activites and home ranges in wild Asian elephants of the Uda Walawe National Park (Sri Lanka)'. In: *Research Update on Elephants and Rhinos* (H.M. Schwammer, T.J. Foose, M.Fouraker, D. Olson, eds.), Schüling Verlag, Münster, 2002, 115–118.

Rensch, B., 'Increase of learning capacity with increase of brain size', *American Naturalist*, 1956, 90: 81–95.

Rensch, B., 'The intelligence of elephants', *Scientific American*, 1957, 196: 44–49.

Rensch, B. & Altevogt, R., 'Zähmung und Dressurleistungen indischer Arbeitselefanten', *Z. Tierpschol*, 1954, 11: 497–510.

Renshaw, E., *Modelling Biological Populations in Space and Time*, Cambridge University Press, Cambridge, 1991.

Riddle, H.S. & Rasmussen, L.E.L., 'Are female African elephants messaging through volatile chemicals? Studies from European, USA, and African groups'. In: *Research Update on Elephants and Rhinos* (H.M. Schwammer, T.J. Foose, M.Fouraker, D. Olson, eds.), Schüling Verlag, Münster, 2002, 119 –120.

Riedman, M.L. 'The evolution of alloparental care and adoption in mammals and birds', *Quaterly Review of Biol*ogy, 1982, 57(4), 405–435.

Rietkerk, F.E., Hiddingh, H. & Van Dijk, S., Hand-rearing an Asian elephant, *Elephas maximus, International Zoo Yearbook*, 1993, 32: 244–232.

Roy, S.D., 'The sad status of elephants in Southern Assam, India', *Gajah*, 1993, 11: 53–54.

Ruckbusch, Y., 'The relevnace of drowsiness in the circadian cycle of farm animals, *Animal Behaviour*, 1972, 20: 637–643.

Rübel, A. & Tanner, R., 'Die Haltung von Elefantenbullen und das Phänomen der Musth', *Bongo*, 1993, 22: 65–72.

Ruthe, H., 'Fussleiden der Elefanten', *Wissenschaftlich Zeitschrift Humbold – Univ. Berlin.*, Math. – Nat. Reihe, 1961, 10: 471–516.

Sachser, N., & Lick, C., 'Social experience, behaviour, and stress in Guinea pigs', *Physiology and Behaviour*, 1991, 50: 83–90.

Sale. J.B., 'Wildlife research in the Indomalayan realm'. In: J.W.Thorsell (ed.), *Conserving Asia's Natural Heritage*, IUCN, Gland, Switzerland, 1990, 137–149.

Salwala, S., 'The role of private orgaizations in elephant conservation'. In: *Giants on our Hands. Proc. Int. Workshop on the Domesticated Asian Elephant* (I. Baker, M. Kashio, eds.), FAO, Bangkok, 2002, 223–226.

Salzert, W., *Elefanten, ihre Pathologie und den Tiergärtner interessierende physiologischen Daten*, 1972, Dissertation, Tierärztlichen Hochschule Hannover.

Salzert, W., Elefanten. In: *Zootierkrankheiten*, H. – G. Klös & E. M. Lang (eds.), Verlag Paul Parey, Berlin & Hamburg: 1976, 133–152.

Sanderson, G.P., *Thirteen Years Among  the Wild Beasts of India*, John Grant, Edinburgh, 1907.

Santiapillai, C., 'News from Sri Lanka', *Gajah*, 1993, 11: 5.

Santiapillai, C., Chambers, M.R. & Ishwaran, N., 'Aspects of the ecology of Asian elephant *Elephas maximus*  L. in the Ruhuna National Park, Sri Lanka', *Biological Conservation*, 1984, 29: 47–61.

Santiapillai, C., & De Silva, M., 'An action plan for the conservation and management of elephant (*Elephas maximus*) in Sri Lanka', *Gajah*, 1994,  13: 1–24.

Santiapillai, C. & Jackson, P., The Asian elephant: an action plan for its conservation, IUCN, WWF, Gland, Switzerland: 1990, 1–62.

Santipillai, C. & Suprahman, H., 'Elephants in Indonesia (Sumatra)', *WWF Monthly Report*, Gland Switzerland, 1985, 287–296.

Sapolsky, R.M., 'Stress in the wild', *Scientific American*, 1990, Jan. 90 : 106–113.

Sapolsky, R.M. *Why Zebras Don't Get Ulcers*, 1998, W.H. Freeman.

Schaaf, C. D. & Stuart, M.D., 'Reproduction of the mongoose lemur in captivity', *Zoo Biology*, 1983, 2: 23–38.

Schaller, G., *The Mountain Gorilla*, 1963, Chicago University Press, Chicago.

Scherer A. & Hill Goldsmith, H. (eds.), *Handbook of affective sciences*, Oxford University Press, 2003.

Scheve, R., Van Dijk, S., Van Der Valk, P., Walstra, W., Engels, F. & Hiddingh, H., Op goed geluuk, *Zoo Informatie van het Noorder Dierenpark*, 1993, 20 (1): 1–32.

Schmid, J., 'Keeping circus elephants temporarily in paddocks – the effects on their behaviour', *Animal Welfare*, 1995, 4: 87–101.

Schmidt, M.J., Schmidt, A.M., Henneous, R.L., Haight J.D., Rutkowski, C. & Sanford, J., 'Progress in the development of the practical artificial insemination technique for elephants'. In: E.G. Silas, M. Krsihna, P.G. Nair & G. Nirmalan (eds.), *The Asian Elephant: Ecology, Biology, Diseases, and Management*, Lumière Printing Works, Trichur: 1992, 123–126.

Schmidt, S., Reimers, M. & Kurt, F., 'Sleep in a herd of captive Asian elephants in the Pinnawela Elephant, Orphanage (Sri Lanka)'. In: *Research Update on Elephants and Rhinos* (H.M. Schwammer, T.J. Foose, M.Fouraker, D. Olson, eds.), Schüling Verlag, Münster, 2002, 124–128.

Schmidt, V., *Augenkrankheiten der Haustiere*, Gustav Fischer, Jena, 1988.

Schmidt–Nielson, K., *Scaling: Why is Animal Size So Important?*, Cambridge University Press, Cambridge, 1984.

Schore, A.N., Affect Regulation and Origin of the Self: the Neurobiology of Emotional Development, Erbaum, Mahwah NJ, 1994.

Schore, A.N., 'The Effects of Relational Trauma on Right Brain Development, Affect Regulation, and Infant Mental Health', *Infant Mental Health Journal*, , 2001, 22, 201–269.

Schore, A.N., Dysregulation of the right brain: a fundamental mechanism of traumatic attachment and the psychopathogenesis of posttraumatic stress disorder, *Australia & New Zealand J. Psychiat.*, 2002, 36, 9–30.

Schore, A.N., 'Attachment, affect regulation, and the developing right brain: Linking developmental neuroscience to pediatrics', *Pediatrics in Review*, 2005, Vol.26. no 6. pp 204–217.

Schreuder, J., 'Memoir of Jan Schreuder'. In: *Selections from the Dutch Records of the Ceylon Government, 5.*, Ceylon Government Press, Colombo: 1946, 1–202.

Schulte, B. A. & Rasmussen, L.E.L.,'Signal-receiver interplay in the communication of male condition by Asian elephants', *Animal Behaviour*, 1999, 57: 1265–1274.

Schwammer, H., 'Elefantenprojekt in Thailand', *Schönbrunner Tiergarten – Journal*, 2000, 9 (2): 1–5.

Schweitzer, C., 'Journal and diary of a six years East Indian journey', *The National Museum of Ceylon Translation Series*, 1953, 1: 37–82.

Shoshani, J., 'Anatomy and physiology'. In: *The Illustrated Encyclopedia of Elephants,* 1991, G. Rodgers & S. Watkinson (eds.), Salamander Books Ltd, London: 30–47.

Shoshani, J., Understanding proboscidean evolution: a formidable task, TREE 13, 1998, (12): 480–487.

Shoshani, J. & Eisenberg, J. F., *Elephas maximus, Mammalian Species*, 1982, 182: 1–8.

Shoshani, J., Golenberg, E.M. & Yang, H., 'Elephantidae phylogeny: morphological versus molecular results', *Acta theriologica*, Supplement 5: 1998, 89–122.

Shoshani, J. & Tassi, P. (eds.) *The Proboscidea: Evolution and Palaeoecology of Elephants and their Relatives*, Oxford University Press, Oxford, 1996.

Sikes, S.K., *The Natural History of the African Elephant*, 1971, Weidenfeld & Nicolson, London.

Sivanagesan, N. & Kumar, A., 'Status of feral elephants in Andamans'. In: *A Week with Elephants* (J. C. Daniel & H. Datye (eds.)), Bombay Natural History Society, Bombay: 1995, 97–119.

Sivanagesan N. & Sathyanarana, M.C., 'Tree mortality caused by elephants in Mudumalai wildlfe sanctuary, South India'. In: *A Week with Elephants* (J. C. Daniel & H. Datye (eds.), Bombay Natural History Society, Bombay: 1995, 314–330.

Slotow, R., Van Dyk, G., Poole, J.C., Page, B. & Klocke A., 'Older bull elephants control young males', *Nature*, 2000, 408: 425–426.

Smith, T.P., Jollie, K.G. & Mohr, J.L., 'Gut protozoans of zoo elephants', Abstract, *Journal of Protozool.*, 1982, 29: 482.

Spinage, C., *Elephants*, T & AD Poyser Ltd, London, 1994.

Spittel, R.L., *Wild Ceylon*, Colombo Book Centre, Colombo, 1951.

Staufacher, M., 15 'Thesen zur Haltungsoptimierung im Zoo', *Der Zoologische Garten* (Neue Folge), 1998, 68: 201–218.

Stevens, V.J., 'Basic operant research in the zoo'. In: *Behaviour of Captive Wild Animals* (H. Markowitz & V. J. Stevens, eds.), Neslon-Hall, Chicago, 1978.

Stevenson, M.F., *Management Guidelines for the Welfare of Elephants*, 2002, The Federation of Zoological Gardens of Great Britain and Ireland, London.

Stolba, A., Baker, N. & Wood–Gush, D.G.M., 'The characterization of stereotyped behaviour in stalled sows by informational redundancy', *Behaviour*, 1983, 87: 157–182.

Stolba, A. & Müllers, B., 'Die Bedeutung von Tierarten, Gehege und Verhalten für den Schauwert im Zoo', *Der Zoologische Garten* (Neue Folge), 1990, 60: 349–368.

Stracey, P.D., *Elephant Gold*, 1963, Weidenfeld and Nicholson, London.

Sukumar, R., *The Asian Elephant – Ecology and Management*, 1992, Cambridge University Press, Cambridge.

Sukumar, R., 'Minimum viable population for elephant conservation', *Gajah*, 1993, 11: 48–52.

Sukumar, R., 'Minimum viable populations for Asian elephant conservation'. In: *A Week with Elephants* (eds.: J. C. Daniel & H. Datye), 1995, Bombay Natural History Society, Bombay:279–288.

Sukumar, R., *The Living Elephants*, 2003, Oxford University Press, New York.

Sukumar, R., Joshi, N. V. & Krishnamurthy V., 'Growth in the Asian elephant', *Proceedings of the Indian Academy of Science (Animal Science)*, 1988, 97: 561–571.

Sukumar, R., Krishnamurthy, V., Wemmer, C. & Rodden, M., 'Demography of captive Asian elephants (*Elephas maximus*) in southern India', *Zoo Biology*, 1997, 16: 263–272.

Tanner, R., *Dicke Haut und zarte Seel. Mein Leben mit Elefanten*, 2000, Tecklenborg Verlag, Steinfurt.

Tashiro, H. & Schwardt, H.H., 'Biological studies of horse flies in New York', *J. Econ. Entomol.*, 1953, 46: 813–822.

Tavernier, J.B., *Reisen zu den Reichtümern Indiens. Abenteuerliche Jahre beim Grossmogul*, 1641–1667, 1984, K. Thienemanns Verlag, Stuttgart.

Taya, K., Komura, H, Kondoh, M., Ogawa, Y., Nakada, K., Watanabe, G., Sasamoto, S., Tanabe, K., Saito, K., Tajima, H. & Narushima, E., 'Concentrations of progesterone, testosterone and estradiol – 17beta in the serum during the estrous cycle of Asian elephants (*Elephas maximus*)', *Zoo Biology*, 1991,10: 299–307.

Tennent, J.E., *Natural History of Ceylon*, 1861, Longman, Green, Longman & Roberts, London.

Tennent, J.E., *The Wild Elephant and the Method of Capturing and Taming it in Ceylon*, 1867, Longman, Green and Co., London.

Tennessen, T., 'Coping with confinement – features of the environment that influence animal's ability to adapt', *Journal of Applied Behaviour Research*, 1989, 22: 139–149.

Terranova, C.J. & Coffmann, B.S., 'Body weight of wild and captive lemurs', *Zoo Biology*, 1997, 16: 17–30.

Tiedemann, R., 'Sexual selection in Asian elephants', *Science*, 1997, 278: 1550–1551.

Tiedemann, R., Hardy, O., Vekemans, X. & Milinkovitch, M.C., 'Higher impact of female than male migration on population structure in large mammals', *Molecular Ecology* 9: 1159–1163.

Tiedemann, R. & Kurt, F., 'A stochastic simulation model for Asian elephant *Elephas maximus* populations and the inheritance of tusks'. In: *Ecological genetics in mammals II*, G. B. Hartl & J. Markowski (eds.), *Acta Theriologica*, Supplement, 1995, 3: 111–124.

Tiedemann, R., Hammer, S., Suchentrunk, F. & Hartl, G.B., 'Allozyme variability in medium-sized and large mammals: determinants, estimators, and significance for conservation', *Biodiversity Letters*, 1996, 3: 81– 91.

Tiedemann, R., Kurt, F., Haberhauer, A. & Hartl, G.B., 'Differences in spatial genetic population structure between African and Asian elephant (*Loxodonta africana, Elephas maximus*) as revealed by sequence analysis of the mitochondrial Cyt *b* gene', *Acta Theriologica*, Supplement 5: 1998, 123–134.

Tiedemann, R., Kurt, F., Goonesekere, N., Gunasekera, M. B.& Ratnasooriya W.D., 'A simualtion study on the viability of Sri Lankan elephant *Elephas maximus* populations: retrospective model validation and future perspectives', *Folia Zoologica* 48, Supplement 1: 95–104, 1999.

Tilakaratne, N. & Santiapillai, C, 'The status of domesticated elephants at the Pinnawela Elephant Orphanage, Sri Lanka', *Gajah*, 2002, 21: 41–52.

Tinbergen, E.A. & Tinbergen, N., 'Early childhood autism – an ethological approach', *Advances in Ethology*, 1972, 10: 1–53.

Tobler, I., 'Behavioral sleep in the Asian elephant', *Sleep*, 1992, 15 (1): 1–12.

Toke Gale U., *Burmese Timber elephants*, 1974, Trade Corporation, Rangoon.

Touma, C. & Kurt, F., Ethologische Untersuchungen zur Musth bei Elefantenbullen. In: *Elefant in Menschenhand* (F. Kurt, ed.), Filander, Fürth, 2001, 185–196.

Trautmann, T.R., 'Elephants and the Mauriyas'. In: *India: History and Thought*, Mukerjee (ed.), Calcutta: 1982, 254–281.

Trivers, R.L., 'The evolution of reciprocal altruism', *Q. Rev. Biol*, 1971, 46: 35–37.

Trivers, R.L., 'Parental investment and sexual selection'. In: *Sexual Selection and the Descent of Man* (B. Champbell, ed.) Aldine, Chicago, 1972, 139–179.

Tudge, C., *Letzte Zuflucht Zoo – Die Erhaltung bedrohter Arten in Zoologischen Gärten*, Spektrum Akademischer Verlag, Berlin, Heidelberg, Oxford, 1993.

Tu Yin, *Wild Animals of Burma*, Rangoon Gazette Ltd., Rangoon, 1967.

Uragoda, C.G., *Wildlife Conservation in Sri Lanka*, Wildlife & Nature Protection Society of Sri Lanka, Colombo, 1994.

Vancuylenberg, B.W.B., *The Feeding Behaviour of Asiatic Elephants in Southeastern Ceylon*, M.Sc. thesis, University of Sri Lanka, Peradeniya Campus, 1974.

Vancuylenberg, B.W.B., 'Feeding behaviour of the Asiatic elephant in Southeast Sri Lanka in relation to conservation', *Biological Conservation*, 1977, 12: 33–54.

Van Goens, R, Memories of Rychloff van Goens (1663–1665). In: *Selections from the Dutch Records of the Ceylon Government, 3*, Ceylon Government Press: 1–112, 1932.

Walker, S. & Cheeran, J. 'Mahout training programme', *Zoo's Print*, 1996, 11 (1): 51–55.

Watve, M.G. & Sukumar, R., 'Asian elephants with longer tusks have lower parasite loads', *Current Science*, 1997, 72 (11): 885–888.

Webster, F., *Calf Husbandry, Health and Welfare*, Collins, London, 1984.

Wechsler, B., 'Stereotypies in polar bears', *Zoo Biology*, 1991, 10: 177–188.

Wechsler, B., 'Zur Genese von Verhaltensstörungen'. In: *Aktuelle Arbeiten zur artgemässen Tierhaltung* 1991, 1992, 23. International Arbeitstagung Angewandte Ethologie bei Nutztieren.: 9–17.

Wechsler, B., 'Coping and coping strategies: a behavioural review', *Journal of Applied Behaviour Research*, 1995, 43: 123–134.

Weihs, W., 'Molar growth and chewing frequencies as age indicators in Asian elephants'. In: *Research Update on Elephants and Rhinos* (H.M. Schwammer, T.J. Foose, M.Fouraker, D. Olson, eds), Schüling Verlag, Münster, 2002, 294–296.

Weihs, W., Weisz, I., Wüstenhagen, A. & Kurt, F., 'Body growth and food intake in a herd of captive Asian elephants at the Pinnawela Elephant Orphanage (Sri Lanka)', In: *A Research Update on Elephants and Rhinos* (H.M. Schwammer, T.J.Foose, M. Fouraker & D. Olson, eds.), Schüling, Münster, 2002, 141–145.

Wemmer, C. & Mishra, H.R., 'Observational learning by an Asiatic elephant of an unusual sound producing method', *Mammalia*, 1982, 46 (4): 556–557.

Wemmer, C. & Krishnamurthy V, 'Methods for taking standard measurements of live domestic elephants'. In: *The Asian Elephant, Ecology, Biology, Diseases, Conservation and Management*, E.G. Silas, M. K. Nair & G. Nirmalan (eds.), Lumière Printing Works, Trichur: 1992, 34–37.

Wickler, W. & Seibt, U., 'Aimed object – throwing by a wild African elephant in an interspecific encounter', *Ethology*, 1997, 103: 365–368.

Wiepkema, P.R., 'Behavioural aspects of stress'. In: *The Biology of Stress in Farm Animals, an Integrated Approach* (P. W. M. van Andrichem & P. R. Wiepkema, eds.), Martinus Nijhoff, Dordrecht, 1987, 113–134.

Wiese, R.J., 'Asian elephants are not self-sustaining in North America', *Zoo Biology*, 2000, 19: 299–309.

Wiesner, H., 'Zum Vorkommen des Arcus scleralis beim Elefanten', *Der Zoologische Garten* (Neue Folge), 1992, 62 (4): 287–293.

Williams, J.H., *Elephant Bill*, Rupert Hart–Davis, London, 1950.

Wilson, E.O., *Sociobiology: The New Synthesis*, Belknap / Harvard University Press, Bambridge Mass, 1971.

Wilson, J.F., Mahajan, U., Wainwright, S.A. & Croner, L.J., 'A continuum model of elephant trunks', *Journal of Biomechanical Engineering*, 1991, 113: 79–84.

Wingate, L. & Lasley, B., 'Is musth a reproductive event: An examination of arguments for and against this view'. In: *Research Update on Elephants and Rhinos* (H.M. Schwammer, T.J. Foose, M.Fouraker, D. Olson, eds.), Schüling Verlag, Münster, 2002, 150 –156.

Wirz, P., *Kataragama, the Holiest Place in Ceylon*, Lake House, Colombo, 1966.

Wyatt, J.R. & Eltringham, S.K., 'The daily activity of the elephants in the Rwenzori National Park, Uganda', *East African, Wildlife Journal*, 1974, 12: 273–289.

Zeppelin, H. & Rechtschaffen, A., Mammalian sleep, longevity, and energy metabolism, *Brain Behaviour and Evolution*, 1974, 10: 425–470.

Zimmer, H., Spiel um den Elefanten, Düsseldorf & Köln.

PHOTO GALLERY

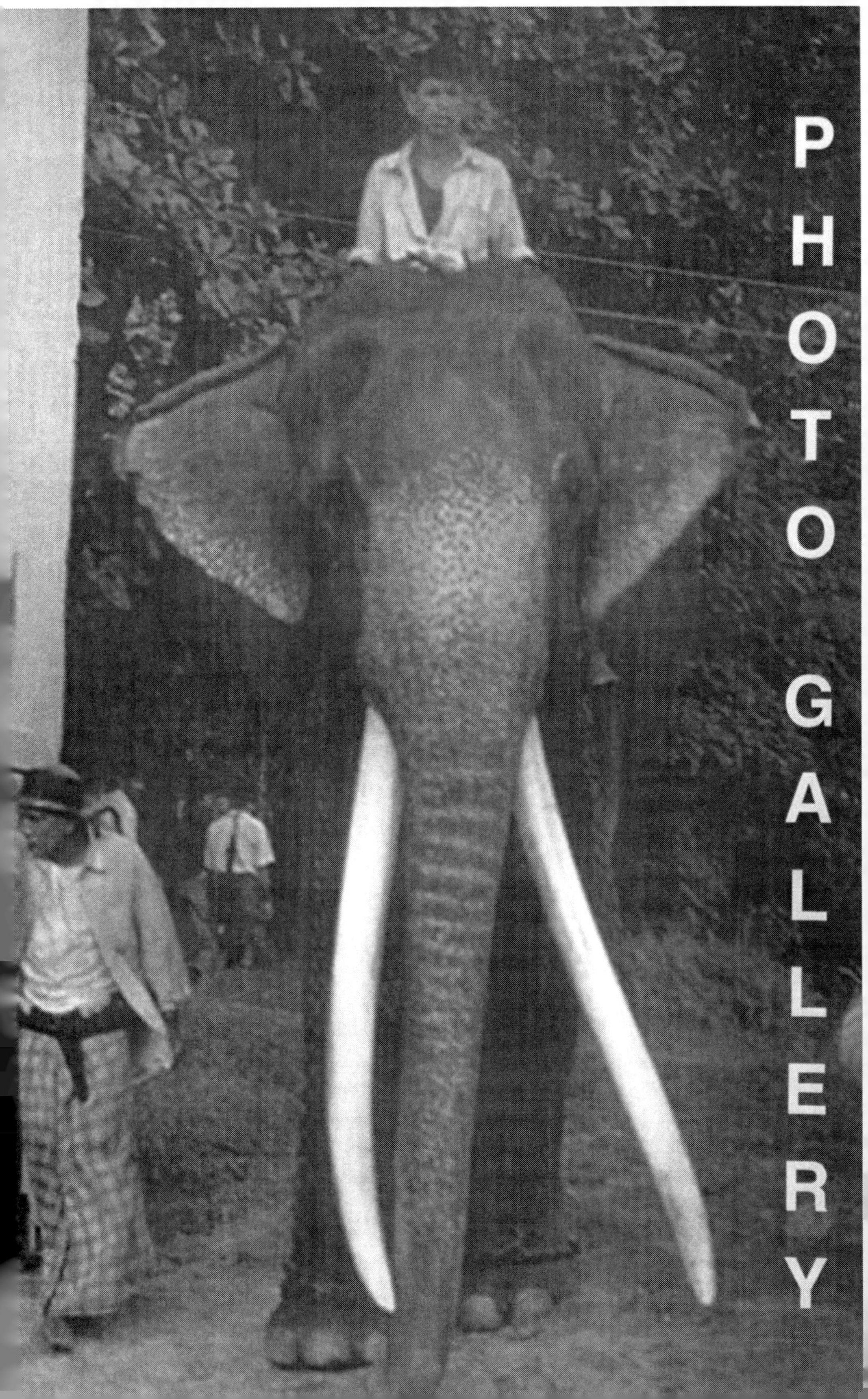

1. A young elephant caught in a pit trap. Kerala, South India.
(Photo: Jacob V. Cheeran)

2. A wild makna in the Ruhuna National Park. The tail was bitten off in a fight.
(Photo: Fred Kurt)

3. Hand-noosing of wild elephants in Sri Lanka in the 19th century.
(Archive: Fred Kurt)

4. *Mela-shikar* in Southeast Asia in the 18th century.
(Archive: Fred Kurt)

5. *Khedda* in Kakanakota, Karnataka, in 1906.
(Archive: Fred Kurt)

6. A captured elephant collapses due to exertion in Kakanakota, Karnataka.
(Photo: Fred Kurt)

7. Extensively kept elephants in the jungle village of Nardi, Tamil Nadu.
(Photo: Fred Kurt)

8. Many intensively kept elephants are used in religious festivals. Kerala.
(Photo: Fred Kurt)

9. Extensively kept elephants are guided with a thin rod. Mudumalai, Tamil Nadu. (Photo: Fred Kurt)

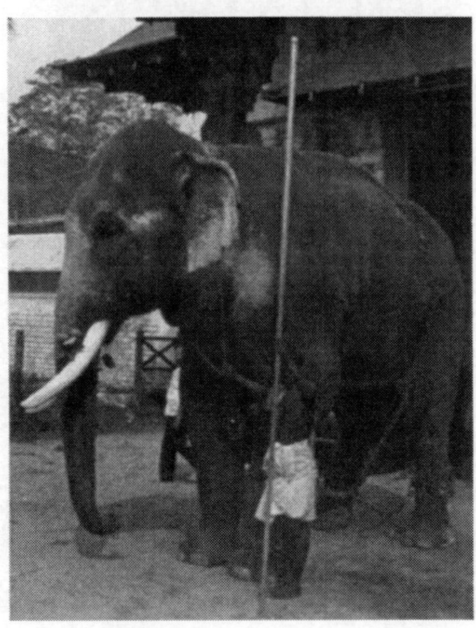

10. Intensively kept elephants are controlled with an elaborate chain-work. Kerala. (Photo: Fred Kurt)

11. Intensively kept older elephants are often seriously wounded by ankus or chains. Kandy. (Photo: Fred Kurt)

12. Extensively kept elephants often work in teams. Mudumalai, Tamil Nadu. (Photo: Fred Kurt)

13. An extensively kept bull receives a portion of energy rich food. Nardi, Tamil Nadu. (Photo: Fred Kurt)

14. The stems of the Kitul palm are one of the basic foods of captive elephants in Sri Lanka. (Photo: Fred Kurt)

15. A tusker brings a portion of Kitul palm leaves to the night quarter. Kandy.
(Photo: Fred Kurt)

16. Branches of the Jackfruit tree belong to the basic food plants of
elephants in Sri Lanka. (Photo: Fred Kurt)

17. Measuring captive elephants in Sri Lanka. (Photo: Fred Kurt)

18. Weighing a captive elephant at the railway station in Kandy.
(Photo: Fred Kurt)

19. The herd at the Pinnawela Elephant Orphanage.
(Photo: Marion E. Garaï)

20. Juvenile elephants take sand bath in Pinnawela.
(Photo: Marion E. Garaï)

21. In wild herds, young elephants normally stay in the centre.
Ruhuna National Park. (Photo: Fred Kurt)

22. Captive born elephant in the centre of a group at Pinnawela.
(Photo: Marion E. Garaï)

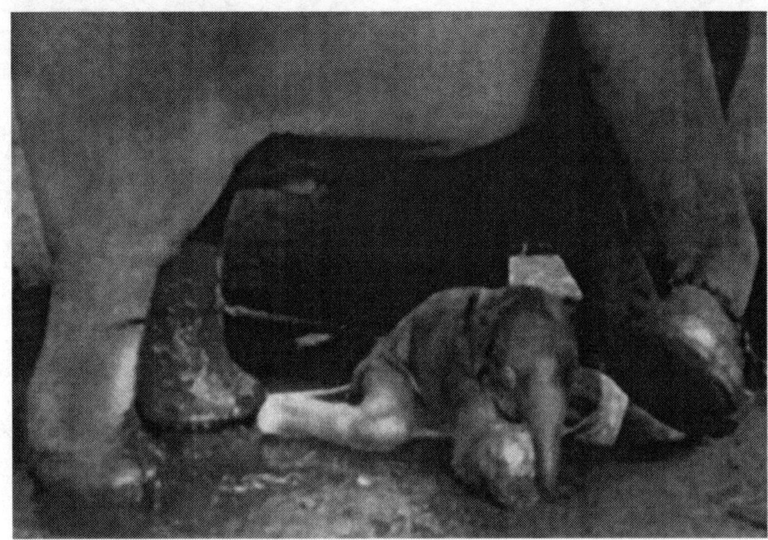

23. Birth of an elephant at the Pinnawela Elephant Orphanage.
(Photo: Birgit Dastig)

24. In modern European zoos offspring are born in groups of unchained
females. (Photo: Stephan Hering-Hagenbeck)

25. During the hottest hours of the day wild elephants rest and sleep.
Mudumalai. (Photo: Ajay Desai)

26. A wild bull sleeps in a mud hole. Ruhuna National Park.
(Photo: Fred Kurt)

27. Musth bulls are found in vicinity or among the female groups. Mudumalai, Tamil Nadu. (Photo: Ajay Desai)

28. After bathing a musth bull smears the fluid of his temporal glands on a tree stem. Ruhuna National Park. (Photo: Fred Kurt)

# Index

Lightning Source UK Ltd.
Milton Keynes UK
22 August 2010

158818UK00001B/40/P